When predicting the effects of changing climate and carbon dioxide on plants at the global scale there is a major stumbling block: we have very little information, in many cases none, about how plants will respond in the future. In order to circumvent this problem, and until more information on species accumulates, we reduce the diversity of species to a diversity of function and structures. The structures may be trees, shrubs, herbs and grasses. The functions may be types of photosynthetic process, the capacity to minimize water loss, and varying the timing of growth.

This book brings together a first-rate team of authors to describe approaches and methods for defining these functional types in ways that maximize our potential to predict accurately the responses of real vegetation with real species diversity.

Plant Functional Types

The International Geosphere–Biosphere Programme was established in 1986 by the International Council of Scientific Unions, with the stated aim

To describe and understand the interactive physical, chemical and biological processes that regulate the total Earth system, the unique environment that it provides for life, the changes that are occurring in this system, and the manner in which they are influenced by human activities.

A wide-ranging and multi-disciplinary project of this kind is unlikely to be effective unless it identifies priorities and goals, and the IGBP defined six key questions that it seeks to answer. These are:

- How is the chemistry of the global atmosphere regulated, and what is the role of biological processes in producing and consuming trace gases?
- How will global changes affect terrestrial ecosystems?
- How does vegetation interact with physical processes of the hydrological cycle?
- How will changes in land-use, sea level and climate alter coastal ecosystems, and what are the wider consequences?
- How do ocean biogeochemical processes influence and respond to climate change?
- What significant climatic and environmental changes occurred in the past, and what were their causes?

The **International Geosphere–Biosphere Programme Book Series** brings new work on topics within these themes to the attention of the wider scientific audience.

INTERNATIONAL GEOSPHERE–BIOSPHERE PROGRAMME BOOK SERIES

Plant Functional Types

Their relevance to ecosystem properties
and global change

Edited by

T. M. Smith
University of Virginia

H. H. Shugart
University of Virginia

F. I. Woodward
University of Sheffield

CAMBRIDGE
UNIVERSITY PRESS

1997

PUBLISHED BY THE PRESS SYNDICATE OF THE UNIVERSITY OF CAMBRIDGE
The Pitt Building, Trumpington Street, Cambridge CB2 1RP, United Kingdom

CAMBRIDGE UNIVERSITY PRESS
The Edinburgh Building, Cambridge CB2 2RU, United Kingdom
40 West 20th Street, New York, NY 10011-4211, USA
10 Stamford Road, Oakleigh, Melbourne 3166, Australia

First published 1997

Printed in the United Kingdom at the University Press, Cambridge

Typeset in Ehrhardt 10.5/14pt

A catalogue record for this book is available from the British Library

Library of Congress Cataloguing in Publication data

Plant functional types : their relevance to ecosystem properties and
   global change / edited by T. M. Smith, H. H. Shugart, and
   F. I. Woodward.
        p. cm. – (International Geosphere–Biosphere Programme book series)
     Includes bibliographical references.
     ISBN 0 521 48231 3 (hc). – ISBN 0 521 56643 6 (pbk.)
        1. Plant ecophysiology.   2. Vegetation and climate.   3. Climatic
   changes.   4. International Geosphere–Biosphere Program 'Global
   Changes'.   I. Smith, T. M. (Thomas Michael), 1955–
   II. Shugart, H. H.   III. Woodward, F. I.   IV International
   Geophere-Biosphere Program 'Global Changes'.   V. Series.
   QK905.P565   1996
   581.5–dc20   96–1582 CIP

ISBN 0 521 48231 3 *hardback*
ISBN 0 521 56643 6 *paperback*

Contents

Contributors

A. C. Blackmore
Natal Parks Board, St Lucia, South Africa

W. J. Bond
Botany Department, University of Cape Town, Private Bag, Rondebosch, South Africa

I. C. Burke
Department of Forest Science and Natural Resource Ecology Laboratory, Colorado State University, Fort Collins, Colorado 80523, USA

M. J. W. Burke
NERC Unit of Comparative Plant Ecology, University of Sheffield, Sheffield S10 2TN, United Kingdom

B. D. Campbell
NERC Unit of Comparative Plant Ecology, University of Sheffield, Sheffield S10 2TN, United Kingdom

D. P. Coffin
Department of Range Science and Natural Resource Ecology Laboratory, Colorado State University, Fort Collins, Colorado 80523, USA

W. Cramer
Department of Global Change and Natural Systems, Potsdam Institute of Climate Impact Research, PO Box 60 12 03, D-14412 Potsdam, Germany

S. Diaz
NERC Unit of Comparative Plant Ecology, University of Sheffield, Sheffield S10 2TN, United Kingdom

W. N. Ellery
Department of Geology, University of the Witwatersrand, Johannesburg 2050, South Africa

A. E. Giblin
The Ecosystems Center, Marine Biological Laboratory, Woods Hole, Massachusetts 02543, USA

H. Gitay
Ecosystem Dynamics Group, Research School of Biological Sciences, Australian National University, Canberra ACT 0200, Australia

R. A. Golluscio
Department of Ecology, University of Buenos Aires, Av San Martin,
Buenos Aires 1417, Argentina

J. P. Grime
NERC Unit of Comparative Plant Ecology, University of Sheffield, Sheffield s10 2TN,
United Kingdom

G. A. F. Hendry
NERC Unit of Comparative Plant Ecology, University of Sheffield, Sheffield s10 2TN,
United Kingdom

S. H. Hillier
NERC Unit of Comparative Plant Ecology, University of Sheffield, Sheffield s10 2TN,
United Kingdom

R. J. Hobbs
CSIRO Division of Wildlife & Ecology, LMB 4, PO Midland, Western Australia 6056

J. G. Hodgson
NERC Unit of Comparative Plant Ecology, University of Sheffield, Sheffield s10 2TN,
United Kingdom

R. Hunt
NERC Unit of Comparative Plant Ecology, University of Sheffield, Sheffield s10 2TN,
United Kingdom

A. Jalili
NERC Unit of Comparative Plant Ecology, University of Sheffield, Sheffield s10 2TN,
United Kingdom

C. K. Kelly
Department of Biological Sciences, University of North Texas, PO Box 5218, Denton,
Texas 76203-52218, USA

W. K. Lauenroth
Department of Range Science and Natural Resource Ecology Laboratory, Colorado
State University, Fort Collins, Colorado 80523, USA

R. Leemans
Department of Terrestrial Ecology & Global Change, National Institute of Public
Health & Environmental Protection, PO Box 1, 3720 BA Bilthoven, The Netherlands

M. Leishman
School of Biological Sciences, Macquarie University, North Ryde, NSW 2109, Australia

H. A. Mooney
Department of Biological Sciences, Stanford University, Stanford,
California 94305, USA

K. J. Nadelhoffer
The Ecosystems Center, Marine Biological Laboratory, Woods Hole,
Massachusetts 02543, USA

I. R. Noble
Ecosystem Dynamics Group, Research School of Biological Sciences, Australian
National University, Canberra ACT 0200, Australia

G. Pickett
Department of Botany, University of the Witwatersrand, Johannesburg 2050,
South Africa

E. B. Rastetter
The Ecosystems Center, Marine Biological Laboratory, Woods Hole,
Massachusetts 02543, USA

J. F. Reynolds
Department of Botany, Phytotron Building, Box 90340, Durham,
North Carolina 27708-0340, USA

O. E. Sala
Department of Ecology, University of Buenos Aires, Av San Martin,
Buenos Aires 1417, Argentina

W. H. Schlesinger
Department of Botany, Phytotron Building, Box 90340, Durham,
North Carolina 27708-0340, USA

R. J. Scholes
Forestek, CSIR, Box 395, Pretoria 0001, South Africa

G. R. Shaver
The Ecosystems Center, Marine Biological Laboratory, Woods Hole,
Massachusetts 02543, USA

H. H. Shugart
Department of Environmental Sciences, University of Virginia, Charlottesville,
Virginia 22903, USA

T. M. Smith
Department of Environmental Sciences, University of Virginia, Charlottesville,
Virginia 22903, USA

K. Thompson
NERC Unit of Comparative Plant Ecology, University of Sheffield, Sheffield S10 2TN,
United Kingdom

R. A. Virginia
Environmental Studies Program, Murdough Center, Dartmouth College, Hanover,
New Hampshire 03744-3560, USA

B. H. Walker

CSIRO Division of Wildlife & Ecology, PO Box 84, Lyneham, ACT 2602, Australia

M. Westoby

School of Biological Sciences, Macquarie University, North Ryde, NSW 2109, Australia

F. I. Woodward

Department of Animal and Plant Sciences, University of Sheffield, PO Box 601, Sheffield S10 2UQ, United Kingdom

Preface

This volume is the first in a series of International Geosphere–Biosphere Programme (IGBP) reports to be published by Cambridge University Press and is specifically concerned with an activity within the core project Global Change and Terrestrial Ecosystems (GCTE). The objectives of GCTE are:

1. To predict the effects of changes in climate, atmospheric composition, and land use on terrestrial ecosystems, including agricultural and production forest systems.
2. To determine how these effects lead to feedbacks to the atmosphere and the physical climate system.

The scientific agenda outlined by GCTE to meet these objectives is organized into four foci: (1) Ecosystem Physiology, (2) Change in Ecosystem Structure, (3) Global Change Impact on Agriculture and Forestry, and (4) Global Change and Ecological Complexity. At the heart of Focus 2 is the goal of modelling the patterns of change in ecosystem composition and structure resulting from such factors as increasing atmospheric concentrations of CO_2, climate change, and changes in land use patterns. These predictions are important in understanding not only the potential impacts on terrestrial ecosystems, but also how changes in the terrestrial surface may influence atmospheric dynamics (i.e. feedback effects).

The potential impacts of global change on terrestrial ecosystems are being explored at local to regional scales, using a variety of models, each suited to the ecosystem of interest. In contrast, the question of feedbacks to the global climate system is a global issue and must be approached using global-scale models. The necessity for global-scale models presents many problems, one of the greatest being the inability to characterize the functional and structural diversity of the organisms that make up the earth's ecosystems.

It is not feasible to develop models for every ecosystem of the globe nor to represent every species within those ecosystems. Thus, the concept that the complexity of nature can be reduced in models by treating a smaller number of 'functional types' (FTs) is central to the work of Focus 2.

The idea that a functionally oriented (as opposed to a phenetic or phylogenetic) classification of organisms may be an effective way of reducing the complexity of modelling ecosystem processes has a long history in ecology. It has often been argued that the essential dynamics of ecosystems can be captured by grouping species into a restricted number of groups or functional types. Experience indicates that this grouping can work well for specific ecosystems but that the groups often have characteristics unique to the ecosystem under consideration. At present there appears to be neither congruity between FT schemes nor agreement about the capacity of FT schemes to predict dynamic responses of FT groups or ecosystems to environmental change.

The chapters in this volume address the major issues concerned with the recognition, definition and operation of FT schemes in the context of environmental change impacts on terrestrial ecosystems. The papers are grouped into five sections. The first section provides an introduction and historical background to the concept of functional classifications of species for ecological research (Chapters 1 and 2).

The second section addresses the theoretical and practical aspects of developing functional classifications of species for global change research. The authors approach the development of functional classifications from a number of perspectives. Chapters 3–5 focus on the theoretical and objective oriented process of classifying species into functional groupings, and Chapters 6 and 7 provide a practical application of the concept to field studies.

The third section focuses on the development of functional classifications for various ecosystems of the world (tundra, fynbos, shrublands, grasslands and savanna); the fourth section focuses on the global application of the functional type concept (Chapters 14 and 15).

The fifth and last section examines the consequences and limitations of modelling vegetation dynamics using a functional types approach (Chapter 16) and the role of biodiversity in ecosystem function (Chapter 17).

The papers in this volume were first presented at a workshop on plant functional types held at the University of Virginia, Charlottesville, Virginia, USA in 1993. We are grateful to the John D. and Catherine T. MacArthur Foundation, LTER, NSF, UNEP and the Global Systems Analysis Program at the University of Virginia for the financial assistance to run the workshop. In addition we would like to express our thanks to Will Steffen and Lyndele Von Schill for helping with the administration of the workshop and to Miss Erica Schwarz for very careful editing of the individual chapters.

T. M. Smith
H. H. Shugart
F. I. Woodward

Part one

① What are functional types and how should we seek them?

H. Gitay and I. R. Noble

Introduction

In recent years ecologists have placed increasing emphasis on the use of non-phylogenetic classifications of organisms when describing the structure and functioning of ecosystems. Some (e.g. Heal & Grime 1991) feel that 'classical taxonomy will have to give way to functional classifications'. Hawkins & MacMahon (1989) and Simberloff & Dayan (1991) have reviewed the increasing body of literature on functional classifications and their wide-ranging and often confusing use in ecology.

GCTE has adopted the concept of a functional classification of organisms as a fundamental part of its operational plan. It concluded that

> *It will not be feasible to develop models for every ecosystem of the globe nor represent every species within those ecosystems. Thus, the concept that the complexity of the models can be reduced by treating a smaller number of 'functional types' (FTs) is central to the work of Focus 2; it has often been argued that the essential dynamics of ecosystems can be captured by grouping species into a limited number of FTs.* (Steffen et al. 1992).

Thus GCTE is committed to the use of functional types*, yet it is not clear what functional types are and how we should seek them.

Here, we briefly review past work related to functional types, concentrating on those issues most relevant to the objectives of GCTE. We examine the various definitions put forward, discuss how functional types might be identified, and ask what success there has been to date in seeking them. Finally we discuss the successes and failures in the application of functional classifications and the issues relevant to GCTE.

---

* Throughout this paper we use the terms functional type and functional group synonymously. We do not know of any authors who make a clear distinction between the terms 'group' and 'type'. We have used the term used by authors when closely quoting their work and the most appropriate term in other situations.

Definitions

Although the idea of functional classification can be traced back at least to Theophrastus (about 300 BC), the modern debate really goes back to suggestions made in the 1960s. Root (1967) introduced the ecological concept of *guilds*, defining them as 'a group of species that exploit the same class of environmental resources in a similar way' or species 'that overlap significantly in their niche requirements'. Cummins (1974) discussed a *functional grouping* of organisms in which he wanted '. . . to stress important process-oriented ecological questions'. Botkin (1975) suggested that species can be grouped into a much smaller number of functional types and that these new groupings would 'still allow consideration of the important population interactions'. Related terms are *character* [or adaptive] *syndromes* to describe certain characters that cannot be decoupled because they contribute to a common functional role (e.g. Swaine & Whitmore 1988) and sometimes may have a common phylogenetic origin (Stebbins 1974), and *strategy* as used by Grime *et al.* (1988) to define 'a grouping of similar or analogous genetic characteristics which re-occurs widely amongst species or populations and causes them to exhibit similar ecology'. Paine (1980) used the term *modules* to define groups of closely interacting species.

In addition, functional types have often been described as those biotic components of ecosystems that perform the same function or set of functions within the ecosystem. Friedel *et al.* (1988) defined them as groups that respond similarly to the same perturbation. Noble (1989) discussed a classification based on a set of physiological, reproductive and life history characters where variation in each character has specific ecologically predictive (rather than descriptive) value. Keddy (1992) stated that species can be aggregated into [functional] groups sharing similar traits.

Other terms have been introduced as extensions of the guild concept. Jaksic (1981) highlighted the confusion over the use of the term guild pointing out that some are based on taxonomic classification (*assemblage guilds*) and others on the use of the same resource (*community guilds*). Verner (1984) defined *management guilds* as a group of species that respond in a similar way to a variety of changes likely to affect their environment. Szaro (1986) used the term *functional guilds* with essentially the same meaning as Root's guilds, and, in addition, introduced two more terms: *structural guilds* for a group of species that use the same resource, although not necessarily in the same manner or for the same purpose, and *response guilds* for species that respond in a similar manner to a habitat perturbation. Menge *et al.* (1986) described the same dichotomy as Szaro's structural and response guilds but called them guilds and functional types.

Various subdivisions and amalgamations of basic guild concepts have
been proposed. Faber (1991) suggested the use of the term *league* to
describe a group of organisms that use more than one resource in a similar
way. The need for such a term depends on one's definition of a 'resource'.
Walker (1992) described functional groups as the result of further subdividing guilds based on functional attributes. Barbault *et al.* (1991) argued
that functional types should be defined in terms of morphology and physiology 'particularly as these properties relate to resources and species interactions' and thus represent 'feeding guilds' or 'plant growth forms'. They
defined functionally similar taxa within a functional type to be *functional
analogues*. Yodzis (1982) used the term *clique* to describe a set of species
that have some food resource in common (although they do not have to use
it in a similar way) and a *dominant clique* to describe a clique that is
contained in no other clique. Thus each guild in Root's sense is a clique,
but a dominant clique may include more species than a guild or even
several guilds. Yodzis suggested that a dominant clique may be called a
trophic guild. Bahr (1982) also suggested higher-order categories called an
ecological sector for 'broad trophic groups of organisms in common vertical
habitat zones' and *ecological species* for groups defined by a binomial
consisting of the ecological sector they occupy and their guild. Swaine &
Whitmore (1988) also used the term *ecological species groups*, but to refer to
groupings similar to Szaro's (1986) response guilds.

In summary, we see several common ideas in the various definitions of
guilds and functional types. The main differences are that in one set of
ideas, the species can be grouped on the basis that they use the same
resource (i.e. *guilds*) and in others by their response to a specified perturbation. The ideas can be further subdivided according to whether the
species use the shared resource in the same way, or respond to the perturbation by the same mechanism. We might expect that a group of species
that use resources, or respond to disturbances, by similar mechanisms might
behave similarly under a range of circumstances and perturbations, and that
the classification would have greater extrapolative power. We present a
summary of the definitions and suggested names in Tables 1.1 and 1.2
along with the terminology of other authors.

It will be difficult to achieve an agreed and precise definition of a functional group given the wide range of ways the term has been, and probably
will continue to be, used. However, we suggest that its use should be
confined to groupings based on the response of organisms to perturbations.
Thus, a functional group is a non-phylogenetic classification leading to a
grouping of organisms that respond in a similar way to a syndrome of
environmental factors. Used in this way, a functional group is the basis for

Table 1.1 *Differences between guilds, response groups and functional groups*

	Resource use		Response to perturbation	
	same resource	same way	same response	same mechanism
Structural guild	Yes	No	—	—
Functional (cf. Root) guild	Yes	Yes	—	—
Response group	—	—	Yes	No
Functional group	—	—	Yes	Yes

Table 1.2 *A comparison of the definitions given by various workers*

We have used the best match to our definition from their definitions. Asterisks indicate cases where we know the match is poor.

Authors	Term used	Authors	Term used
our term	**Structural guild**	*our term*	**Response group**
Szaro (1986)	structural guild	Verner (1984)	management guild
Menge *et al.* (1986)	guild	Szaro (1986)	response guild
Yodzis (1982)*	clique	Friedel *et al.* (1988)	functional group
		Swaine & Whitmore (1988)	ecological species group
our term	**Functional guild**	*our term*	**Functional group**
Root (1967)	guild	Noble (1989)	functional type
Szaro (1986)	functional guild		
Menge *et al.* (1986)	functional type		
Walker (1992)	functional type		
Faber (1991)*	league		
Barbault *et al.* (1991)*	functional analogue		

a context-specific simplification of the real world to deal with predictions of the dynamics of the systems or any of their components. We refer to 'organisms' to emphasize that the classification is not restricted to species but can apply to different phylogenetic levels and different life-history stages. A syndrome of environmental factors is a combination of biotic and abiotic processes that change as a result of a perturbation to the system, for example as a consequence of the passage of a fire. We also point out that similarity must be defined by the user in the context of the syndrome of environmental factors.

We further suggest that the term *response group* (or type) should be used to describe groups of organisms based simply on their behaviour in response to a particular perturbation and the term *functional group* be confined to groups where the response is mediated through the same mechanism.

GCTE and functional types

Given that GCTE wants to predict the response of functional types (FTs) under a wide range of environmental syndromes, then it has to seek a grouping of species that obeys the following mathematical relation:

$$\int(\{x_1 + x_2 + x_3\}, \{x_{4} + x_5\}, \ldots \{\ldots + x_n\}, E) \approx \int(x_1, x_2, x_3, x_4, \ldots x_n, E)$$

where the x_i are species (taxa), the {bracketed} x_i are functional types, and E is the residual biotic and abiotic environment not incorporated in the x_i. Thus GCTE's interest in functional types is pragmatic. It needs a simplification tool, as originally suggested by Botkin. GCTE also requires that this equation apply not only to current combinations of x_i, but to future combinations and, more importantly, to new environmental scenarios, E.

Thus, GCTE requires a strong extrapolative property of its functional classification and thus seeks functional groups as defined above (see Tables 1.1 and 1.2). However, in terms of the papers in this volume most authors, except for Scholes, have used the 'functional type' in the sense of our structural guilds (see Table 1.1), without considering the response of the organisms involved.

What is the relationship between guilds, response groups and functional groups as we have defined them? If two species are in the same structural guild they use the same resource but in a different way and there is no *a priori* reason to expect them to respond to a perturbation in a similar way. Similarly, the observation that two species are in the same response group conveys no information about the mechanisms leading to the response and thus no information about whether they might be in the same guild. However, if two species are in the same functional guild, they use the same resources in the same way, and thus it is more likely that they will respond to a perturbation in a similar way and therefore fall within the same response group and possibly the same functional group.

How can we recognize functional types?

There are three main approaches that have been used to identify functional types and guilds: we will call them subjective, deductive and data-defined (a form of pattern analysis).

A subjective approach is based on observations of one or more ecosystems in which it is taken for granted that functional types exist and that these can be defined inductively. Most ecological writing, as with most of our everyday description of our biological environment, is based on such groups (e.g. trees and shrubs, tree feeders, etc.). Various workers have used such a classification for wide-ranging groups of organisms. Examples of these include the work of Baker (1971), MacMahon (1976), Mannan *et al.* (1984), De Graaf *et al.* (1985), Shorrocks & Rosewell (1986), Adserson (1988), DuBowy (1988), Rader & Ward (1988), Wilson (1989), Croonquist & Brooks (1991), Faber (1991), Simberloff & Dayan (1991) and Root & Cappuccino (1992).

Some very comprehensive subjective, and mostly untested, classifications have been proposed. Examples include those of Johnson (1981), Verner (1984) and Newsome & Noble (1986). Johnson has attempted a classification of terrestrial vegetation of the USA for environmental impact assessment. He defined 95 hierarchically arranged guilds, based on dispersal, establishment and growth characters. The main purpose was to provide a non-subjective classification suitable for use by non-specialists. Verner (1984) produced an inductive two-dimensional classification of nesting and feeding zones of forest birds for management purposes. Newsome & Noble (1986) classified 86 weed species by multivariate techniques using 17 traits. They then compared the classes found in this way with subjective classes defined for invasive birds and described four groups applicable to both.

In a deductive approach, such as that of Noble & Slatyer (1980), van der Valk (1981), Huston & Smith (1987), Keddy (1992) and Scholes *et al.* (Chapter 13, this volume), a functional classification is derived from an *a priori* statement (or model) of the importance of particular processes or properties in the functioning of an ecosystem. The feasible set of functional categories are then deduced from these premises. The keystone species classification of Paine (1980) can also be considered to be such an approach. The proposal by Walker (Chapter 5, this volume) is also a deductive approach based on assumptions about which characters might be most important in terms of climate change and can be obtained at a global scale. This is the most direct way of achieving a functional classification for a specific purpose; however, it is not clear whether it will lead to a workable number of functional types.

A data-defined approach uses multivariate techniques to seek clusters of species based on a set of characters. Again this has been used for a wide range of organisms. Examples include the work of Hagmeier & Dexter Stults (1964), Crome (1978), Holmes *et al.* (1978), Landres & MacMahon (1980), Pianka (1980), Folse (1981), Joern & Lawlor (1981), Bowers & Brown (1982), Newsome & Noble (1986), Niemi (1985), Szaro (1986), Cornell & Kahn (1989), Hawkins & MacMahon (1989), Jaksic & Medel (1990), Keddy (1990), Maddox & Root (1990), Winemiller & Pianka (1990), Finch (1991), Simberloff & Dayan (1991) and Leishman & Westoby (1992). In most cases the analyses are based on morphological and growth characters and might best be described as searching for a set of character syndromes (Stebbins 1974) or strategies (Grime *et al.* 1988). Thus, with respect to functional types, they are usually based on a set of characters that are surrogates for the response data. We discuss this point later in this chapter.

The three approaches can be used at different scales. The subjective and deductive classifications can be applied at local to global scales. The data-defined approach is appropriate to the local or regional scale and is unlikely to be used at the global scale because it would be difficult to gather appropriate data sets. There is also the question of whether the species properties and thus the functional types are discrete or continuous. If they are continuous then the data-defined approach would not produce clear groups and there will also be difficulty with a subjective approach. We discuss the general questions that can be raised with regards to this elsewhere (Noble & Gitay 1995).

Formal tests of functional classifications

One of the commonest approaches in attempting to find guilds or functional types has been to use multivariate analysis of character sets describing a set of species. This approach may not be suitable to the goals of GCTE as the detailed information required for each of the species to produce a classification may not be achievable even at a regional scale. It may be that a deductive approach is more appropriate for GCTE's objectives. However, since the data-defined approach has been popular and some see the quantitative methodology as providing a sense of objectivity, despite the subjectivity in choosing character sets, distance or similarity measures and linkage methods, we will now consider the likely problems with the data-defined approach in producing a functional classification.

Functional classifications assume that there is an inherent structure in the world and that guilds and functional types are a revelation of that structure. Is this the case, and if it is, then how do we find that structure? If we fail to find consistent structure, does this imply that it does not exist, or is it a problem of inadequate selection of characters or inappropriate analytical techniques?

Ideally, a functional classification should be applicable to that suite of species wherever they occur and under a prescribed, but wide, range of environmental conditions and perturbations. Realistically, we might expect that classifications will vary in their robustness. Here we present a formal ranking of the degrees of robustness and then examine the performance of some existing classifications. We assume the following scenario.

- That we have collected data for a given pool of organisms. These data might include any characters or properties that reflect the function of the organisms in an ecosystem or their responses to perturbations.
- That we have also independently collected the data set at several different times and/or in several different locations; i.e. we have replicates of the data set.
- That several different researchers have been involved, each with the same purpose of the classification in mind, but each having collected different suites of characters that describe these organisms.
- That other researchers have also described the same pool of organisms, but this time with a different purpose in mind in seeking a functional classification.

We could analyse these data sets in several ways, and recognize the following outcomes.

1. *Uniqueness*: when similar classifications of a given data set are achieved by using *different analytical techniques*.
2. *Repeatability*: when similar classifications are achieved by using the same character sets collected at *different sites or different times*. By implication, character syndromes exist.
3. *Congruency*: when similar classifications are achieved by using *different character sets*, i.e. there are correlations between character syndromes. A hypothetical example is given in Table 1.3.
4. *Convergence*: when similar classifications of a pool of organisms are

Table 1.3 *Possible distribution of characters in organisms that would result in congruency*

The rows represent taxa and the columns characters. The taxa can be grouped based on the first five characters, which represent two character syndromes. They can also be grouped on the remaining six characters, which represent another two character syndromes. In this case the groupings of the taxa are identical and thus the classifications are congruent.

achieved by using data collected and analysed for *different purposes*. This implies that there is a limited number of feasible suites of physiological and life-history characters that must cover responses to all environmental syndromes; for example, groups with a similar response to drought will have a similar response to fire.

Some test results

Uniqueness

Anyone with experience of using multivariate techniques knows that different results can be obtained by using different similarity indices and linkage methods, and so this will not be discussed further. However, if classifications are to be compared, then it is important that the analytical techniques should be as similar as possible to ensure that any difference in the outcome reflects biological differences and not analytical ones.

Repeatability

We found some examples in the literature which can be used to test the criterion of repeatability. Friedel *et al.* (1988) gathered data sets on plants from two semi-arid sites. A cluster analysis produced ecologically interpretable groups, but the structure of the dendrogram was different (as illustrated in Fig. 1.1), although some of the groups were the same, implying that there is some repeatability.

Joern & Lawlor (1981) obtained a diet list of grasshoppers at several sites and concluded that there is reasonable agreement between the groups derived from their data. We have reanalysed one pair of their sites with 75% species overlap using a χ^2 test and found a significant repeatability of the groups ($p = 0.01$).

Szaro (1986) tested Verner's (1984) classification by counting bird densities in defined habitats in ponderosa pine forests eight times per year for three years. He found no consistent patterns in the variation of densities of individual species within a guild compared with the overall density of that guild. When he clustered species based on their density variation between habitats he found a different number of clusters in each of the three years with little consistency in the membership of the groups.

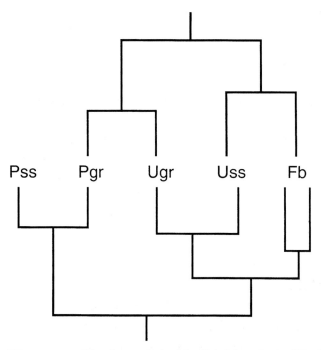

Figure 1.1   *The clusters derived by Friedel* et al. *(1988). Fb, forbs; Pss, palatable subshrubs; Pgr, palatable grasses; Ugr, unpalatable grasses; Uss, unpalatable subshrubs.*

Congruency and convergence

Grime *et al.* (1988) measured two data sets based on characters describing the regenerative and established phases of the same suite of plant species. The established phase showed three groups (reflecting mostly life history and life form), and the regenerative phase showed ten groups with no simple ecological (or phylogenetic) interpretation. The congruency between the classifications appears to be low. Shipley *et al.* (1989) collected morphological data for two life stages, juvenile and adult, where the characters were selected to reflect an adaptive syndrome to contrasting shore conditions. They found no association between the syndromes based on juvenile and adult life stages.

Joern & Lawlor (1981) used data sets based on diet characters and micro-habitat variables to classify grasshoppers. They concluded that there is a reasonable degree of agreement between the groups; this is supported by our reanalysis of their data using a χ^2 test in which we found that there was almost a significant level of congruency ($p = 0.06$).

We have analysed two data sets for rainforest species in Queensland, Australia (H. Gitay & I. R. Noble, unpublished), one representing morpho-logical and life-history traits and the other representing the dynamic behaviour of the species over 30 years of observation made by J. H. Connell and others (Connell *et al.* 1984). We found that there was no relationship between the cluster membership in the two classifications (see Table 1.4). Thus there was no congruency.

We have found no suitable data sets in the literature to test for convergence.

In summary, there is some evidence of repeatability but little evidence of congruency, except within a restricted taxon (see Table 1.5) and we have not come across any data to test for convergence. We conclude that it is feasible to group species based on character syndromes (Stebbins 1974) and that these groupings are repeatable to some extent when based on the same character set measured in different locations or different times. There appears to be little evidence of congruency of the groups based on different character syndromes, such as groups based on juvenile and adult traits.

Tests of the utility of functional classifications

Few specific tests of the utility of functional classifications exist. Mannan *et al.* (1984) defined *a priori* members of bird guilds and found that these guilds did not respond consistently to a series of forest perturbations. They concluded that more guild categories were needed, approaching a species classification. Szaro (1986) also found that individual species within *a priori* defined guilds responded inconsistently between forest treatments.

Table 1.4 *A comparison of the congruency between classifications based on morphological and dynamic character sets*

Information statistics = 100.00, 56 d.f., $p \approx 0.55$.

Morphology	Dynamic								Total
	1	2	3	4	5	6	7	8	
2	1	3	0	0	0	1	0	1	6
3	0	1	0	1	1	1	0	1	5
4	5	4	0	0	1	1	0	0	11
5	0	2	0	0	1	0	0	0	3
6	1	2	1	1	1	0	0	0	6
7	3	5	1	2	0	0	0	0	11
8	1	1	0	0	1	0	1	0	4
9	1	0	0	0	2	0	0	0	3
10	0	0	0	0	1	0	0	0	1
Total	12	18	2	4	8	3	1	2	50

Source: H. Gitay & I. R. Noble, unpublished.

Sale & Guy (1992) observed that species assemblages of tropical fish are highly variable. Classifying the fish into subjectively defined guilds reduced the variability, but no more than with a null classification. They concluded that there is no evidence of an underlying organization of these assemblages at the guild level.

Against these three direct tests of functional classifications we must set the many examples where users have found classifications (usually subjective) to be satisfactory for their purposes (e.g. Diamond 1975, Noble & Slatyer 1980, Johnson 1981, Verner 1984, Keddy 1992, and many other modellers).

Conclusion

The use of the terms *guilds* and *functional types* has been inconsistent and has become confusing. We suggest that the term guild be used for classifications based on whether species use the same resources, and functional groups for classifications based on whether species respond in a similar way to a specified perturbation. Thus, functional types are a context dependent classification. For the purposes of GCTE, there is a more severe constraint since, in addition to being a means of simplification of the system that

Table 1.5 *Summary of case studies testing repeatability and congruency*

The table shows the result of the test, whether different life forms were incorporated in the range of organisms used in the analysis, whether a restricted range of taxa were included, and whether contrasting habitats were used.

Author	Test	Result a success?	Different life forms	Restricted taxa	Habitat contrast	Organisms
Friedel et al. (1988)	Repeatability, site	?Yes	Yes	No	?	Semi-arid plants
Joern & Lawlor (1981)	Repeatability, site	Yes	No	Yes	No	Grasshoppers
Szaro (1986)	Repeatability, time	No	Yes	No	No	Birds
Joern & Lawlor (1981)	Congruency	?Yes	No	Yes	No	Grasshoppers
H. Gitay & I. R. Noble (unpublished)	Congruency	No	No	No	No	Rainforest trees
Grime et al. (1988)	Congruency	No	Yes	No	?	Various UK plants
Shipley et al. (1989)	Congruency	No	Yes	No	Yes	Shoreline plants

describes similarity in responses to perturbations, the responses should also be due to the same mechanism if extrapolations to new circumstances are likely to be valid.

There is no doubt as to the utility of the concept of functional classifications, but as suggested by Hawkins & MacMahon (1989) and Landres (1983), this does not mean that functional groups exist. Unlike Heal & Grime (1991), we doubt that there is 'a universal [functional] classification of organisms in an ecosystem' since context is important. However, it is likely that functional types can be defined for specific purposes.

There has been some success in developing functional classifications that lead to ecologically interpretable groups. However, many classifications do not seem to get past the empirically obvious morphological and life-history-based groups. When we apply formal tests of robustness to the functional classifications of past work, we find some evidence that classifications are repeatable in space and time but little evidence of congruency: classifications carried out for the same purpose but with different character sets result in different groups and thus do not support the existence of an inherent structure in communities.

The apparent lack of congruency is a major warning of the difficulties still facing us in defining functional types on a broad scale. It is also important to know whether we seek a structural guild type of classification, where we are only interested in the resource use, or whether we want a classification based on the response of the species to perturbations (response groups) and have extrapolative power as GCTE will require (functional groups).

Acknowledgements

Thanks to Jacqui de Chazal and Phil Dunne for their help during the preparation of the manuscript and to Nugget Coombs, Sandra Lavorel and Bastow Wilson for useful discussions.

References

Adserson, H. (1988) Null hypothesis and species composition in the Galápagos Islands. In During, H. J., Werger, M. J. A. & Williems, J. H. (eds), *Diversity and Pattern in Plant Communities*, pp. 37–46. The Hague, The Netherlands: SPB Publishing.

Bahr, L. M. (1982) Functional taxonomy: an immodest proposal. *Ecological Modelling*, **15**, 211–33.

Baker, R. H. (1971) Nutritional strategies of Myomorph rodents in North American grasslands. *Journal of Mammalogy*, **52**, 800–5.

Barbault, R., Colwell, R. K., Dias, B., Hawksworth, D. L., Huston, M., Laserre, P., Stone, D. & Younes, T. (1991) B. Conceptual framework and research issues for species diversity at the community level. In O. T. Solbrig (ed.), *From Genes to Ecosystems: a Research Agenda for Biodiversity*, pp. 37–71. Cambridge, Massachusetts: IUBS.

Botkin, D. B. (1975) Functional groups of organisms in model ecosystems. In Levin, S. A. (ed.), *Ecosystem Analysis and Prediction*, pp. 98–102. Philadelphia: Society for Industrial and Applied Mathematics.

Bowers, M. A. & Brown, J. H. (1982) Body size and coexistence in desert rodents: chance or community structure? *Ecology*, 63, 391–400.

Connell, J. H., Tracey, J. G. & Webb, L. J. (1984) Compensatory recruitment, growth, and mortality as factors maintaining rain forest tree diversity. *Ecological Monographs*, 54, 141–64.

Cornell, H. V. & Kahn, D. M. (1989) Guild structure in the British arboreal arthropods: is it stable and predictable? *Journal of Animal Ecology*, 58, 1003–20.

Crome, F. H. J. (1978) Foraging ecology of an assemblage of birds in lowland rainforest in northern Queensland. *Australian Journal of Ecology*, 3, 195–212.

Croonquist, M. J. & Brooks, R. P. (1991) Use of avian and mammalian guilds·as indicators of cumulative impacts in Riparian-Wetland Areas. *Environmental Management*, 15, 701–14.

Cummins, K. W. (1974) Structure and function of stream ecosystems. *Bioscience*, 24, 631–41.

De Graaf, R. M., Lilghman, N. G. & Anderson, S. H. (1985) Foraging guilds of North American birds. *Environmental Management*, 9, 493–536.

Diamond, J. M. (1975) Assembly of species communities. In Cody, M. L. & Diamond, J. M. (eds), *Ecology and Evolution of Communities*, pp. 342–444. Cambridge, Massachusetts: Harvard University Press.

DuBowy, P. J. (1988) Waterfowl communities and seasonal environments: temporal variability in interspecific competition. *Ecology*, 69, 1439–53.

Faber, J. H. (1991) Functional classification of soil fauna: a new approach. *Oikos*, 62, 110–17.

Finch, D. M. (1991) Positive associations among riparian bird species correspond to elevational changes in plant communities. *Canadian Journal of Zoology*, 69, 951–63.

Folse, L. T. Jr. (1981) Ecological relationships of grasslands birds to habitat and food supply in east Africa. In Capen, D. E. (ed), *The Use of Multivariate Statistics in Studies of Wildlife Habitat*, pp. 160–6. Ft. Collins, Colorado: Rocky Mountain Forest and Range Experiment Station.

Friedel, M. H., Bastin, G. N. & Griffin, G. F. (1988) Range assessment and monitoring of arid lands: the derivation of functional groups to simplify vegetation data. *Journal of Environmental Management*, 27, 85–97.

Grime, J. P., Hodgson, J. G. & Hunt, R. (1988) *Comparative Plant Ecology: A Functional Approach to Common British Species*. London: Unwin Hyman.

Hagmeier, E. M. & Dexter Stults, C. (1964) A numerical analysis of the distributional patterns of North American mammals. *Systematic Zoology*, 13, 125–55.

Hawkins, C. P. & MacMahon, J. A. (1989) Guilds: the multiple meanings of a concept. *Annual Review of Entomology*, 34, 423–51.

Heal, O. W. & Grime, J. P. (1991) Comparative analysis of ecosystems: past lessons and future directions. In Cole, J., Lovett, G. & Findlay, S. (eds), *Comparative Analysis of Ecosystems: Patterns, Mechanisms and Theories*, pp. 7–23. New York: Springer-Verlag.

Holmes, R. T. Jr., Bonney, R. E. & Pacala, S. W. (1978) Guild structure of the Hubbard Brook bird community: a multivariate approach. *Ecology*, 60, 512–20.

Huston, M. & Smith, T. (1987) Plant succession: life history and competition. *American Naturalist*, 130, 168–98.

Jaksic, F. M. (1981) Abuse and misuse of the term 'guild' in ecological studies. *Oikos*, 37, 397–400.

Jaksic, F. M. & Medel, R. G. (1990) Objective

recognition of guilds: testing for statisically significant species clusters. *Oecologia*, **82**, 87–92.

Joern, A. & Lawlor, L. R. (1981) Guild structure in grasshopper assemblages based on food and micro-habitat resources. *Oikos*, **37**, 93–104.

Johnson, R. A. (1981) Application of the guild concept to environmental impact analysis of terrestrial vegetation. *Journal of Environmental Management*, **13**, 205–22.

Keddy, P. A. (1990) The use of functional as opposed to phylogenetic systematics: a first step in predictive community ecology. In Kawano, S. (ed.), *Biological Approaches and Evolutionary Trends in Plants*, pp. 387–406. London: Harcourt Brace Jovanovich.

Keddy, P. A. (1992) Assembly and response rules: two goals for predictive community ecology. *Journal of Vegetation Science*, **3**, 157–64.

Landres, P. B. (1983) Use of the guild concept in environmental impact assessment. *Environmental Management*, **7**, 393–8.

Landres, P. B. & MacMahon, J. A. (1980) Guilds and community organisation: analysis of an oak woodland avifauna in Sonora, Mexico. *Auk*, **97**, 351–65.

Leishman, M. R. & Westoby, M. (1992) Classifying plants into groups on the basis of associations of individual traits – evidence from Australian semi-arid woodlands. *Journal of Ecology*, **80**, 417–24.

MacMahon, J. A. (1976) Species and guilds similarity of North American desert mammal faunas: a functional analysis of communities. In Goodall, D. W. (ed.), *Evolution of Desert Biota*, pp. 133–48. Austin and London: University of Texas Press.

Maddox, G. D. & Root, R. B. (1990) Structure of the encounter between goldenrod (*Solidago altissima*) and its diverse insect fauna. *Ecology*, **71**, 2115–24.

Mannan, R. W., Morrison, M. L. and Meslow, E. C. (1984) The use of guilds in forest bird management. *Wildlife Society Bulletin*, **12**, 426–30.

Menge, B. A., Lubchenco, J., Ashkenas, L. R. & Ramsey, F. (1986) Experimental separation of effects of consumers on sessile

prey in the lower zone of a rocky shore in the Bay of Panama: direct and indirect consequences of food web complexity. *Journal of Experimental Marine Biology and Ecology*, **100**, 225–69.

Newsome, A. E. & Noble, I. R. (1986) Ecological and physiological characters of invading species. In Groves, R. H. & Burdon, J. J. (eds), *Ecology of Biological Invasions: An Australian Perspective*, pp. 1–20. Canberra: Australian Academy of Science.

Niemi, G. J. (1985) Patterns of morphological evolution in bird genera of New World and Old World peatlands. *Ecology*, **4**, 1215–28.

Noble, I. R. (1989) Attributes of invaders and the invading process: terrestrial and vascular plants. In Drake, J. A., Mooney, H. A., di Castri, F., Groves, R. H., Kruger, F. J., Rejmánek, M. & Williamson, M. (eds), *Biological Invasions: a Global Perspective*, pp. 301–11. Scientific Committee on Problems of the Environment. Chichester: John Wiley & Sons Ltd.

Noble, I. R. & Gitay, H. (1995) Functional classifications for predicting the dynamics of landscapes. *Journal of Vegetation Science*, **6**, (in press).

Noble, I. R. & Slatyer, R. O. (1980) The use of vital attributes to predict successional changes in plant communities subject to recurrent disturbance. *Vegetatio*, **43**, 5–21.

Paine, R. T. (1980) Food webs: linkage interaction strength and community infrastructure. *Journal of Animal Ecology*, **49**, 667–85.

Pianka, E. R. (1980) Guild structure in desert lizards. *Oikos*, **35**, 194–201.

Rader, R. B. & Ward, J. V. (1988) The influence of environmental predictability/disturbance characterstics on the structure of a guild of mountain stream insects. *Oikos*, **54**, 107–16.

Root, R. B. (1967) The niche exploration pattern of a blue grey gnatcatcher. *Ecological Monographs*, **37**, 317–50.

Root, R. B. & Cappuccino, N. (1992) Patterns in population change and the organisation of the insect community associated with

Goldenrod. *Ecological Monographs*, **62**, 393–420.

Sale, P. F. & Guy, J. A. (1992) Persistence of community structure–what happens when you change taxonomic scale. *Coral Reefs*, **11**, 147–54.

Shipley, B., Keddy, P. A., Moore, D. R. J. & Lemky, K. (1989) Regeneration and establishment strategies of emergent macrophytes. *Journal of Ecology*, **77**, 1093–110.

Shorrocks, B. & Rosewell, J. (1986) Guild size in drosophilids: a simulation model. *Journal of Animal Ecology*, **55**, 527–41.

Simberloff, D. & Dayan, T. (1991) The guild concept and the structure of ecological communities. *Annual Review of Ecology and Systematics*, **22**, 115–43.

Stebbins, G. L. (1974) *Flowering Plants: Evolution Above the Species Level*. Massachusetts: The Belknap Press of Harvard University Press.

Steffen, W. L., Walker, B. H., Ingram, J. S. I. & Koch, G. W. (eds) (1992) *Global Change and Terrestrial Ecosystems: the Operational Plan*. Stockholm: IGBP and ICSU.

Swaine, M. D. & Whitmore, T. C. (1988) On the definition of ecological species groups in tropical rain forests. *Vegetatio*, **75**, 81–6.

Szaro, R. C. (1986) Guild management: an evaluation of avian guilds. *Environmental Management*, **10**, 681–8.

Valk, A. G. van der (1981) Succession in wetlands: a Gleasonian approach. *Ecology*, **62**, 688–96.

Verner, J. (1984) The guild concept applied to management of bird populations. *Environmental Management*, **8**, 1–14.

Walker, B. H. (1992) Biodiversity and ecological redundancy. *Conservation Biology*, **6**, 18–23.

Wilson, J. B. (1989) Mechanisms for species co-existence: twelve explanations for Hutchinson's 'paradox of the plankton': evidence from New Zealand plant communities. *New Zealand Journal of Ecology*, **13**, 17–42.

Winemiller, K. O. & Pianka, E. R. (1990) Organisation in natural assemblages of desert lizards and tropical fishes. *Ecological Monographs*, **60**, 27–55.

Yodzis, P. (1982) The compartmentation of real and assembled ecosystems. *American Naturalist*, **120**, 551–70.

❷ Plant and ecosystem functional types

H. H. Shugart

Introduction

It is an essential aspect of science to simplify the systems under consideration. In experiments, the usual intent is to control, to as great a degree as possible, all extraneous factors with respect to a given experimental objective. In developing models of dynamic systems, the underlying assumptions are often complex hypotheses involving which factors can be left out of one's accounting of system dynamics. It may be true that, as Francis Thompson related, 'thou canst not stir a flower without troubling a star,' but, in computing the motions of stars and planets, the effects of flowers do not loom large. It is the disregarding of the effects of flowers on stars that allows progress in astronomy. The regarding of the reciprocal (the effects of the motions of stars on flowers and other things) is the grist of astrology. Appropriate abstraction is critical to progress in science.

Ecologists deal with systems of great complexity, and the science is strongly oriented towards description of detail and the production of counter-examples to generalities. Ecological studies are often at relatively small spatial scales. Karieva & Anderson (1988) found that 95% of a recent survey of papers in the ecological literature were conducted on study plots of less than 1 ha. Half of these studies used plots of 1 m^2 or smaller. Given that the focus in our field is towards description often at relatively fine scale, the synthesis of this detail toward abstractions aimed at depicting larger-scale responses of ecological systems will regularly be thwarted by cases that do not fit. It is important to appreciate when counter-examples need consideration and when they should be ignored with respect to the goals at hand.

This chapter will focus on species and ecosystem functional types, drawing from examples in forests. In this chapter, 'plant functional types' will be used to connote species or groups of species that have similar responses to a suite of environmental conditions. Related or analogous terms include: life-forms (Raunkiaer 1934), guilds (see Simberloff & Dayan 1991), plant forms (Box 1981), plant strategies (Grime 1974,1977; Tilman

1988) and temperament (Oldeman & van Dijk (1991). 'Ecosystem functional types' are aggregated components of ecosystems whose interactions with one another and with the environment produce differences in patterns of ecosystem structure and dynamics. Because ecosystems are abstractions, these functional types are also abstractions. The distinction between plant and ecosystem functional types is largely in the emphasis on plant adaptations in the former and on ecosystem responses in the latter. A group of plants could be in the same ecosystem functional type but different plant functional types. Also, different plant functional types could be equivalent ecosystem functional types but in different ecosystems.

Functional types and diversity/stability theory

The development of ecosystem functional types has a direct relationship with a set of important and still largely unresolved theoretical questions in ecology from the 1960s and 1970s. Questions involving the effects on ecosystem behaviour of eliminating species ('the relationship between diversity and stability') often had at their root an underlying assumption that, if species' function contributed to the dynamical behaviour of an ecosystem and if enough species were eliminated, then the functioning of the ecosystem would be disrupted to some degree. In some explanations, sets of species were thought of as having parallel functions. Thus, the presence of several species provided a redundancy that might allow ecosystems to maintain their function even in the face of species extinctions. This latter view inspired information-theory-based indices of diversity and has at its basis a view of aggregation of species into functional types.

Interest in relating diversity and stability often stumbled over which of the two relatively loosely defined attributes were causal on the other. Was diversity a consequence of stability, so that relatively unchanged systems over time evolved diverse assemblages of plants and animals? Was diversity itself a stabilizing factor, allowing diverse systems to harbour species of great antiquity that indicated long-term stability? Did diversity and stability reinforce one another so that the disruption to one might harm the other? The influential aspect of this time in ecology was a rich development of theory about the manner in which ecosystems were structured. It is far from appropriate to criticize these antecedent developments here. It is important to realize that the issues that were treated in the diversity and stability discussions are to some degree revisited in the discussions of ecosystem functional types.

What additional insights do we bring to the problem that we did not have 20 years ago? Perhaps, none. Perhaps, the retrospection that comes with time. Perhaps, the increased understanding of the role of disturbance in ecosystem dynamics; the large-area characterizations of ecosystems in past environments from palaeoecological studies; the relative improvement in our ability to model complex systems; increased computational power; the rich recent experience in understanding the physiological ecology of important species: all may contribute to a better understanding. One development that should prove useful is a better appreciation of ecological scale.

Scale and importance of processes

The temporal and spatial scale at which one chooses to consider a dynamic system has a great degree of influence on which environmental factors will appear important and on which state variables of the system will respond. That scale is an integral part of the process of defining an ecosystem for a given purpose seems apparent in the initial definition of the term 'ecosystem'. Tansley (1935), in the sentence that defined and first used this term, declared that ecosystems were 'of the most various kinds and sizes' and sounded a theme that has recently been reiterated in modern ecology as hierarchy theory (Allen & Starr 1982; Allen & Hoekstra 1984; O'Neill *et al.* 1986). It is my own inclination in considering ecosystem functional types to be concerned with relatively long timescales (usually longer than one year and typically several generations for dynamic responses) and landscape or larger spatial scales (see Smith *et al.* 1993). Regardless of the particular scale of one's application, it is important to realize that otherwise equivalent ecosystems inspected at different temporal and spatial scales are likely to have different control and response variables.

Historical considerations

Box (1981) noted that the interest in relating vegetation structure with climate and other environmental factors that was a hallmark of early plant ecologists and plant geographers gave way in the past decade or two to the sharpened interest in plant demography. The rich insights of the early ecologists should not be overlooked in developing functional types for large-scale interrelating of vegetation and climate. One of the challenges in developing plant functional types is interweaving the physiognomic adaptations emphasized in the early systems with population and community dynamics. This interweaving seems essential in any functional type categories that can deal with dynamic changes.

Approaches, including *r*, *K* and adversity strategies (MacArthur & Wilson 1967; Southwood 1977; Greenslade 1983) early and late-successional species (Budowski 1965, 1970; Whittaker 1975; Bazzaz 1979; Finegan 1984), exploitative and conservative species (Bormann & Likens 1979a,b), ruderal, stress-tolerant and competitive strategies (Grime 1977, 1979), gap and non-gap species (Hartshorn 1978; Brokaw 1985a,b), structural classifications (Raunkiaer 1934; Hallé 1974; Hallé & Oldeman 1975; Webb *et al.* 1970; Walker *et al.* 1981), and vital attributes (Noble & Slatyer 1980), are based on the constraints imposed on plants by different environmental conditions, such as resource availability or disturbance regime. After classifying environmental conditions, most schemes categorize plant strategies based on responses to those environmental conditions and the associated life-history characteristics, either explicitly or implicitly considering the consequences of resource allocation.

Form and function, pattern and process

Biology has a central tenet: that the form or shape of entities is both modified by and a creator of function. The three-dimensional shape of a protein determines its function as an enzyme; the length of a bird's first primary feather indicates the lift the wing will generate; the geometry of a flower restricts the potential pollinators and increases the efficiency of cross-pollination; grazers with chambered stomachs can shelter symbionts that digest cellulose. The relationship between form and function, or pattern and process, is also one of the classic ecological themes. Lindeman (1942) laid a basis to relate the functional efficiency of ecological energetics to the structure of Eltonian trophic pyramids. Watt (1947) and Whittaker & Levin (1977) associated ecosystem pattern with generating processes. Bormann & Likens (1979a,b) pointed out the effects of changes in forest structure on processes such as productivity and nutrient cycling. Many ecologists recognize that pattern and process are mutually causal, with changes in ecosystem processes causing changes in pattern, and modifications in ecosystem pattern changing processes.

Form and function are interwoven to the degree that each can be thought of as causing the other. It is difficult to investigate directly the feedback relation between pattern and process. Such feedback can potentially produce a rich array of system behaviour, which is difficult to assess without a coordination between theory and experimentation. In the development of functional types, it is important to realize the mutual cause between form and function.

Effect of function on form

The clearest cases of the effect of function on the form of plants are in morphological adaptations to environmental conditions. The shape of various organs is often characterized with respect to the degree with which they adapt the plant to survive and prosper in a particular environmental regime. Harper (1982) sounded a cautionary note regarding such discourses in discussing what he called 'eureka ecology': the ascribing of morphological (or other) similarities among plants in a particular environment as examples of parallel adaptations to the environment and the ascribing of dissimilarities as examples of niche differentiation. Beyond Harper's caution, it is clear that rather subtle differences in the interpretation of the function of plant tissues can lead to different interpretations of appropriate morphologies.

For example, in two classic papers, Parkhurst & Loucks (1972) and Givnish & Vermelj (1976) attempted to derive the optimal shape of a leaf under the assumption that leaf shape adaptation to environmental conditions could be considered the consequence of optimizing the carbon, water and heat budgets of the leaves. In the most general sense each of these studies grappled with the interaction among leaf size and the heat balance equations, the evaporative flux equations (which are strongly tied to the heat equations through the evaporation terms), and resistance-based CO_2 flux equations. Parkhurst & Loucks (1972) assumed that leaves under natural selection would maximize their water-use efficiency. In their study, Givnish & Vermelj (1976) made an alternative assumption that leaves optimized the pay-off of having leaves of a given size vs. the costs of maintaining leaves of a given size. Either assumption appears reasonable. Each study produced a diagram of optimal leaf size and environment, and these differed in their predictions of leaf size in low-light, high-moisture conditions (Fig. 2.1a,b).

This example case is illustrative of the sorts of problems that may arise from applying classic optimization techniques to develop a theoretical basis for relating form to function. In many cases, it is not clear what natural selection is optimizing. There are several obvious choices ranging from optimizing (or maximizing) survival, fitness, efficacious use of energy, water or nutrients, occupancy of space, etc. Further, it is possible that two sympatric species functioning in the same environment may, owing to their past evolutionary and biogeographical history, be optimized through natural selection towards different goals.

It is clear that studies of optimal form with respect to the functioning of a plant in a particular environment provide considerable insight into what ranges of adaptations are possible and the manner in which particular plants perform.

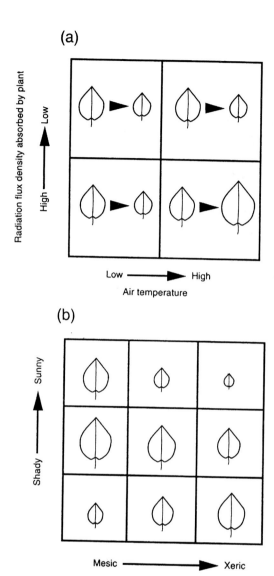

Figure 2.1 *The optimal sizes of leaves in different environmental conditions derived by assuming that leaves are in some sense optimized to the environment. (a) Parkhurst & Loucks' (1972) prediction of changes in leaf shape under natural selection. Leaf shape changes are based on the assumption that leaves tend to optimize their water-use efficiency. (b) Givnish & Vermelj's (1976) prediction of leaf shape based on the assumption that leaves tend to optimize the difference between gains associated with leaf size and costs (e.g., increased respiration, need for structure to support large leaves) associated with leaf size. While the axes of the two diagrams are different, they clearly illustrate that reasonable assumptions about the optimization goal can produce rather different expected outcomes.*

Effect of form on function

The geometry of plants can have a profound effect on the manner in which they function. Figure 2.2 illustrates typical responses of net photosynthesis of plants to changes in light level. Bazzaz (1979) illustrated the tendency for plants found in early succession to have high rates of carbon fixation at high light intensities and lower rates of carbon fixation at low light intensities, relative to late-successional plants (Fig. 2.2a). This pattern, manifested at the leaf tissue level, has been implicated as a mechanism that gives plants different levels of shade tolerance and thus can be used to explain patterns in the successional patterns of ecological systems (e.g. Bazzaz 1979; Bazzaz & Pickett 1980; Kozlowski *et al.* 1991).

Horn (1971) investigated trees that had equivalent tissue-level responses to changes in light intensity but different geometries with respect to leaf layering, and produced the whole-plant net photosynthesis curves illustrated in Figure 2.2b. These curves demonstrate the same pattern as summarized by Bazzaz (Fig. 2.2a). Plants differ in their ability to utilize light, owing to their geometry (Horn 1971) as well as their physiology (Bazzaz 1979). It seems likely that combinations of physiology and geometry might amplify the effect of an apparent trade-off in the ability of a plant to use light at different intensities.

Oldeman and van Dijk (1991) developed a scheme for tropical trees,

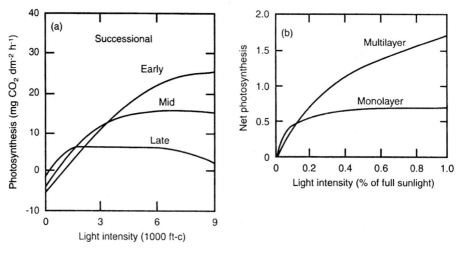

Figure 2.2   *Typical responses of net photosynthesis in plants as a function of light intensity. Note that at the tissue and canopy levels, there is a reversal of photosynthetic efficiency at high vs. low light intensities. (a) Idealized light saturation curves for early-, mid- and late-successional plants (from Bazzaz 1979). (b) The effect of light on net photosynthesis in multilayered and monolayered trees (from Horn 1971).*

based on combinations of three size categories and two responses to light intensity, to develop tree temperaments: 'Gamblers', having many of the attributes associated with early-successional plants in the sense of Bazzaz (1979), such as fast growth rates and intolerance to shading; 'Strugglers', having slower growth rates and relatively greater shade tolerance. These two strategies with respect to light were combined with three life stages of the tree (seedling, pole and mature) to produce a categorization of tree strategies depending on whether the tree was a 'gambler' when small, a 'struggler' when somewhat larger, etc. (Fig. 2.3). Although Oldeman & van Dijk (1991) did not explicitly draw on Horn's multilayer–monolayer dichotomy of tree geometry (Fig. 2.2b) to ascribe response curves for the strategies they developed, the functional consequences of form are evident in the development of the scheme.

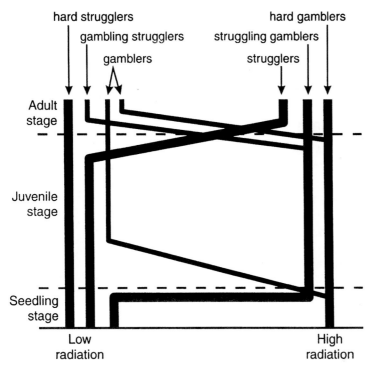

Figure 2.3 *Six tree 'temperaments' according to Oldeman & van Dijk (1991). 'Hard gamblers' and 'hard strugglers' remain constant in shade tolerance through their entire lives. Other 'temperaments' change in their shade tolerance at some stage in the life cycle. Strugglers correspond to classic 'shade-tolerant trees' or 'climax species'. Hard gamblers are the classic 'light-demanding trees' or 'pioneer species'. The relative frequency of a temperament in tropical rainforest is indicated by the thickness of the bar.*

A simple system of plant functional types for forests

To illustrate how one might formulate a system of ecosystem functional types, it is useful to develop a set of ordering rules that can be used to develop plant functional types for a particular ecosystem. This example applies in environments in which the vegetation response would favour trees, and within the part of Grime's triangle (Grime 1977, 1979) in which one would expect the interactions to be controlled by competition for space. The example is in the domain of gap dynamics in forests. Considering the development of a minimum number of categories of plant functional types with respect to gap competition in trees, one can divide tree species into:

1. species requiring a canopy gap for successful regeneration or otherwise;
2. species typically generating a gap with the death of a mature individual of typical size or otherwise.

As a product of this pair of dichotomies, there are implied four resultant strategies, categorized as species roles (Fig. 2.4). Life-history traits associated with each role are intuitive adaptive syndromes: gap-forming trees are large, whereas species that do not create gaps are typically smaller; gap-

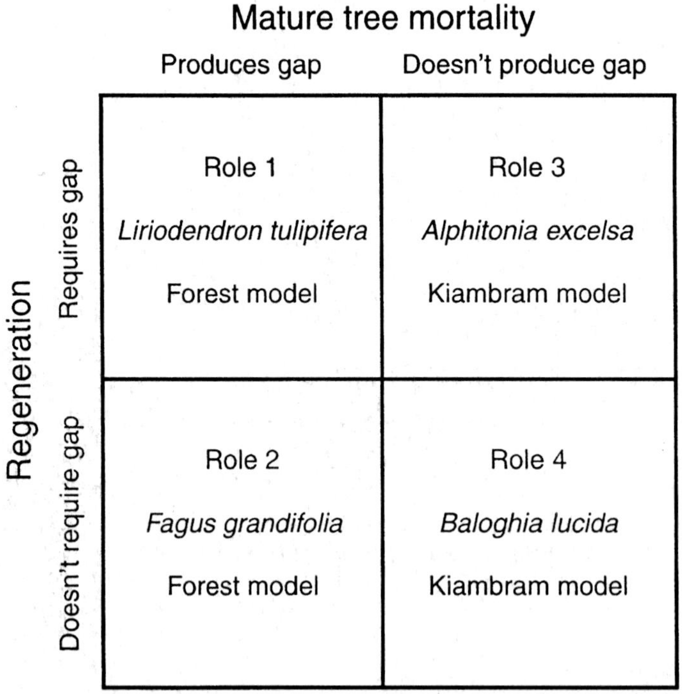

Mature tree mortality

	Produces gap	Doesn't produce gap
Requires gap	Role 1 *Liriodendron tulipifera* Forest model	Role 3 *Alphitonia excelsa* Kiambram model
Doesn't require gap	Role 2 *Fagus grandifolia* Forest model	Role 4 *Baloghia lucida* Kiambram model

(left axis label: **Regeneration**)

Figure 2.4   *The role of gap-requiring and gap-producing tree life-history traits with example species for each of the four roles. See Fig. 2.6 for dynamics of forests consisting solely of each of the example species (from Shugart 1984).*

regenerating species are shade-intolerant, whereas species that do not
require gaps are shade-tolerant (Shugart 1984, 1987).

This classification is intentionally simplistic, and it could be elaborated
by subdividing the four roles or by considering other criteria or resources.
For example, Grubb's (1977) regeneration niches consider one axis of this
dichotomy, regeneration, in detail. As another example, several authors have
further divided the 'needs gap' axis to account for subtle differences in
species' shade tolerance (e.g. gap-size partitioning by tropical trees):
(Denslow 1980; Brokaw 1985a,b). Yet even this two-by-two categorization
can produce a rich array of forest gap dynamics.

Some qualitative features follow from interactions among the four roles
(Fig. 2.5). Two of the roles are self-reinforcing: role 1 species can create
the gaps they need to regenerate; role 4 species can regenerate in the shade
and do not drastically open the canopy when they die. Some combinations

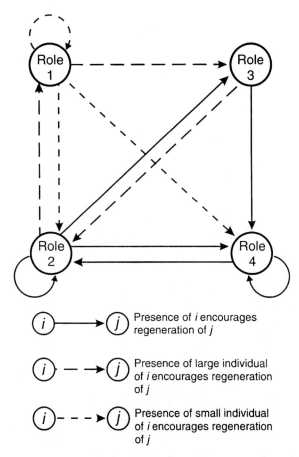

Figure 2.5   *The directions of positive or strongly positive interactions*
among roles of tree species with respect to their influence on regeneration
(from Shugart 1984).

of roles can replace each other reciprocally: role 1 species can take over a gap created by role 2, and vice versa. Other roles tend to give away the space they occupy when they die: role 3 species cannot create the conditions they need to regenerate, but instead favour roles 2 and 4 in the subsequent cohort of trees. Of course, if trees of the larger roles (1 and 2) die at a smaller size, they influence the ecosystem dynamics in the same ways as their smaller counterparts (roles 3 and 4, respectively). A role 1 tree that dies when small is equivalent to a role 3 species, and hence does no favour to its progeny in its early death.

These qualitative patterns imply a set of rules governing patterns of gap dynamics (and species replacement). Like any set of rules, these may be followed strictly, loosely, or not at all. In the case of forests patterned by gap dynamics (i.e. by competition for and temporary occupancy of space by individual trees), qualitatively different modes of competition may occur. Trees may compete within the rules implied by the foregoing discussion. Under these rules, role 3 is a loser in local competition involving gap dynamics, but can persist by not playing by this set of rules and competing in a different, larger arena. This is the case in fugitive (ruderal) species, which concede gap-scale competitive ability in order to persist regionally (Marks 1974; Urban *et al.* 1987). Finally, trees have a sufficiently large local effect on the environment that species can alter the rules that govern competitive outcomes. Certain fire-adapted trees (often role 3 species) may alter mortality by interacting with fires in ways that take the advantage away from longer-lived (larger) roles 1 and 2 (Heinselman 1981; Foster & King 1986). In general, frequent disturbances can shift the competitive advantage to role 3 in many forests. Environmental factors can also change the rules of competition. At high latitudes, low temperatures, a short growing season, and low sun angles create a condition under which gap formation by the death of a single tree does not occur. This rules out roles 1 and 2 as possible strategies. For this reason, upper-latitude boreal forests should be dominated by roles 3 and 4.

These simple notions of functional roles of trees have clear implications in interpreting forest dynamics. The basic premise is that life-history traits effectively couple demographic processes of mortality and regeneration. Morphological and physiological consideration impose a correlation structure among life-history traits, such that only certain combinations of traits may be realized in any single species: shade-tolerant species tend to be intolerant of moisture stress; long-lived trees tend to grow more slowly than short-lived species; and so on (Huston & Smith 1987). Thus, however functional roles might be defined, there is a limited number of such roles, and these will interact to imply a limited set of rules constraining forest dynamics. These roles, and the implications of their interactions, can be conveniently explored by implementing them in simulation models.

Roles of trees and forest temporal dynamics

Having established a simple categorization of species functional types, this chapter will elaborate on the interactions among these types and with the environment to provide insight into the development of ecosystem functional types: categories of plants that help us to understand ecosystem dynamics.

The different roles of trees produce essentially different biomass and numbers dynamics when monospecific stands are simulated with small model plots (*ca*. 0.1 ha). The long-term behaviour of numbers and biomass for example species are shown in Fig. 2.6. Each of the roles has a fundamentally different signature of numbers and biomass dynamics in monospecific stands. The dynamics of a role 1 stand feature explosive increases in biomass and numbers following the death of a large canopy tree, and an even-aged character to the stand, reflecting the tendency for episodic recruitment following the death of a large tree. Role 2-dominated stands are mixed-age and mixed-size structured even at small scales (e.g. see Fig. 2.6b), with episodes of enhanced recruitment associated with drops in biomass from large tree mortality. Role 3 dynamics feature episodes of recruitment that are preceded by the senescence of the previous cohort and an associated drop in biomass. Role 4 dynamics are of mixed-aged and mixed-sized forest with little variance in either numbers or biomass dynamics (Fig. 2.6d), except when simultaneous death of several trees produces a transient response (as occurs at around year 1200 in Fig. 2.6d).

Although these modes of stand dynamics are intended as examples of functional types, there are forests that display similar kinetics. For example, Shugart (1987) found the role 3 dynamics of numbers and mass to be in agreement with the dynamics of *Pinus sylvestris* stands at high latitude reported by Zyabchenko (1982). It is clear from Fig. 2.6 that the numbers and mass dynamics of forests composed of only one role are different one from another. Through the gap creation–gap colonization process, each of these cases represents a different temporal pattern of environmental variables that co-vary locally with the gap processes.

Factors affecting differential abundance can be coupled to the environment or to one another. In the model experiments that have been conducted by creating forests with more than one species-role, the interactions among the species of different roles can produce forests with a variety of successional pathways. Huston and Smith (1987), in a series of

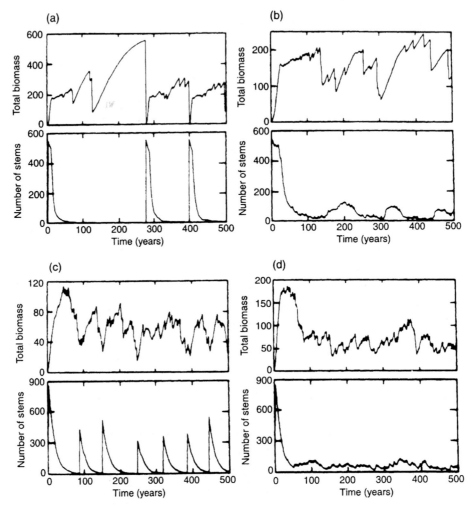

Figure 2.6 *Number and biomass dynamics of hypothetical patches in forests of a single species that conform to the roles shown in Fig. 2.4 (from Shugart 1984). (a) Role 1:* Liriodendron tulipifera *on a 1/12 ha plot as simulated by the FORET model. (b) Role 2:* Fagus grandifolia *on a 1/12 ha plot as simulated by the FORET model. (c) Role 3:* Alphitonia excelsa *on a 1/30 ha plot as simulated by the KIAMBRAM model. (d) Role 4:* Baloghia lucida *on a 1/30 ha plot as simulated by the KIAMBRAM model.*

model experiments, used the ZELIG model to allow species with a spectrum of life-history attributes to compete against one another. They found that the different combinations of species (and the life-history strategies these species represented) could produce a rich array of successional pathways. The relative success of species with different roles depended on environmental context. A species with a given set of life-history attributes might be judged successful under one set of environmental conditions and competitors, but a loser under different conditions. The logical trade-offs in life-history and physiological features of the species render impossible the emergence of a set of attributes that is always successful.

Tilman (1982, 1988) has used biological trade-offs as a basis for understanding the diversity of plant species with regard to the division of resource axes. It is refreshing that his results, based on models that seem appropriate to populations that are well-mixed and without significant size structure, are in qualitative agreement with those obtained with individual-based models that are not subject to these restrictions.

One feature that emerges from model experiments is the potential importance of multi-level competition in forests. The species roles can produce patterns of inter-role interactions are that are asymmetric, competitive or mutualistic, even though the fundamental individual-level interactions are based on competition for space and resources (Shugart 1984). Species can compete within the context of a set of rules. For example, role 3 species may be extremely competitive for safe sites in forest gaps given a particular forest condition. Species of different roles can in some cases compete to determine the rules that apply in the forest. For example, a successful role 1 species would tend to reinforce gap dynamics in a given forest, whereas a successful role 4 species would reduce the importance of gap regeneration in the same forest. Competition of the first sort (competition within the rules) is the usual case for competitive interactions and one would expect species that were the most similar to be most competitive. In the second sort of competition (competition to change the rules), one would expect species that are the most dissimilar to be the most competitive. This type of competition is more difficult to study under field conditions. Model experiments involving these sorts of interactions have produced ecosystem dynamics featuring multiple stable states and hysteresis (Shugart *et al.* 1980). In these instances, the dynamics associated with one role dominate and install a self-reinforcing mode of overall system behaviour at one extreme of an environmental gradient; a second role similarly controls the system behaviour at the other extreme of the gradient. In the intermediate range of the gradient, the ecosystem has two quasi-stable modes of behaviour. In this range the state of the system is a consequence of the prior history of conditions to which the ecosystem has been exposed (Shugart *et al.* 1980).

Ecosystem functional types

If species 'strategies' emerge from an interplay of form and function at the individual-plant level, they are part of a spectrum of form–function, pattern–process interactions in a nested hierarchy of system rules that can be expressed as theories. In practice, such theories are constrained (either explicitly, or implicitly by the domain of their successful applications) to particular spatial and temporal scales. One theory can 'work' in producing successful predictions at one scale, while another theory, with different fundamental assumptions, can also 'work', albeit at a different scale. This latter feature is consistent with a hierarchical concept of vegetation pattern. For example, Grime's triangle of plant response strategies (Grime 1977, 1979) is a set of rules that can be used to predict a pattern of strategies (and associated life-forms) expected under a particular environmental regime. There are other, relatively higher-level ordering rules that could be used to determine the fundamental properties of the vegetation that attend the environmental conditions at larger spatial scales (e.g. Holdridge 1967; Box 1981; Woodward 1987).

Features of ecosystem functional types

That the interaction between two functional types could transform a forest system from one with a single stable state to one with a multiplicity of stable states can be taken as a basis for ecosystem functional types. A related basis is the concept that plants may compete with one another either within a relatively fixed set of rules or by rules altered through feedbacks onto the environment. Both bases emphasize interactions of ecosystem functional types as altering ecosystem function and ecosystem–environment interactions.

To provide a simple case of ecosystem functional types, consider the example plant functional types discussed above. The four roles of trees (Shugart 1984) can also be used as examples of ecosystem functional types. The dimensions of ecosystem functional types involve alterations of ecosystem dynamics. There are three domains of ecosystem responses: (1) the time domain; (2) the frequency domain; (3) the spatial domain. Each of these cases will be discussed using the example plant functional types (and their interactions) for illustration.

The time domain: fast versus slow responders

Perhaps the most straightforward way in which species can alter the dynamic response of a ecosystem is by changing the rate at which important processes occur. Species with rapid growth rates can shorten the time needed for a forest to recover a closed canopy following a disturbance. An example of a possible change in the eastern deciduous forest of North America might be the demise of American chestnut (*Castanea dentata*) following the introduction of the fungal disease chestnut blight (*Endothia parasitica*). American chestnut once accounted for as much as 50% of the biomass in some widespread forest types and was eliminated as a canopy tree during the early part of the twentieth century. This species had, among other attributes, leaves with high decomposition rates compared with the oak species (notably white oak, *Quercus alba*, and chestnut oak, *Q. prinus*) that have tended to replace it. One would expect that the demise of American chestnut would result in forests with a slower rate of decomposition and an attendant slowing down of the rate of natural recycling of essential plant nutrients. American chestnut also sustained a large wildlife population that fed on its regularly produced, sweet seeds. The species that have replaced American chestnut either have a tendency to mast (irregular seed production) or to have tannin-laden acorns. Thus, the elimination of this one dominant species, even though it was replaced by other species, may have resulted in a forest that was different from that before with respect to element cycling rates and seed predation rates.

The frequency domain: attenuators versus amplifiers

There is a direct relationship between the temporal response of a dynamic system and the time responses of the system. Systems with rapidly responding components are affected by relatively high frequency variation that is attenuated by systems dominated by components with slower responses. Forests have a considerable number of internal feedbacks among processes and should be expected to attenuate certain frequencies of disturbances or climate cycles, and also to amplify other frequencies. Based on experiments with simulation models, Shugart & Urban (1989) found that forests could be expected to amplify variations in the range of 100–200 year periodicities. Such amplifications would be manifested as cohort dominance in mature forests and periodic synchronous diebacks in all forests. These features have been reported for a variety of natural forests.

Emanuel *et al.* (1978) conducted a series of experiments using simulation models in which they inspected the frequency response (the response of the dynamics of the system to inputs of different frequencies) with the addition and removal of certain species and under conditions of simulated stress. The results showed that certain species, as a consequence of their life-history strategies in interaction with the environment, have a tendency to amplify certain periodicities, whereas other species have the opposite tendency and attenuate periodic inputs.

Amplifiers were often what were categorized as role 1 species (species that require and generate canopy gaps) and tended to form forests that locally were even-aged. Attenuators were role 2 and role 4 species and tended to form forests with a J-shaped diameter frequency distribution (even at relatively small spatial scales).

Although these results are based on model experiments and involve inspections of actual data that would require hundreds of years of detailed observations, there is considerable evidence for cohort-dominance in role 1 trees in a variety of forests. A second result of the model investigations of forest responses in the frequency domain was that model forests under stress conditions tended to display a reduction in the richness of the frequency response, and this has been demonstrated experimentally in microcosm-based studies (Van Voris *et al.* 1980).

The spatial domain: homogenizers versus heterogenizers

The earlier discussions of species' roles with respect to gap replacement indicated that two of the roles (1 and 4) were self-reinforcing. Role 1 species require and generate canopy gaps; role 4 species are the logical converse. Each of the roles competes with the other by changing the spatial grain of the forest's light environment. A plenitude of role 1 individuals favours a forest with a high degree of variation in the spatial pattern of light on the forest floor. An abundance of role 4 individuals would tend to result in forests with a greater degree of homogeneity.

Because the competitiveness of an individual tree is strongly related to its size, the interactions among trees are complex. Size is not typically a state variable used in developing competition theory in animal populations, so that comments regarding tree competition in relation to traditional competition theory are a special case. Competition among trees at the scale of a small patch involves two sorts of interactions. First, trees can compete with one another with a right to determine the 'rules of the game'. For example, role 1 and role 3 species prosper when the typical tree in the forest is large. However, if role 3 species are successful in their competition for sites with role 1 species, they tend to install a small-canopy forest, which in turn disadvantages the role 1 species. Thus we have a situation in which the presence of a role 1 species is advantageous to a role 3 species, but the presence of a role 3 species is disadvantageous to role 1 species. These relationships, if represented by a zoologist, would resemble predation in some cases, parasitism in others, and mutualism in still others (Fig. 2.5).

The interactions among the roles of trees include several that are competitive between roles. Some interactions are mutualistic in that roles tend to help one another. The important point is that interactions among plants are as complex as the interactions among entire food webs of animals competing with and feeding upon one another. In this case, rather than competing for resources such as food energy, the trees compete for space. Presumably, within a role, competition would occur much as it does in animal communities. That is, role 1 species would compete strongly against other role 1 species for the opportunity to be the dominant role 1 species in the forest. It is not clear at this point whether competition among trees within roles is as strong as the variety of interactions across roles. It is clear that species that can change the spatial grain of the ecosystem to either more or less heterogeneous could potentially manifest an extremely large effect on ecosystem function.

The environment and functional types

Forest ecosystems and trees will be used as examples to illustrate some of the considerations that might arise in developing functional-type categorizations for different systems. Note that the environment acts to constrain the expression of these interactions to a great degree. The relative similarity of vegetation under similar climates is *prima facie* evidence for such control. The development of improved global predictions of vegetation type from climate could be achieved by incorporating more detailed biological information about the vegetation using process-based (biophysical) models that describe the envelope within which the vegetation is constrained under various environmental conditions. For example, a general photosynthesis and growth model (Woodward 1993) has been applied for global prediction of vegetation in response to climate variables. Similar biophysical models have been used to couple surface processes to atmospheric models (Dickinson 1984; Jarvis & McNaughton 1986; Martin 1993). One challenge in the global-scale application of these models is the estimation of plant-process-related coefficients over a large range of conditions for a wide variety of different plant species. These coefficients (for example, those which define the influence of the environment on stomatal conductance) can be considered a dimension or characteristic of a classification of plant functional types. Such coefficients are *intensive* in that they incorporate detailed and precise information about a single process (Smith *et al.* 1993). For the model described by Woodward (1993), a physiologically based model of global leaf area was enriched with information including the length of the growing season (for tundra) and the lowest annual temperature (for temperate deciduous forest). Smith *et al.* (1993) saw the determination of such vegetation properties as a conceptual basis for the development of descriptors or dimensions of plant functional types. They divided these descriptors into two types: intensive and extensive.

Intensive descriptors incorporate detailed and precise information about a single process. In the Woodward (1993) model of global leaf area (see also Smith *et al.* 1993), the intensive characteristics are as follows.

1. Leaf longevity.
2. Threshold temperature for low-temperature mortality.

3. Threshold temperature for growth.
4. Daylength for growth.
5. Seed longevity.
6. Photosynthate allocation to leaves, stems and roots.
7. Environmental coefficients for stomatal conductance.
8. Environmental coefficients for mesophyll conductance.
9. Environmental coefficients for leaf growth.
10. Environmental coefficients for respiration.

These intensive functional-type dimensions are contrasted by *extensive* functional-type characteristics. In this latter case, the characteristic is not a model parameter but is a descriptor that implies extensive properties. Such a descriptor is the term *arboreal*. Plant types that are arboreal have woody stems, are long-lived and tall and, when mature, can control the environmental responses of subordinates. The arboreal descriptor may be divided into trees or shrubs and then further subdivided into angiosperm or gymnosperm. A gymnosperm tree further implies a tracheidal xylem, (generally) narrow or needle leaves, low stomatal and mesophyll conductance and a low wood density. In some cases, at least, the extensive descriptors imply estimates of intensive coefficients (e.g. photosynthetic maxima).

Important descriptors for the development of a extensive descriptor-based classification of functional types (from Smith *et al.* 1993) are:

1. Physiognomy.
2. Desiccation features.
3. Life span.
4. Pollination.
5. Seed dispersal.
6. Photosynthetic pathway.
7. Shade tolerance.
8. Fire tolerance.
9. Nutrient tolerance.

Extensive dimensions for developing functional types are primarily, but not entirely, related to either life cycle or structural and morphological properties. In this sense, these dimensions are like those at the basis of several of the classic classifications of global vegetation.

Intensive and extensive dimensions of classifications of plant functional types imply a knowledge of the underlying factors that determine the envelope within which plants operate to form vegetation structure. One sees this explicitly in Box's (1981) 'eco-sieve' approach of matching types of plants against the range of environmental variables to determine which plant types are possible elements of the vegetation in a given environment.

In the Box (1981) example, the range of a given type of plant along an environmental variable appears to have been fitted to the current distributions of the plant types in some cases. The Box (1981) system could be greatly enriched using basic information on the physiological constraints in plants. Problems in the approach of interpreting adaptations according to the costs and benefits associated with a given set of environmental conditions were mentioned above. None the less, this basic approach has proven valuable in understanding the consequences for leaf size and shape (Parkhurst & Loucks 1972; Givnish 1978, 1979), leaf type in arid environments (Orians & Solbrig 1977), plant height (Givnish 1982; Chazdon 1986; Friend 1993), photosynthesis (Mooney & Gulmon 1979; Cowan 1986), the use of multiple resources (Chapin *et al.* 1987), herbivore defence and nutrient use (Mooney & Gulmon 1982; Bryant *et al.* 1983); and the effect of light response and life history on plant succession (Huston & Smith 1987).

Conclusions

As was discussed at the beginning of this chapter, the relationship between form and function (or pattern and process) is a venerable concept in biology. The interactions of plants, during the formation of vegetation, to alter the structure of the vegetation canopy (its height, depth, layering and spatial heterogeneity) represent a focus for the development of functional types. Simberloff & Dayan (1991) have recently reviewed the 'guild concept' in the structuring ecological communities and note that the concept has a wide range of uses in ecology. They further note that the guild concept was specifically meant to relate to species using the same resources in similar ways and that the concept is most elaborated with respect to animal communities. In discussing guilds in plants, they state, 'Also, perhaps because researchers are animals, differences in the ways that plants use resources do not seem as obvious as they do for animals.'

The basis for ecosystem functional types discussed in this chapter arises from the manner in which plants modify their environment, and hence the ecological systems in which they are found, by changing the temporal, spectral or spatial fabric of the system response. The basis for plant functional types is in the manner in which the environment interacts with the plants to produce consistent patterns of physiological, morphological and life-history response in the plant species. In the earlier sections a simple plant functional type categorization (the four roles of trees) was used in discussing different ecosystem functional types (attenuators, amplifiers, etc.) that were associated with these plant functional types.

The relationship between plant and ecosystem functional types may not be a one-to-one mapping. Species that are quite different at the plant functional type level might alter ecosystem response in similar ways and thus be considered similar ecosystem functional types. Strong reinforcing feedback between alterations in the environment caused by the success of a group of species of the same ecosystem functional type that also favours the plant functional types found in the group represents an ordering principle for ecosystems. The implication is that not only do species get sorted by the environment (e.g. Box 1981) but that ecosystems, in a sense, also sort their components. Such a principle could explain not only why the structure and life history of plants with different floristic origins are often convergent at the species level in different environments, but also why the ecosystems themselves appear similar.

Acknowledgements

The author thanks W. R. Emanuel for long discussions leading to the idea of ecosystem functional types: T. M. Smith, F. I. Woodward, P. Martin, A. Friend and W. Cramer for discussions about functional types while at the International Institute of Applied Systems Analysis in Laxenburg, Austria. Finally, the support of Brian Walker and Will Steffen in discussing ideas connected with this project is greatly appreciated.

References

Allen, T. F. H. & Starr, T. B. (1982) *Hierarchy: Perspectives for Ecological Complexity.* Chicago, Illinois: University of Chicago Press.

Allen, T. F. H. & Hoekstra, T. W. (1984) Nested and non-nested hierarchies: a significant distinction for ecological systems. In Smith, A. W. (ed.), *Proceedings of the Society for General Systems Research. I. Systems Methodologies and Isomorphies,* pp. 175–80. Lewiston, New York: Intersystems Publications, Coutts Library Service.

Bazzaz, F. A. (1979) The physiological ecology of plant succession. *Annual Review of Ecology and Systematics,* **10**, 351–71.

Bazzaz, F. A. & Pickett, S. T. A. (1980) Physiological ecology of tropical succession:

A comparative review. *Annual Review of Ecology and Systematics,* **11**, 287–310.

Bormann, F. H. & Likens, G. E. (1979a) *Pattern and Process in a Forested Ecosystem.* New York: Springer-Verlag.

Bormann, F. H. & Likens, G. E. (1979b) Catastrophic disturbance and the steady state in northern hardwood forests. *American Scientist,* **67**, 660–9.

Box, E. O. (1981) *Macroclimate and Plant Forms: An Introduction in Predictive Modeling in Phytogeography.* The Hague: Dr. W. Junk.

Brokaw, N. V. L. (1985a) Gap-phase regeneration in a tropical forest. *Ecology,* **66**, 682–7.

Brokaw, N. V. L. (1985b) Treefalls, regrowth, and community structure in tropical forests. In Pickett, S. T. A. & White, P. S. (eds), *The Ecology of Natural Disturbance and Patch Dynamics*. New York: Academic Press.

Bryant, J. P., Chapin, F. S. & Klein, D. R. (1983) Carbon/nutrient balance of boreal plants in relation to vertebrate herbivory. *Oikos*, **40**, 357–68.

Budowski, G. (1965) Distribution of tropical American trees in the light of successional process. *Turrialba*, **15**, 40–2.

Budowski, G. (1970) The distinction between old secondary and climax species in tropical Central American lowland forests. *Tropical Ecology*, **11**, 44–8.

Chapin, F. S., Bloom, A. J., Field, C. B. & Waring, R. H. (1987) Plant responses to multiple environmental factors. *BioScience*, **37**, 49–57.

Chazdon, R. L. (1986). The costs of leaf support in understory palms: economy versus safety. *American Naturalist*, **127**, 9–30.

Cowan, I. R. (1986). Economics of carbon fixation in higher plants. In Givnish, T. J., (ed.), *On the Economy of Plant Form and Function*, pp. 133–70. Cambridge University Press.

Denslow, J. S. (1980) Gap partitioning among tropical rainforest trees. *Biotropica*, **12**, 47–55 (suppl.).

Dickinson, R. E. (1984) Modeling evapotranspiration for three-dimensional global climate models. In Hansen, J. E. & Takahashi, T. (eds), *Climate Processes and Climate Sensitivity*, pp. 58–72. Geophysical Monograph 29. Washington, D.C.: American Geophysical Union.

Emanuel, W. R., Shugart, H. H. & West, D. C. (1978) Spectral analysis and forest dynamics: The effects of perturbations on long-term dynamics. In Shugart, H. H. (ed.), *Time Series and Ecological Processes*, pp. 295–310. Philadelphia, Pennsylvania: Society for Industrial and Applied Mathematics.

Finegan, B. (1984) Forest Succession. *Nature*, **312**, 109–14.

Foster, D. H. & King, G. A. (1986) Vegetation pattern and diversity in SE Labrador, Canada: *Betula papyrifera* (birch) forest development in relation to fire history and physiography. *Journal of Ecology*, **74**, 465–83.

Friend, A. D. (1993) The prediction and physiological significance of tree height. In Solomon, A. M. & Shugart, H. H. (eds), *Vegetation Dynamics and Global Change*, pp. 101–15. New York: Chapman and Hall.

Givnish, T. J. (1978) On the adaptive significance of compound leaves, with particular reference to tropical trees. In Tomlinson, P. B. & Zimmermann, M. H. (eds), *Tropical Trees as Living Systems*, pp. 351–80. Cambridge University Press.

Givnish, T. J. (1979) On the adaptive significance of leaf form. In Solbrig, O. T., Jain, S., Johnson, G. B. & Raven, P. H. (eds), *Topics in Plant Population Biology*, pp. 375–407. New York: Columbia University Press.

Givnish, T. J. (1982) On the adaptive significance of leaf height in forest herbs. *American Naturalist*, **120**, 353–81.

Givnish, T. J. & Vermelj, G. V. (1976) Sizes and shapes of liane leaves. *American Naturalist*, **110**, 743–78.

Greenslade, P. J. M. (1983) Adversity selection and the habitat template. *American Naturalist*, **122**, 352–65.

Grime, J. P. (1974) Vegetation classification by reference to strategy. *Nature*, **250**, 26–31.

Grime, J. P. (1977) Evidence for the existence of three primary strategies in plants and its relevance to ecological and evolutionary theory. *American Naturalist*, **111**, 1169–94.

Grime, J. P. (1979) *Plant Strategies and Vegetation Processes*. New York: John Wiley.

Grubb, P. J. (1977) The maintenance of species-richness in plant communities: The importance of the regeneration niche. *Biological Review*, **52**, 107–45.

Hallé, F. (1974) Architecture of trees in the rain forest of Morobe District, New Guinea. *Biotropica*, **6**, 43–50.

Hallé, F. & Oldemann, R. A. A. (1975) *Essay on the Architecture and Dynamics of Growth of Tropical Trees*. Kuala Lumpur, Malaysia: Penerbit University.

Harper, J. L. (1982) After description. In Newman, E. I. (ed.), *The Plant Community as a Working Mechanism*, pp. 11–25. British Ecological Society Special Publications Series 1. Oxford: Blackwell Scientific Publications.

Hartshorn, G. S. (1978) Tree falls and tropical forest dynamics: In Tomlinson, P. B. and Zimmermann, M. H. (eds), *Tropical Trees as Living Systems*, pp. 617–38. Cambridge University Press.

Heinselman, M. L. (1981) Fire and succession in the conifer forests of northern North America. In West, D. L., Shugart, H. H. & West, D. C. (eds), *Forest Succession: Concepts and Application*, pp. 374–405. New York: Springer-Verlag.

Holdridge, L. R. (1967) *Life Zone Ecology*. San Jose, Costa Rica: Tropical Science Center.

Horn, H. S. (1971) *The Adaptive Geometry of Trees*. Princeton University Press. 144 pp.

Huston, M. A. & Smith, T. M. (1987) Plant succession: life history and competition. *American Naturalist*, 130, 168–98.

Jarvis, P. G. & McNaughton, K. G. (1986) Stomatal control of transpiration: Scaling up from leaf to region. *Advances in Ecological Research*, 15, 1–49.

Karieva, P. & Anderson, M. (1988) Spatial aspects of species interactions: the wedding of models and experiments. In Hastings, A. (ed.), *Community Ecology*, pp. 35–50. New York: Springer-Verlag.

Kozlowski, T. T., Kramer, P. J. & Pallardy, S. G. (1991) *The Physiological Ecology of Woody Plants*. San Diego: Academic Press.

Lindeman, R. L. (1942) The trophic-dynamic aspect of ecology. *Ecology*, 23, 399–418.

MacArthur, R. H. & Wilson, E. O. (1967) *The Theory of Island Biogeography*. Princeton, New Jersey: Princeton University Press.

Marks, P. L. (1974) The role of pin cherry (*Prunus pensylvanica*) in the maintenance of stability in northern hardwood ecosystems. *Ecological Monographs*, 44, 73–88.

Martin, P. (1993) Coupling atmosphere with vegetation. In Solomon, A. M. & Shugart, H. H. (eds), *Vegetation Dynamics and Global Change*, pp. 133–49. New York: Chapman and Hall.

Mooney, H. A. & Gulmon, S. L. (1979) Environmental and evolutionary constraints on the photosynthetic characteristics of higher plants. In Solbrig, O. T., Jain, S., Johnson, G. B. & Raven, P. H. (eds), *Topics in Plant Population Biology*, pp. 316–37. New York: Columbia University Press.

Mooney, H. A. & Gulmon, S. L. (1982) Constraints on leaf structure and function in reference to herbivory. *BioScience*, 32, 198–206.

Noble, I. R. & Slatyer, R. O. (1980) The use of vital attributes to predict successional changes in plant communities subject to recurrent disturbances. *Vegetatio*, 43, 5–21.

O'Neill, R. V., DeAngelis, D. L., Waide, J. B. & Allen, T. F. H. (1986) *A Hierarchical Concept of the Ecosystem*. Princeton, New Jersey: Princeton University Press.

Oldeman, R. A. A. & van Dijk, J. (1991) Diagnosis of the temperament of tropical rain forest trees. In Gómez-Pompa, A., Whitmore, T. C. & Hadley, M. (eds), *Rain Forest Regeneration and Management*, pp. 21–66. Man and the Biosphere Series. Volume 6. Paris: UNESCO.

Orians, G. H. & Solbrig, O. T. (1977) A cost-income model of leaves and roots with special reference to arid and semiarid areas. *American Naturalist*, 111, 677–90.

Parkhurst, D. G. & Loucks, O. L. (1972) Optimal leaf size in relation to environment. *Journal of Ecology*, 60, 505–37.

Raunkiaer, C. (1934) *The Life-forms of Plants and Statistical Plant Geography*. Oxford University Press.

Shugart, H. H. (1984) *A Theory of Forest Dynamics*. New York: Springer-Verlag.

Shugart, H. H. (1987) Dynamic ecosystem consequences of tree birth and death patterns. *BioScience*, 37, 596–602.

Shugart, H. H., Emanuel, W. R., West, D. C. & DeAngelis, D. L. (1980) Environmental gradients in a beech–yellow poplar stand simulation model. *Mathematical Biosciences*, 50, 163–70.

Shugart, H. H. & Urban, D. L. (1989) Factors affecting the relative abundance of forest tree species. In Grubb, P. J. & Whittaker,

J. B. (eds), *Toward a More Exact Ecology*, pp. 249–74. Jubilee Symposium of the British Ecological Society. Oxford: Blackwell.

Simberloff, D. & Dayan, T. (1991) The guild concept and the structure of ecological communities. *Annual Review of Ecology and Systematics*, **22**, 115–43.

Smith, T. M., Shugart, H. H., Woodward, F. I. & Burton, P. J. (1993) Plant functional types. In Solomon, A. M. & Shugart, H. H. (eds), *Vegetation Dynamics and Global Change*. pp. 272–92. New York: Chapman and Hall.

Southwood, T. R. E. (1977) Habitat, the template for ecological strategies. *Journal of Animal Ecology*, **46**, 337–65.

Tansley, A. G. (1935) The use and abuse of vegetational concepts and terms. *Ecology*, **16**, 284–307.

Tilman, D. (1982) *Resource Competition and Community Structure*. Princeton, New Jersey: Princeton University Press.

Tilman, D. (1988) *Plant Strategies and the Dynamics and Structure of Plant Communities*. Princeton, New Jersey: Princeton University Press.

Urban, D., O'Neill, R. V. & Shugart, H. H. (1987) Landscape Ecology. *BioScience*, **37**, 119–27.

Van Voris, P., O'Neill, R. V., Emanuel, W. R. & Shugart, H. H. (1980) Functional complexity and ecosystem stability. *Ecology*, **61**, 1352–60.

Walker, B. H., Ludwig, D., Holling, C. S. & Peterman, R. M. (1981) Stability of semi-arid savanna grazing systems. *Journal of Ecology*, **69**, 473–98.

Watt, A. S. (1947) Pattern and process in the plant community. *Journal of Ecology*, **35**, 1–22.

Webb, L. J., Tracey, J. G., Williams, W.T & Lance, G. N. (1970) Studies in the numerical analysis of complex rain-forest communities. V. A comparison of the properties of floristic and physiognomic-structural data. *Journal of Ecology*, **58**, 203–32.

Whittaker, R. H. (1975) *Communities and Ecosystems*. New York: MacMillan.

Whittaker, R. H. & Levin, S. I. (1977) The role of mosaic phenomena in natural communities. *Theoretical Population Biology*, **12**, 117–39.

Woodward, F. I. (1987) *Climate and Plant Distribution*. Cambridge University Press.

Woodward, F. I. (1993) Leaf responses to the environment and extrapolation to larger scales. In Solomon, A. M. & Shugart, H. H. (eds), *Vegetation Dynamics and Global Change*, pp. 71–100. New York: Chapman and Hall.

Zyabchenko, S. S. (1982) Age dynamics of scotch pine forests in the European North. *Lesovedenie*, **2**, 3–10.

Part two

❸ Plant functional types: towards a definition by environmental constraints

F. I. Woodward and C. K. Kelly

Introduction

There has always been a need for a global classification of plants in terms of plant ecophysiological properties. These properties can be described as functional attributes and here include, for example, photosynthetic capacity, survival, water relations and growth. The first requirements, tracing back to the height of the ancient Greek civilization, were for defining and understanding the climatic and edaphic controls of plant distribution (Woodward 1987). Subsequently the need was for defining appropriate crop and garden plants for a specified climate and soil. A functional classification is emphasized because this can predict development and yield more accurately than a typical phylogenetically-based taxonomic classification, and this is the underlying *raison d'être* of this chapter.

Functional definitions may cut across the traditional taxonomic classification of plants, often with taxonomically closely related species showing more marked differences in environmental sensitivity, e.g. low-temperature tolerance, than unrelated species with very similar ecology (Box 1981; Woodward 1987). The realization that there is a possible but undefined ecological classification has encouraged a number of generations of ecologists to define a workable scheme. Raunkiaer (1934) described a scheme in which the functional capabilities of plants were defined in terms of the disposition and seasonality of meristems (buds). This scheme attempted to define the functional characteristics of plants on the basis of easily observed structural properties. Although a useful classification, it does not have significant resolving power to differentiate between plants with the same structural properties but occurring in very different ecological situations.

Box (1981) significantly expanded the geographical scope and diversity of the Raunkiaer philosophy of attempting to define functional attributes from structural properties. The approach provided an extensive global coverage; however, plants were defined in terms of obvious structural rather than functional attributes. It has proved difficult, to date, to convert these characteristics into functional attributes.

This chapter aims to define functional attributes of plants at the global scale and in terms of resource acquisition, growth and survival. Plant taxonomy has described more species than can ever be studied in terms of functional attributes. Therefore any scheme of classification that aims to be all-embracing globally, in terms of its applicability to all plant species, must have different levels of resolving power. The approach adopted here is to follow an initially broad classification, which will apply to all species, by 'finer-tuned' classifications which, at each step, define smaller units of classification.

General description of the approach

The classificatory approach to be adopted here has temperature as the central controller (Fig. 3.1). The absolute maximum (Gauslaa 1984) and minimum (Woodward 1987) temperature tolerances of plants define the high and low temperature limits of plant survival. Within these broad limits temperature influences resource capture, through effects on light capture by photosynthesis, i.e. C3, C4 and CAM photosynthetic pathways (Jones 1992). Temperature also influences shade tolerance through different respiratory responses of shade-tolerant and -intolerant species (Björkman *et al.* 1972; Jones 1992). However, shade tolerance is not considered further here, as it is a globally universal phenomenon that defines small-scale – within-ecosystem – structure and not broad-scale distributional patterns (Woodward 1987). The uptake of water and nutrients is temperature-sensitive, through changes in the permeability in the plasma membrane of

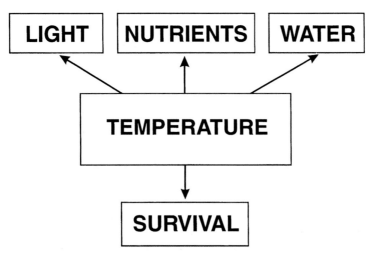

Figure 3.1 *General schematic of functional classification of plants, based around temperature.*

plant cells and through the temperature sensitivity of enzymatic reactions
(Long & Woodward 1988).

Woodward & Diament (1991) integrated the notion of functional types
with that of environmental filters. In this union (Fig. 3.2) species are amal-
gamated into functional types on the basis of similarity of responses to
environmental limitations, such as extremes of temperature and modes of
nutrient uptake. At any particular location, all knowable functional types
have the chance to occur; these are represented at the top of the figure
(Fig. 3.2). All locations have particular environments. Certain features of
these environments, such as extremes of temperature, will cause the
exclusion of (filter out) certain functional types. The functional types that
can survive the first filter will then be subjected to further filters, until a
small subset of all the knowable functional types has the potential to occur
in the selected location. Clearly a major problem with this approach is
associated with first recognizing the nature of the environmental filter and
then accruing adequate knowledge about species responses to define a

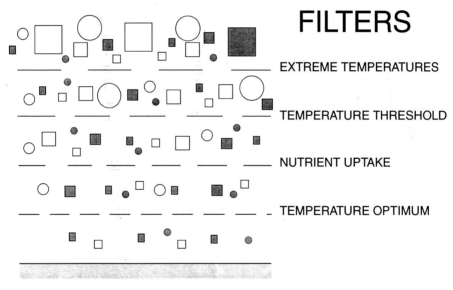

FILTERS

EXTREME TEMPERATURES

TEMPERATURE THRESHOLD

NUTRIENT UPTAKE

TEMPERATURE OPTIMUM

Figure 3.2 *A scheme of environmental filters for defining functional types
(from Woodward & Diament 1991). Each symbol represents a set of species
with the same response to the environment. Four environmental filters – extreme
temperatures, temperature thresholds for growth and hydraulic conductivity,
nutrient uptake and the temperature optimum for photosynthesis – are shown
here. Details of these filters are given in the text.*

*The filter scheme assumes that all functional types are available at a
particular location. These are represented at the top of the figure. Each environmental filter selects out functional types that cannot endure the filter conditions.
The functional types that can survive all of the environmental filters are
represented at the bottom of the figure.*

particular functional type. This chapter aims to define a small set of filters and functional types which have the capacity to delimit the distribution and functioning of major large-scale biomes, or vegetation types (Woodward 1987). The particular application of this approach is for predicting the responses of vegetation to future climatic change.

Absolute temperature limits

The absolute maximum and minimum temperatures for North and Central America are presented in Fig. 3.3 for major meteorological stations (from Müller 1982). Except for the stations in the extreme north, the absolute maximum air temperature at screen height ranges between 30 and 45 °C. For all stations, the absolute minimum temperature shows a much greater range, from −55 to +15 °C. Woodward (1987) described experimental and mechanistic evidence for the occurrence of cardinal temperatures associated with low-temperature resistance. Four basic limits were recognized; these are shown in Fig. 3.4. These species limits were integrated (Woodward 1987) into physiognomic classes of vegetation, i.e. evergreen broadleaf and chilling-sensitive (10 °C limit), or frost-sensitive (0 °C limit); limit of evergreen broadleaf survival (−15 °C); limit of broadleaf deciduous survival

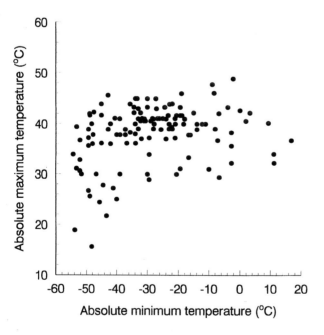

Figure 3.3   *Absolute minimum and maximum temperatures for meteorological stations in north and central America (from Müller 1982).*

(−40 °C) and boreal species with no minimum temperature limits. This classification splits North and Central America into five functional classes (Fig. 3.4).

Gauslaa (1984) provided a global-scale analysis of the absolute maximum temperature limits for mature plant survival. The majority of species had similar survival thresholds, averaging 49 °C (Fig. 3.5). A small group of species could survive to about 63 °C (Nobel & Smith 1983; Nobel 1988). These species are succulent species, generally with CAM photosynthetic metabolism and found in arid or very arid environments. The temperature thresholds for survival in these species (Fig. 3.5) appear to significantly exceed the observed air temperature maxima. However, this is an important point and it must be emphasized that the plant temperature maximum may greatly exceed the air temperature through variations in the plant's energy balance (Woodward & Sheehy 1983; Woodward 1987; Jones 1992). This effect has been modelled (Fig. 3.6) at a day temperature of 50 °C and an irradiance of 800 W m^{-2}, for a succulent plant with closed stomata and with a cylinder-like structure characteristic of barrel and pad cacti (Nobel 1988). The lethal high-temperature threshold of 63 °C can be reached for plants with low boundary-layer conductances to heat transfer (i.e. large barrel-like structures) and with high absorptivity of solar radiation (Fig. 3.6). It is

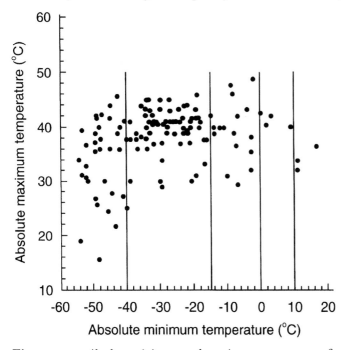

Figure 3.4 *Absolute minimum and maximum temperatures for north and central America, with lines demarcating different characterized low temperature limits of survival for different physiognomic classes of plants (Woodward 1987).*

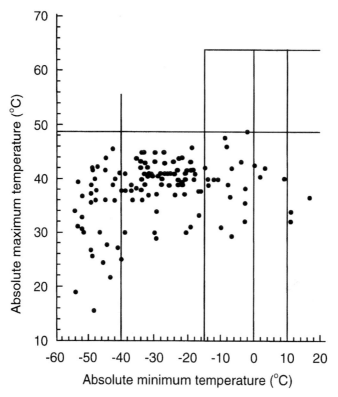

Figure 3.5   *Data and limits as for Fig. 3.4 with the addition of absolute maximum temperature limits (from Gauslaa 1984 and Nobel 1988).*

interesting to note that barrel cacti are typically covered in a dense layer of highly reflective spines, so that such plants with low boundary-layer conductances are dependent on reflectance for survival at high air temperatures. This example indicates a simple but important interplay between structure (boundary-layer conductance, plant absorptivity) and function (high-temperature survival). Although the absolute limit to temperature survival for succulents is reached in the high-absorptivity and low-conductance corner of the relationship (Fig. 3.6), it is also clear that in the conditions defined in Figure 3.6, the maximum temperature for plant survival will exceed that which is typical for the majority of terrestrial plants (Fig. 3.5).

Temperature limits of photosynthesis

Larcher (1983) described a clear association between the maximum temperature for photosynthesis and growing season temperature. Woodward & Smith (1994) have expanded on these observations and described a general model of the responses of C_3 and C_4 pathways of photosynthesis to

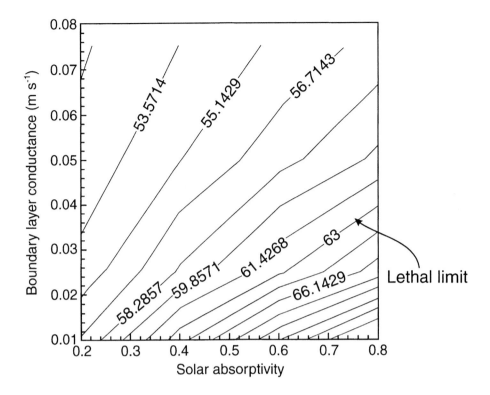

Figure 3.6 *Modelled responses (by method described in Woodward & Sheehy 1983) of plant temperatures (numbers drawn on isotherms in °C) to variations in plant solar absorptivity and boundary layer conductance to heat. The lethal temperature limit for succulent plants (Nobel 1988) is shown.*

temperature. C_4 photosynthesis is characteristic of plants in arid to moist climates of warm temperate, subtropical and tropical environments in which trees are not the major dominants of the vegetation. The optimum and low temperature thresholds for photosynthesis are higher for C_4 photosynthesis than for C_3 photosynthesis (reviewed in Jones 1992; Woodward & Smith 1994). At a global scale, the majority of plant species and the majority of the vegetation dominants of the world's terrestrial vegetation exhibit C_3 photosynthesis, therefore only aspects of the functional types of C_3 species will be discussed here.

The general response of the maximum rate of photosynthesis (A_{max}) to temperature is shown in Figure 3.7 and has been derived (Woodward & Smith 1994) from the temperature responses of CO_2 fixation, O_2 fixation (photorespiration) and enzyme activation at high temperature. The responses of the maximum rates of photosynthesis by major biome types are defined within the envelope of maximum values defined by the curve of A_{max}. In all biomes there is an optimum temperature for photosynthesis,

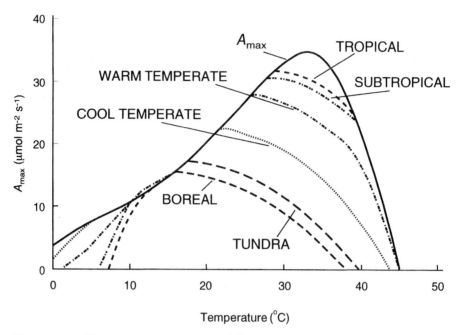

Figure 3.7   *The temperature responses of the maximum photosynthetic rate by biome (from Woodward & Smith 1994). A$_{max}$ is a general maximum photosynthetic rate which cannot be exceeded in any climate and has been defined from the temperature responses of CO_2 fixation, O_2 fixation and high-temperature enzyme inactivation (Woodward & Smith 1994). The temperature responses of the maximum photosynthetic rates, by biome, have been derived from data referenced in Woodward & Smith (1994).*

which is correlated with the temperature of the growing season (Larcher 1983; Woodward & Smith 1994). The optimum temperatures and absolute rates of maximum photosynthesis all increase along a trend towards the equator. Plants from tundra tend to show somewhat higher rates of photosynthesis than plants from boreal forest. This is probably related to the dwarf stature of the tundra plants and the higher temperatures experienced by leaves during periods of high irradiance, compared with cooler leaves in the tall forest canopy of the boreal forest (Larcher 1983; Woodward 1987).

The nitrogen uptake limits of photosynthesis

Leaf litter accumulates more under boreal forests and tundra vegetation than under any other biome types (Post *et al.* 1985). This is due to two major factors: the low temperatures of the soil layer and the consequent low rate of litter decomposition and the poor nutritional quality (high $C : N$ ratio) of the leaf litter (Read 1991).

Woodward & Smith (1994) have analysed the relations between biome type, litter and soil nutritional quality, mycorrhizal association, and the rate of nitrogen uptake from the soil. A clear relationship has emerged between the rates of uptake of the nitrogen resource and the maximum rate of photosynthesis (Fig. 3.8). Low rates of nitrogen uptake occur from the highly organic soils of the boreal forests and tundra. In these biomes the host plants are incapable of absorbing nitrogen (and phosphorus) from the complex organic forms found in the soil, and this function is replaced by a symbiotic mycorrhizal association (Read 1991). For the tundra vegetation the major mycorrhizal type is ericoid, whereas ectomycorrhizal fungi form the symbiosis under boreal and some of the broadleaved forests of the world (Read 1991).

In ecosystems with lower concentrations of organic N and higher rates of decomposition, the dominant mycorrhizal association is vesicular–arbuscular. Plants with this association and on these soils, usually with high concentrations of available nitrate (Read 1991), have higher rates of N uptake and photosynthesis (Fig. 3.8). Non-mycorrhizal species have the highest rates of N uptake and photosynthesis; these species are characteristically found on soils which are rich in available N.

In terms of the uptake of the N (and P) resource from the soil, temperature plays a more indirect role than its direct effect on photosynthesis (Fig. 3.7). Nevertheless, there is a close tie between the biome–

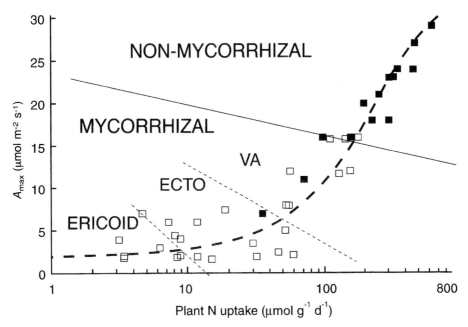

Figure 3.8 *The relationship between the uptake rate of nitrogen, the maximum rate of photosynthesis (from data reviewed and presented in Woodward & Smith 1994) and mycorrhizal association (VA, vesicular–arbuscular mycorrhiza; Ecto, ectomycorrhizal; Ericoid, ericoid endomycorrhiza).*

ecosystem property of soil characteristics and temperatures (Woodward & Smith 1994) which, in turn, influence photosynthetic rate.

Temperature, water and growth

Woodward & Smith (1994) describe a model of photosynthesis, based on the rates of N uptake shown in Figure 3.8, which also predicts an associated stomatal conductance. That is, the rate of photosynthesis is closely tied to the stomatal conductance (Jones 1992). The stomata control the rate of water loss through and out of a plant to the surrounding atmosphere. The synthetic products of photosynthesis are used, in turn, to construct growing cells and also to provide osmotically active solutes, which influence the rate of water supply into cells. Both water flow and the osmotic control of cell water potential determine the rate of cell and plant growth (Cosgrove 1986; Passioura & Fry 1992). This set of information therefore indicates the processes by which both photosynthesis and the water balance of a plant will influence growth and development.

In general the temperature responses of growth and development have been considered in isolation from the temperature responses of photosynthesis and nutrient uptake (Woodward 1987). While recognizing this problem, Woodward (1987) nevertheless defined plant functional types on the basis of their growth responses to the length and warmth of the growing season. The integral of temperature with time in days has the unit of day-degrees. It was noted that the day-degree total for completing the annual cycle of growth decreased towards the poles (Woodward 1987). This relationship is, therefore, a useful heuristic with moderate accuracy and repeatability, but plant growth responses may be modified in changed environments such as with warming, increased CO_2, or changed soil nutrient status. A consequence would be different rates of development for the same day-degree total. What is needed, therefore, is a better mechanistic coupling of the environmental controls of photosynthesis with the environmental controls on growth. It is here proposed that such a coupling can be realized through the mechanisms of growth, based on water flow and osmotic relations of cells as described earlier.

The flow rate of water or water vapour ($\mathcal{J}$) through a plant can be defined (Nobel 1991) as:

$$\mathcal{J} = L\,(\Psi_1 - \Psi_2) \tag{3.1}$$

where L is the hydraulic conductivity of the pathway under consideration. When diffusion of water vapour from the leaf is considered, then L is a

formulation of the stomatal conductance; in the liquid phase of water movement through the plant, L is a measure of liquid water conductivity. Ψ_1 is
the water potential at the start of the pathway under consideration (e.g. the
soil) and Ψ_2 is the water potential at the end of the pathway (e.g. the leaf).

The growth of cells or leaves (G) can be defined by the following equations (Cosgrove 1986; Passioura & Fry 1992).

$$G = \frac{mL}{m+L} (\Delta\pi - Y) \qquad\qquad (3.2)$$

$$G = m (P - Y) \qquad\qquad (3.3)$$

where L is the hydraulic conductance, as in Eqn 3.1, m is the extensibility
of the cell wall, $\Delta\pi$ is the osmotic potential difference between the inside
and outside of the growing cell, or the average difference between the
expanding and non-expanding (photosynthetically active) region of the leaf.
The turgor potential of the leaf is P, and Y is the yield threshold, the
threshold turgor potential below which no expansion is possible.

When the hydraulic conductivity L is not limiting water flow then Eqn
3.2 reduces to Eqn 3.3. During water supply limitation, the water potential
at the end of the pathway (Ψ_2 in Eqn 3.1) is likely to fall, as also may the
turgor potential, P, of the growing tissue. Unless further photosynthetically
derived carbohydrates are imported to the growing tissues, the turgor potential will fall and the growth rate, G, will be reduced.

Under low temperatures, when growth rates are typically low (Woodward
1987), it is often observed that the osmotic potential of the growing region
decreases, through the accumulation of osmotically active solutes (Woodward & Friend 1988). This response indicates that, at low temperatures, it
is not the photosynthetic supply of carbohydrates which limits growth but
rather one or more other aspects of the process of growth, as described in
Eqns 3.2 and 3.3.

The temperature responses of those components of the growth equations
that are readily measured are now considered for plants with different
geographical distributions. The aim is to tease out any general response of
plant growth which relates to temperature and which can be understood in
terms of the equations of growth (Eqns 3.2 and 3.3).

Temperature and leaf growth

The responses of leaf growth to temperature for species of Gramineae from
different temperature zones tend to fall within one overall pattern
(Fig. 3.9). No species appear able to grow quickly at low temperatures; the
fastest growth rates occur at temperatures of about 35 °C, a temperature

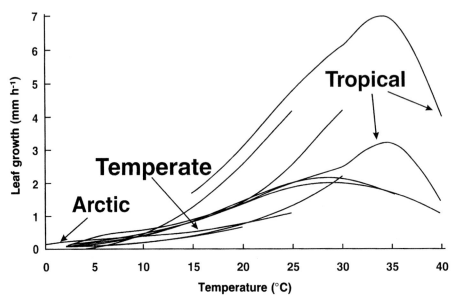

Figure 3.9   *Leaf growth rates from species of arctic (Woodward & Friend 1988), temperate (Barlow 1986; Stoddart et al. 1986; Woodward & Friend 1988; Thomas et al. 1989) and tropical regions (Watts 1971; Ong 1983; Barlow 1986).*

that corresponds with the highest photosynthetic rate (Fig. 3.7).

More resolution between the species from different geographical regions can be achieved by extracting the Q_{10} (the coefficient of the growth rates at two temperatures 10 °K apart) of leaf growth over the range of experienced temperatures. The Q_{10} values of all species increase towards the low temperature threshold for leaf growth (Fig. 3.10). The species from the different climatic zones therefore differentiate on the threshold temperatures for leaf growth. In addition, it appears that a low temperature threshold for growth (e.g. the arctic species in Fig. 3.10) is generally associated with a low Q_{10} for the leaf growth response to increasing temperatures, i.e. there is a trade-off in which low temperature growth potential is at the cost of a low growth rate at higher temperatures.

Not all of the growth studies were able to differentiate between the different components of the growth equations (Eqns 3.2 and 3.3). However, for both the arctic (Woodward & Friend 1988) and temperate species (Thomas *et al.* 1989) the temperature sensitivity of wall extensibility, *m*, was the major controller of the growth responses to temperature.

Temperature and water

Woodward & Friend (1988) pointed out the negative correlation between the capacity of plants to grow at low temperatures and the growth rate at high temperatures. Arctic plants, for example, are able to grow at low temperatures but their growth at high temperatures is much less than, for example, temperate species. However, the temperate species have a higher threshold temperature for growth (Fig. 3.10). Further investigation of the arctic species at high temperatures indicated a poor control of leaf turgor, which decreased and led to low growth rates.

It appears, therefore, that the control of leaf growth changes from a wall extensibility limitation at low temperatures to a water balance limitation at high temperatures. Thomas *et al.* (1989) demonstrated a similar trend in control and were also able to demonstrate matching temperature responses of the cell hydraulic conductivity (L in Eqns 3.1 and 3.2). Therefore it will be interesting to investigate the responses of L to temperature, in order to search for evidence of a match with the responses of leaf growth (Fig. 3.10).

The Q_{10} of the plant hydraulic conductivity (Fig. 3.11), L, matches quite closely the Q_{10} of leaf growth for plants originating from similar climatic zones. The rising Q_{10} at the low temperature response indicates a rapidly approaching low temperature limit and threshold to water movement. This response does not match that of the viscosity of water (Kaufmann 1975), indicating that it is a property – functional type – of the species under study. It is clear that there will be a dual low temperature limit for plants. The first is the low temperature threshold for growth (Fig. 3.10), which is

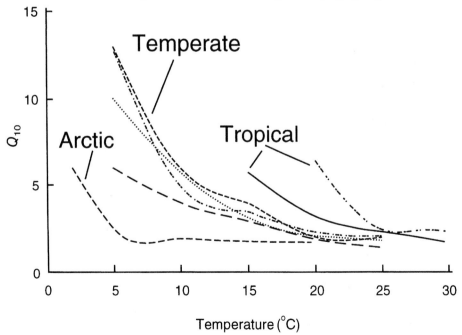

Figure 3.10 *The Q_{10} of leaf growth calculated from the data presented in Fig. 3.8.*

dominated by the temperature sensitivity of cell wall extensibility. The second is the low hydraulic conductivity for the same species at low temperatures. Towards the low temperature limits, plants will be less efficient at conducting water. This is likely to cause a drop in leaf water potential (Eqn 3.1) and turgor potential (Eqns 3.2 and 3.3), leading to reduced growth and the possibility of low-temperature drought (Larcher 1983).

For both leaf growth (Fig. 3.10) and hydraulic conductivity (Fig. 3.11) there is also evidence that the Q_{10} is lowest for the arctic species as temperatures exceed 5–10 °C. Therefore these species will be less able to conduct water and increase leaf growth rate as temperature increases than, for example, temperate species. This will lead to the likelihood of poorer competitive ability (and also greater susceptibility to drought) and the chance of competitive exclusion from warmer climates by functional types with higher Q_{10} values (Woodward 1987).

Sampling schemes

Although taxonomic categories may not provide sufficient information to determine functional types as they are defined here, taxonomic data are

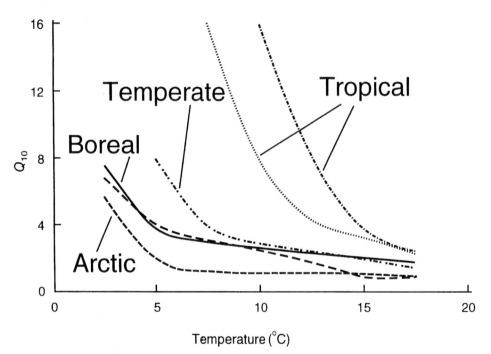

Figure 3.11   *The Q_{10} of plant hydraulic conductivity for plants from different climatic regions. Sources of data: arctic (F. I. Woodward & A. Friend, unpublished); boreal (Kaufmann 1975; Running & Reid 1980); temperate (Thomas et al. 1989); tropical (Kaufmann 1975; Markhart et al. 1979; Steudle et al. 1987).*

essential in designing efficient sampling schemes for precise definitions and recognition of functional types. A large data set is available for defining the absolute minimum temperature limits for survival (Fig. 3.4; Woodward 1987). By contrast the proposals that cell wall extensibility (CWE) and/or hydraulic conductivity (HC) define the 'filters' (*sensu* Woodward & Diament 1991) of low-temperature growth tolerance in plants are backed by very suggestive but limited data. These hypotheses need further support before any declaration can be made on their generality, and before they can be used as functions that define current distribution and potential response to change. Demonstrating support for these hypotheses involves two questions: 'Do these processes indeed work in the ways suggested?' and 'How frequently are these likely to be the limiting mechanisms?' Both questions must be considered in the experimental design.

The question of whether CWE and/or HC has the capacity to limit plant distribution can be tested with a comparison of closely related species pairs, one of which has a lower-temperature distribution than the other (*sensu* Felsenstein 1985; Harvey & Pagel 1991). CWE/HC would be quantified for each member of the pair, and support of the prediction would be that the species found in colder habitats have CWE/HC function at lower temperatures than the species found in warmer habitats. In theory, only five such pairs need be examined to test the hypothesis (the appropriate signs test will have a significant p value with five comparisons, if all comparisons vary in the same direction) (Siegal 1956). However, other factors may confound the issue, requiring more comparisons. For example, CWE/HC processes may allow a plant to grow in areas of low temperature but the species may be excluded for other reasons such as current competition, or historical distribution patterns and the disadvantage of the late-comer. If such a species were chosen as the warmer-habitat representative in one of six comparisons, a signs test would not support the hypothesis of an association with habitat, even though these may indeed be qualities necessary for cold tolerance.

Similarly, the hypothesized association between cold tolerance and optimum temperature for photosynthesis (Fig. 3.7) could be tested with a relatively small data set. The prediction here is that the species of the pair with greater cold tolerance (categorized either by distributions or by lower temperature limits on CWE/HC, depending on the results of the above test) would have lower maximum photosynthetic rates. Exclusion from a cold habitat for reasons other than cold tolerance could confound the chosen comparisons here, as above. Additionally, plants may be restricted to low photosynthetic rates by causes other than the mechanisms that allow low-temperature tolerance, and inclusion of these species could further obscure the relationship between photosynthesis and cold tolerance. Worth noting at this point is that the explicit relationships among nitrogen uptake, low temperatures and photosynthetic rates have yet to be understood, and would also be best tested with the above methods.

Paired-species comparisons, however, can only establish the function of a biological mechanism; they do not necessarily define the generality of the relationship between the mechanism and the function. If comparisons are made exclusively within a genus, or within a family, one can conclude that the function is performed by that mechanism within that genus, or within that family, but not that it is the exclusive or even predominant means of achieving the function. Greater generality in ecological patterns can only come with a more general sampling schedule, which does not depend merely on the number of species examined, but rather on the number of comparisons that can be made. Species should be chosen to maximize the number and robustness of comparisons. The desire to generalize recommends sampling widely, in a taxonomic sense, but there is a trade-off between breadth of sampling and realism of the comparison that demands forethought in constructing the sampling scheme. For more depth on this topic, we suggest Harvey & Pagel (1991) and Kelly & Purvis (1993); the latter specifically addresses the use of such comparisons in establishing ecological patterns.

A larger data set may allow application of more powerful parametric nested ANOVAs or ANCOVAs. With a nested design, the value of a character is compared to the environmental variable for species within genera, genera within families, families within orders, etc. The variance can be examined for the relative contribution of membership in a taxonomic group (i.e. to what extent it can be said that a species has a particular character because it is a member of taxon *x*) vs. a species having a character because that character is associated with a particular environmental condition (i.e. an ecological pattern). For these tests, one must consider the requirements of nested ANOVA or ANCOVA, such as replicates within factor levels and the extent to which a balanced design is necessary, when designing a sampling scheme. Such requirements are well treated in basic statistical texts (e.g. Sokal & Rohlf 1981; Zar 1984) and will not be dealt with here.

Conclusion

The benefit of new techniques for studying plants, as described here and elsewhere (e.g. Jones 1992), is that the mechanisms that lead directly to defining functional types can also be studied directly, rather than by

Table 3.1 *Functional type groups based on temperature and mycorrhizal infection*

Process		Subdivisions
Minimum temperature for survival	$< -40\,°C$	Boreal
	-15 to $-40\,°C$	Broadleaf (Blf) deciduous
	-1 to $-14\,°C$	Blf evergreen, frost-tolerant
	10 to 0 °C	Blf evergreen, chilling-tolerant
	$>10\,°C$	Blf evergreen, chilling-sensitive
Maximum temperature for survival	$>50\,°C$	
	$\leq 50\,°C$	
Temperature for optimum photosynthesis	18 °C	Arctic tundra
	16 °C	Boreal
	21 °C	Cool temperate
	26 °C	Warm temperate
	28 °C	Subtropical and Tropical
Mycorrhizal infection	None	
	Vesicular–arbuscular	
	Ectomycorrhizal	
	Ericoid	
Temperature threshold for leaf growth	$<2.5\,°C$	Arctic
	2.5 to 5 °C	Boreal
	5 to 7 °C	Temperate
	10 to 15 °C	Subtropical and Tropical
Temperature threshold for hydraulic conductivity	$<3\,°C$	Arctic and Boreal
	5 to 7 °C	Temperate
	8 to 13 °C	Subtropical and Tropical

inference or correlation with other recognizable, typically morphological, features. This increases precision but does so at the cost of the numbers of species that can be studied, at least for intensive studies of properties such as hydraulic conductivity and cell wall extensibility.

The attraction of the approach is its capacity to integrate different photo-synthetic capacities (functional types) with growth. Previously these two areas were often considered in isolation, and exemplified by apparently unlikely observations of species with high photosynthetic rates and low growth rates, and the reverse case of low photosynthetic rates and high rates of growth (Woodward 1986). Such cases could not be readily resolved. However, the conflict can be resolved by recourse to the model of growth described earlier. For example, if the plants with high photosynthetic rates occur in areas of low temperature where leaf growth and hydraulic conduc-tivity are low and perhaps close to the low-temperature threshold, then no matter what the range of photosynthetic rate and supply of carbohydrates for reducing the osmotic potential, growth rates will be low and limited by wall extensibility and perhaps some degree of turgor loss as transpiration exceeds water supply by the roots.

The case of plants with low photosynthetic rates but high growth rates is commonly found in shaded habitats (Taylor & Davies 1986; Woodward 1987). In this situation, growth and water supply may not be limited by low temperatures. The cell wall extensibility may be very high in the shade and growth is primarily a function of turgor potential (Eqn 3.3). As long as the plants in the shade are well watered then growth will be rapid.

The process-oriented discussion outlined in this chapter has aimed to define functional types on the basis of mechanisms and to apply these mechanisms across the global suite of terrestrial plants, gradually narrowing down the large initial classes of functional types to a smaller subset. The different definitions of global-scale functional types discussed in this chapter are outlined and quantified in Table 3.1. There are 23 types, with 253 possible paired combinations. Some of these combinations may be unlikely to occur naturally. For example a functional type with growth less than 2.5 °C is unlikely to be capable of enduring plant maximum temperatures over 50 °C. However, many of the other combinations appear, theoretically at least, to be possible. What is clearly required is a large global survey of plant responses to improve the quantification of the thresholds that have been defined in this chapter.

References

Barlow, E. W. R. (1986) Water relations of expanding leaves. *Australian Journal of Plant Physiology*, **13**, 45–58.

Björkman, O., Ludlow, M. M. & Morrow, P. A. (1972) Photosynthetic performance of two rainforest species in their native habitat and analysis of their gas exchange. *Carnegie Institution Year Book*, **71**, 94–102.

Box, E. O. (1981) *Macroclimate and Plant Forms: An Introduction to Predictive Modeling in Phytogeography*. The Hague: Junk.

Cosgrove, D. J. (1986) Biophysical control of plant cell growth. *Annual Review of Plant Physiology*, **37**, 377–405.

Felsenstein, J. (1985) Phylogenies and the comparative method. *American Naturalist*, **125**, 1–15.

Gauslaa, Y. (1984) Heat resistance and energy budget in different Scandinavian plants. *Holarctic Ecology*, **7**, 1–78.

Harvey, P. H. & Pagel, M. D. (1991) *The Comparative Method in Evolutionary Biology*. Oxford University Press.

Jones, H. G. (1992) *Plants and Microclimate*. 2nd edn. Cambridge University Press.

Kaufmann, M. R. (1975) Leaf water stress in Engelmann Spruce. *Plant Physiology*, **56**, 841–6.

Kelly, C. K. & Purvis, A. (1993) Seed size and establishment conditions in tropical trees: On the use of taxonomic relatedness in determining ecological patterns. *Oecologia*, **94**, 356–60.

Larcher, W. (1983) *Physiological Plant Ecology*. Berlin: Springer-Verlag.

Long, S. P. & Woodward, F. I. (eds) (1988) *Plants and Temperature. Symposia of the Society for Experimental Biology* **XLII**. Cambridge: Company of Biologists.

Markhart, A. H., Fiscus, E. L., Naylor, A. W. & Kramer, P. J. (1979) Effect of temperature on water and ion transport in soybean and broccoli systems. *Plant Physiology*, **64**, 83–7.

Müller, M. J. (1982) *Selected Climatic Data for a Global Set of Standard Stations for Vegetation Science.* The Hague: Junk.

Nobel, P. S. (1988) *Environmental Biology of Agaves and Cacti.* Cambridge University Press.

Nobel, P. S. (1991) *Physicochemical and Environmental Plant Physiology.* San Diego: Academic Press.

Nobel, P. S. & Smith, S. D. (1983) High and low temperature tolerances and their relationships to distribution of agaves. *Plant, Cell and Environment*, **6**, 711–19.

Ong, C. K. (1983) Response to temperature in a stand of pearl millet (*Pennisetum typhoides* S. & H.). 4. Extension of individual leaves. *Journal of Experimental Botany*, **34**, 1731–9.

Passioura, J. B. & Fry, S. C. (1992) Turgor and cell expansion: beyond the Lockhart equation. *Australian Journal of Plant Physiology*, **19**, 565–76.

Post, W. M., Pastor, J., Zinke, P. J. & Stangenberger, A. G. (1985) Global patterns of soil nitrogen storage. *Nature*, **317**, 613–16.

Raunkiaer, C. (1934) *The Life Forms of Plants and Statistical Plant Geography* (translated by H. G. Carter, A. G. Tansley & M. Fansboll). Oxford: Clarendon Press.

Read, D. J. (1991) Mycorrhizas in ecosystems. *Experientia*, **47**, 376–91.

Running, S. W. & Reid, C. P. (1980) Soil temperature influences on root resistance of *Pinus contorta* seedlings. *Plant Physiology*, **65**, 635–40.

Siegal, S. (1956) *Non-parametric Statistics.* New York: McGraw-Hill.

Sokal, R. R. & Rohlf, F. J. (1981) *Biometry.* 2nd edn. New York: W. H. Freeman.

Steudle, E., Oren, R. & Schulze, E. D. (1987) Water transport of maize roots. The measurement of hydraulic conductivity, solute permeability, and of reflection coefficients of excised roots using the root pressure probe. *Plant Physiology*, **84**, 1220–32.

Stoddart, J. L., Thomas, H., Lloyd, E. J. & Pollock, C. J. (1986) The use of a temperature-profiled position transducer for the study of low-temperature growth in graminae. *Planta*, **167**, 359–63.

Taylor, G. & Davies, W. J. (1986) Yield turgor of growing leaves of *Betula* and *Acer*. *New Phytologist*, **104**, 347–53.

Thomas, A., Tomos, A. D., Stoddart, J. L., Thomas, H. & Pollock, C. J. (1989) Cell expansion rate, temperature and turgor pressure in growing leaves of *Lolium temulentum* L. *New Phytologist*, **112**, 1–5.

Watts, W. R. (1971) Role of temperature in the regulation of leaf expansion in *Zea mays*. *Nature*, **229**, 46–7.

Woodward, F. I. (1986) Ecophysiological studies in the shrub *Vaccinium myrtillus* L. taken from a wide altitudinal range. *Oecologia*, **70**, 580–96.

Woodward, F. I. (1987) *Climate and Plant Distribution.* Cambridge University Press.

Woodward, F. I. & Diament, A. D. (1991) Functional approaches to predicting the ecological effects of global change. *Functional Ecology*, **5**, 202–12.

Woodward, F. I. & Friend, A. D. (1988) Controlled environment studies on the temperature responses of leaf extension in species of *Poa* with diverse altitudinal ranges. *Journal of Experimental Botany*, **39**, 411–20.

Woodward, F. I. & Sheehy, J. E. (1983) *Principles and Measurements in Environmental Biology.* London: Butterworth.

Woodward, F. I. & Smith, T. M. (1994) Global photosynthesis and stomatal conductance: modelling the controls by soil and climate. *Advances in Botanical Research* **20**, 1–41.

Zar, J. H. (1984) *Biostatistical Analysis.* 2nd edn. Englewood Cliffs, New Jersey: Prentice-Hall.

4 Can we use plant functional types to describe and predict responses to environmental change?

R. J. Hobbs

Introduction

Gitay & Noble (Chapter 1, this volume) have indicated that Global Change and Terrestrial Ecosystems (GCTE) seeks to use functional types, which reflect similar responses to perturbations via similar mechanisms. They suggest that functional groupings have to be context-specific simplifications of the real world which allow predictions of the dynamics of systems or system components. Mooney (Chapter 17, this volume) also indicates that groupings have, of necessity, to be question-, scale-, system- and time-dependent. Other chapters in this volume explore the derivation of functional groupings, either *de novo* or from *a priori* assumptions. Here I explore whether it is possible to use groupings derived from current knowledge to describe and predict vegetation response to environmental change. I first consider a series of questions that have to be asked before analysis of the response to environmental change can be undertaken, and I then examine in detail the responses of two ecosystem types to a variety of changes.

What scales, functions, groups and responses are important?

Before sensible progress can be made with the question of deriving useful functional groupings of species, a number of questions have to be considered. We need to define the temporal and spatial domains in which we are interested, the ecosystem functions we wish to consider, how we wish to group species in relation to these functions, and which types of response we need to focus on. Without answers to these questions, we will spend valuable time adrift on a sea of unbounded and speculative activity, which may prove to be great fun but end up with little of much use to anyone.

The objective of developing functional groups is to allow an easier interface between the different scales of study currently being considered. If we wish to determine how variations in terrestrial ecosystems affect global-scale climatic processes and, in turn, how global changes affect ecosystem structure and function, then we need to simplify the vast diversity of nature into something more manageable. Ecology has traditionally been carried out at the scale of the quadrat or plot, but the need for larger-scale observations has become increasingly apparent. Plots are placed within patches made up of different plant communities, which make up landscapes, which aggregate to regions, continents, etc. Processes can similarly be sorted hierarchically on a spatial scale (Delcourt & Delcourt 1991). Because this chapter covers the response to environmental change, it is appropriate to consider the scale at which this response should be measured. It could be argued that all levels of the hierarchy are relevant for different reasons. Local (plot- or patch-scale) changes may be very important for the persistence of individual species, or for local human industries; ecosystem- and landscape-level changes aggregate these smaller-scale effects and have implications for regional hydrology, local economies, etc. At a larger scale, changes are important for national economies and resource allocation strategies, and because of potential feedbacks to the global system. Methods of observation and analysis will differ at each level, as will the required level of detail (Botkin *et al.* 1984; Hobbs 1990). Predictions of responses to global change have to be useful at, or be translatable to, local and regional levels. Here I first examine local responses to change, since this is the scale at which most information is available, and it is the scale at which management responses to change have to be made. I then consider the landscape level since this may be an appropriate intermediate level linking small- and large-scale patterns and processes.

There are a variety of ways in which species can be grouped on the basis of various aspects of ecosystem function and/or adaptive responses to environmental variables (Hobbs 1992), for example:

1. Resource use.
2. Ecosystem function: production, consumption, decomposition, N fixation.
3. Response to disturbance.
4. Response to environment: mesophytes, xerophytes, etc.
5. Reproductive strategy, pollination systems.
6. Tolerance to stress: halophytes, etc.
7. Physiognomic types (Raunkiaer, etc.).
8. Physiological types: C_3, C_4, etc.
9. Phenology.

Is it possible, or desirable, to produce sensible functional groupings that incorporate some or all of these various classifications? Leishman & Westoby (1992) have made a start on examining this question. One way to treat the problem is to view the different classifications in a hierarchical perspective; multivariate approaches may indicate the existence of groupings arising from the combination of various classifications. It has been argued that groupings will largely reflect the purpose of the classification (Körner 1993; Solbrig 1993). Walker (1992) argues that the most practical approach may be to group species in relation to limiting or dominant ecosystem processes.

I further argue, however, that groupings have to be made in consideration of the types of responses that are likely to be important. Hobbie *et al.* (1993) proposed a set of 'straw man' functional groups, based primarily on growth form (Table 4.1). However, such a set suffers from the lack of consideration of the regeneration phase, and, in particular, the response to disturbance (as, for instance, discussed by Noble & Slatyer (1980)). Consideration of responses to environmental change has to focus on the mechanisms by which these changes will occur. This has not been explicitly explored in most treatments of the subject; a simple assumption that vegetation types are likely to migrate in response to climatic changes (e.g. Graetz *et al.* 1988) ignores the many complex factors involved in vegetation change, such as: individualistic plant responses, variation in dispersal rates and modes of spread; lag effects, including the inertia of existing vegetation; the importance of habitat fragmentation (Davis 1981, 1986; Davis & Botkin 1985; Graetz 1988; Huntley & Webb 1989; Hobbs & Hopkins 1991). I would argue, as do Leishman *et al.* (1992), that the regeneration phase is critical in considerations of responses to environmental change and hence has to be explicitly considered in any development of functional groupings. Bond (Chapter 9, this volume) further argues that the important features to be considered are those that allow persistence, regeneration and/or spread in the face of changing conditions.

What environmental changes are important?

Palynological studies in North America have been used to trace the history of vegetation in relation to postglacial climatic changes (Davis 1986, 1988, 1989; Delcourt & Delcourt 1991). These studies have indicated how vegetation components have migrated in response to climatic amelioration. In Britain and northern Europe, palynological studies have shown similar climate-related vegetation changes, but from three to four thousand years ago the observed changes could not be attributed to climate alone. The influence of bronze- and iron-age humans on the vegetation was profound, and in many cases obscured or modified any climatically induced changes

	High resource/ low light	**Table 4.1** *'Straw-man' functional groups, based primarily on growth form and secondarily on physiological and growth characteristics, and their relation to a light/resource gradient*
Tree		
Deciduous		
High relative growth rate		
Low relative growth rate		
Evergreen		
Shade-intolerant		
Shade-tolerant		
Shrub		
N-fixing		
Non-fixing		
Herb		
C3		
C4		
Moss		
Feathermoss		
Sphagnum		
Lichens		
	Low resource/ high light	

Source: After Hobbie *et al.* (1993).

(Pennington 1969; Gimingham 1972, 1975). This is also the case in the Mediterranean region, where climatically controlled changes gave way to human-induced change after 6000 BP (Le Houérou 1981; Thirgood 1981; Vernet & Thiebault 1987). Delcourt & Delcourt (1991) list numerous other areas where human influence on vegetation and natural ecosystems is obvious from 3000 to 5000 years ago (e.g. eastern USA, mid-Europe, Central America).

Human activities and use of the land have thus had an important influence on vegetation dynamics in many parts of the world for some time, to the extent that it is difficult to disentangle climatically driven changes from human-induced changes in the palynological record. With increasing human populations, improved technologies, agricultural development and modification of natural ecological processes, the impact of humans on the world's ecosystems has increased dramatically over the past few hundred years. Removal, exploitation and modification of native vegetation have resulted in extensive and often irreversible changes in most parts of the world (Wolman & Fournier 1987; Hobbs & Hopkins 1990; Turner *et al.* 1991; Saunders & Hobbs 1992; Angelstam & Arnold 1993; Green 1993). Wyman *et al.* (1991) illustrated how grasslands and forests have declined from each representing one-third of terrestrial ecosystems in 1800 to each representing one-quarter in 1960 and one-sixth in 1990. At current rates of attrition, by 2000, they will represent between one-eighth and one-tenth, and will disappear sometime next century, *even without major changes in climate*.

Human land use has thus been a major factor leading to change in the earth's ecosystems, and it is likely that continuing and increasing human exploitation, together with regional changes in land-use practices, will have an increasing impact. Indeed, Davis (1988) states that, 'For the last century, and presumably for the next, land use has been more important than climatic change in forcing changes in ecological systems and dynamics'. The US Committee on Global Change (1990) also stated, 'Throughout most of history, the primary way in which humans have effected change in the global environment has been by transforming the earth's land surface'. Yet most attention is undoubtedly directed at the effects of human-induced climate change (e.g. Clark 1991), although the situation is now changing (e.g. Leemans, Chapter 15, this volume).

Analyses of likely changes in climate must, of necessity, possess a high degree of uncertainty since the climatic models used to predict future climate changes with good regional accuracy are still in the development phase (e.g. Pearman 1988; Henderson-Sellers & Blong 1989). In many parts of the world, climatic parameters such as temperature and rainfall fluctuate wildly from year to year, and hence detecting directional change becomes difficult, especially given the relative shortness of most climatic records (e.g. Saunders & Hobbs 1992). Predicting ecological responses to potential climate change adds a further layer of uncertainty to this, since in many cases we have yet to develop a good understanding of ecological conditions under current conditions, far less under changed conditions (Hobbs & Hopkins 1991; Saunders & Hobbs 1992). When changes caused by human land use are added to the picture, it becomes difficult to see how any signal from impacts of climate change will show through the confounding effects of land use.

For these reasons, in this chapter I have chosen to focus on the ecological responses to a variety of direct human impacts on ecosystems. Direct human impacts are already extensive and obvious, and hence their effects can be assessed without recourse to 'what if' scenario-based questions. Many different types of impact are available to be assessed, and hence comparative analyses can be carried out in the search for generalizable conclusions.

I thus focus on two main areas to examine the functional responses to environmental change. Firstly, I consider changes brought about by various human-induced and natural impacts in the sclerophyll vegetation of south-western Australia. Secondly, I examine in detail the effects of management, disturbance and climatic variability on the annual grasslands of California.

Responses to environmental change

Sclerophyll vegetation in south-western Australia

The vegetation of south-western Australia consists of a complex mosaic of vegetation types containing a high diversity of species with a high level of endemism (Groves & Hobbs 1992; Hobbs 1992; Hopper 1992). South-western Australia has also been subjected to dramatic human-induced changes over the past 150 years, with large areas being cleared for agriculture, exploited for timber, or modified by exotic herbivores, plant pests or diseases (Hobbs 1992). In addition the vegetation is subjected to 'natural' disturbances such as fire, drought and above-average rainfall. Here I examine the response of vegetation to some of these impacts. I do this by defining functional groups within particular vegetation types and analysing the response of these groups to change.

Response to Phytophthora cinnamomi
The soil-borne pathogenic fungus *Phytophthora cinnamomi* is an important agent causing vegetation change in south-western Australia. The introduced fungus is present throughout extensive areas of the jarrah (*Eucalyptus marginata*) forest, but is also spreading to other vegetation types. It is thought to have spread into the Stirling Range National Park, one of the region's centres of floristic diversity, during the 1960s and now occurs throughout the park (Wills 1994). The fungus has dramatic effects on the species-rich heath found in the park, but plant species vary in their susceptibility to the fungus. Data from Wills (1994 and personal communication; Table 4.2) illustrate this variation.

Wills' data have been grouped in several ways, reflecting various functional attributes. Overall, 35% of the species recorded in Wills' (1994) survey of the park vegetation showed some degree of susceptibility to the fungus. However, susceptible species were all woody perennials, with no

All species	35%
Growth form	
Woody perennials	47%
Herbaceous perennials	0%
Annuals	0%
Climbers/parasites etc.	0%
Nutrition mode	
Proteaceae	83%
N fixers	47%
Other woody perennials	21%
Pollination mechanism	
Wind-pollination	2%
Insect-pollinated	34%
Vertebrate-pollinated	58%
Response to fire	
Seeders	68%
Resprouters	20%
Combination groups (example)	
Vertebrate-pollinated seeders	87%

Table 4.2 *Percentage of species in heath vegetation in Stirling Range National Park that show susceptibility to* Phytophthora cinnamomi, *sorted into various functional groupings*

Source: Wills (1994); R. T. Wills, unpublished data.

herbaceous species exhibiting any degree of damage. Within the woody perennials, functional types also differed markedly in their susceptibility. Groups with specialized models of nutrition, i.e. the Proteaceae (which possess specialized root structures; Lamont 1982, 1992) and N fixers, were proportionally more susceptible than other woody perennials. The Proteaceae in particular had a high proportion of species susceptible to the fungus. Grouping on mode of pollination indicated that animal-pollinated species were proportionally more susceptible than wind-pollinated ones. Grouping on response to fire indicated a greater incidence of susceptibility among seeder species compared with resprouting species. These different groupings can be combined in a variety of ways, and as an example it can be seen that the group 'vertebrate-pollinated seeders' has a very high incidence of susceptibility.

These differing degrees of susceptibility can be translated into the effect of the fungus on plant community composition. Projections of changes in the relative importance of the various groups discussed above can be produced, assuming that all susceptible species eventually disappear from the area (Fig. 4.1). Clearly, the fungus results in predicted changes in the relative importance of particular groups, whichever characteristics are used to derive groupings. This type of analysis thus allows a consideration of the likely impact of *Phytophthora* on various aspects of ecosystem function. These projections can be compared against real data recorded by Wills (1994) in areas with and without *Phytophthora* (Fig. 4.2). Clearly, the largest influence is the reduction in abundance of proteaceous species. Infestation results in a decline in proteaceous shrubs and a corresponding increase in herbaceous species. Vegetation composition and physiognomy is dramatically affected, with a shift from woody perennials to herbaceous perennials. Nutrient dynamics are likely to be affected by the reduction in abundance of groups with specialized nutrient uptake, and response to disturbance will change because of the reduced proportion of seeders. The reduced importance of vertebrate-pollinated plants may have important flow-on effects for the vertebrate community dependent on the nectar resources provided by this group (Wills 1994).

The question to be asked, however, is whether this type of analysis provides more valuable information than a simple taxonomic assessment. Taxonomically, the group most severely affected by *Phytophthora* is the Proteaceae. All members of this family possess specialized roots, many are seeders after fire, and many are vertebrate-pollinated. So would we arrive at the same conclusion regarding possible effects on ecosystem function by traditional taxonomic survey? The effort required to obtain a taxonomic inventory is probably less than that required to assess the various functional aspects but, of course, the assessment of the relation between taxonomy and function is only possible when such an assessment has been made.

Other types of disturbance

For comparison with the impacts of *Phytophthora*, similar analyses of the effects of other types of disturbance have been conducted. In a study of the effects of drought (2 years of below-average rainfall) on northern sandplain heath, Hnatiuk & Hopkins (1980) found similar patterns of susceptibility to those reported by Wills (1994) (Fig. 4.3). Proteaceae were the group most affected; annuals and herbaceous perennials were least affected. Thus, again, the response to severe drought might be expected to be a reduction in Proteaceae and an increase in herbaceous species. These data go some way to providing a test of meta-congruency as discussed by Gitay & Noble (Chapter 1, this volume) by indicating that the groupings used show similar responses to different types of disturbance. This is an imperfect test, however, because the data were collected in different locations.

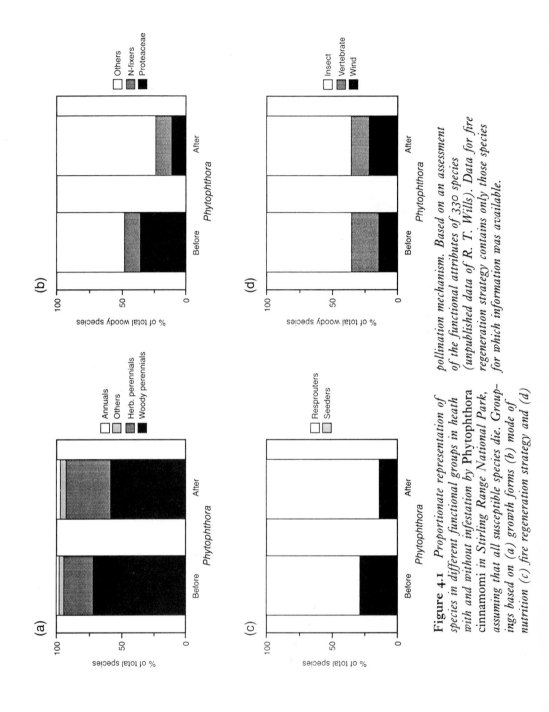

Figure 4.1 *Proportionate representation of species in different functional groups in heath with and without infestation by Phytophthora cinnamomi in Stirling Range National Park, assuming that all susceptible species die. Groupings based on (a) growth forms (b) mode of nutrition (c) fire regeneration strategy and (d) pollination mechanism. Based on an assessment of the functional attributes of 330 species (unpublished data of R. T. Wills). Data for fire regeneration strategy contains only those species for which information was available.*

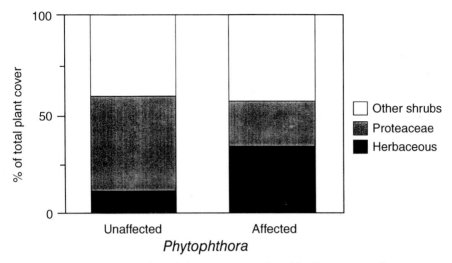

Figure 4.2 *Percentage of total plant cover contributed by Proteaceae, other shrubs and herbaceous perennials in heath with and without infestation by* Phytophthora cinnamomi *on Stirling Range National Park (data from Wills 1994).*

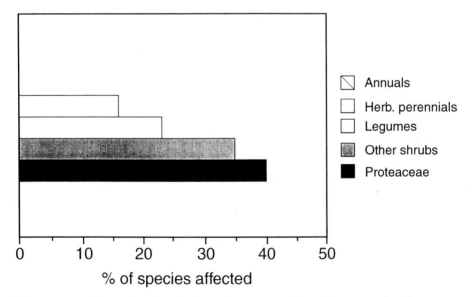

Figure 4.3 *Proportion of species in various groups affected by severe drought in northern sandplain heath (data from Hnatiuk & Hopkins 1980).*

Studies in the central Australian wheatbelt have also indicated that disturbances of various types alter the relative importance of various groups. In sandplain heath cleared for agriculture, cropped for a few years and then abandoned about 40 years ago, there is a clear shift in the regenerated vegetation from woody perennials to annuals (R. J. Hobbs, unpublished data; Fig. 4.4). This shift appears not to be simply a successional stage,

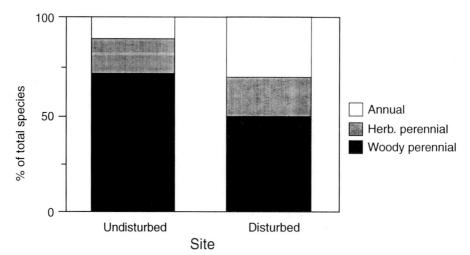

Figure 4.4 *Proportionate representation of species in different groups in undisturbed sandplain heath and equivalent heath which was cleared and briefly cultivated before abandonment in the 1930s. Both sites are in Durokoppin Nature Reserve in the central wheatbelt (R. J. Hobbs, unpublished data).*

since the annual component now successfully prevents further colonization by perennials (Hobbs & Atkins 1991). A similar increase in annuals and herbaceous perennials is observed after fire in the heath community, but in this case it appears to be a transient state, with fire-following annuals and short-lived herbaceous perennials giving way to woody perennials after a few years (R. J. Hobbs, unpublished data). Clearly, a distinction thus has to be made between transient dynamics (Tilman 1988), as observed following fire, and the formation of new metastable states, as in the case of temporary clearance.

Grazing by introduced stock can also alter the representation of various groups, as illustrated by work by Scougall (1991) and Scougall *et al.* (1993) in Figure 4.5. In this case, grazing has completely removed the shrub component and increased the herbaceous component. Because grazing also prevents tree regeneration, in the longer term the tree component would decline too. Another feature of the changes caused by grazing is the pronounced increase in representation of non-native species in the herbaceous component.

The above set of examples serves to indicate that changes in the relative importance of different functional groups can be detected in response to environmental change caused either by climatic extremes or by various forms of disturbance or management. A common feature in all the data sets is a reduction in the woody perennial component (or parts thereof) and an increase in the herbaceous component. In some cases this shift may be temporary, but in others it appears to be more permanent. These findings are, in themselves, fairly trivial; increased herbaceousness has been noted

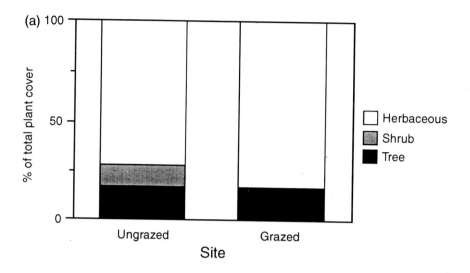

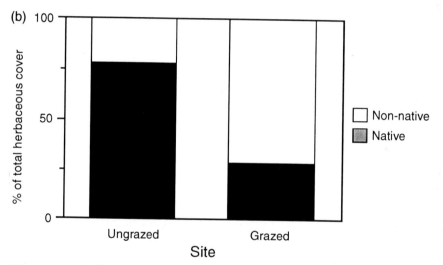

Figure 4.5 *Relative abundance of tree, shrub and herbaceous components in grazed and ungrazed* Eucalpytus loxophleba/Acacia acuminata *woodland in the central wheatbelt (a) and proportions of the herbaceous component made up of native and non-native species (b) (data from Scougall 1991).*

following disturbance in a range of vegetation types (e.g. Delcourt & Delcourt 1991). However, the data at least serve to suggest that some generalities about functional responses to environmental change in south-western Australian sclerophyll vegetation may be possible.

California: annual grasslands

Annual grasslands in California are a relatively well-studied and understood ecosystem (Huenneke & Mooney 1989), and they possess several attributes which make them a useful object of study in the present context. Firstly, they consist mostly of annual and herbaceous perennial species, which means that they respond relatively rapidly to any environmental or management change (cf. Woodward 1992). Secondly, they contain a number of easily defined functional groups, which can be distinguished in terms of their life histories, phenology and physiology (Mooney *et al.* 1986; Chiariello 1989; J. K. Armstrong & R. J. Hobbs, unpublished data). Here I explore the response of the various groups to a variety of disturbances and management changes, and to variation in microclimate and rainfall amounts and distributions. I will concentrate first on grasslands occurring on serpentine soils, but will also examine the more widespread non-serpentine grasslands, which are predominantly composed of non-native annuals.

Serpentine grasslands

The serpentine grassland at Jasper Ridge, near Stanford University, is subject to a high degree of disturbance by pocket gophers, *Thomomys bottae*, and this disturbance produces characteristic patterning of the vegetation (Hobbs & Mooney 1985). Exclosures set up to exclude or reduce gopher disturbance revealed that disturbance reduced the abundance of perennial grasses and geophytes, but increased the abundance of late-growing annuals and legumes (Fig. 4.6) (Hobbs & Mooney 1991). Fertilization studies revealed that increased nutrient availability increased the proportion of grasses and decreased the forb component (Fig. 4.7) (Hobbs *et al.* 1988); more detailed studies supported this for complete nutrient supplementation (i.e. N, P and others) but found complex responses to the addition of

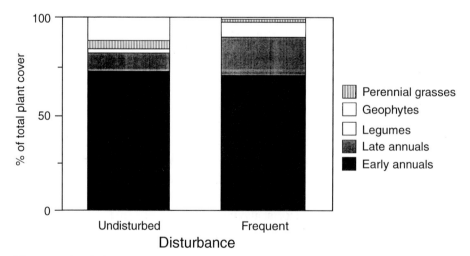

Figure 4.6   *Relative abundance of different groups in areas free of gopher disturbance compared with areas subjected to frequent disturbance in serpentine grassland at Jasper Ridge, California (data from Hobbs & Mooney 1991).*

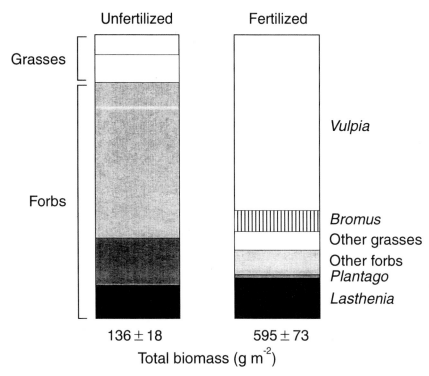

Figure 4.7 *Responses of grasses and forbs to fertilization with NPK in serpentine grassland at Jasper Ridge, California (data from Hobbs* et al. *1988).*

nutrients singly or in different combinations (Huenneke *et al.* 1990). In a study of revegetation operations, Koide & Mooney (1987) found that the addition of P alone increased the abundance of legumes at the expenses of grasses. Exclosures indicated that removal of grazing and/or trampling by cattle increased the abundance of grasses and reduced the forb component (L. F. Huenneke, personal communication).

In areas of diverse topography, slope and aspect have an important influence on grassland microclimate, with south-facing slopes receiving more insolation and being hotter and drier than north-facing slopes (Weiss *et al.* 1988; Weiss & Murphy 1993; J. K. Armstrong & R. J. Hobbs, unpublished data). This is translated into marked differences in the grassland vegetation on different slopes and aspects (Fig. 4.8) (J. K. Armstrong & R. J. Hobbs, unpublished data). Functional groups vary in abundance along a microclimate gradient formed by the different slopes, with, for instance, geophytes being most abundant on flat or shallow slopes and perennial forbs and grasses being most abundant on north-facing slopes. Annual grasses are also most abundant on north slopes.

Examination of vegetation responses to annual rainfall variation also indicates that some groups may show large responses to this variation (Hobbs &

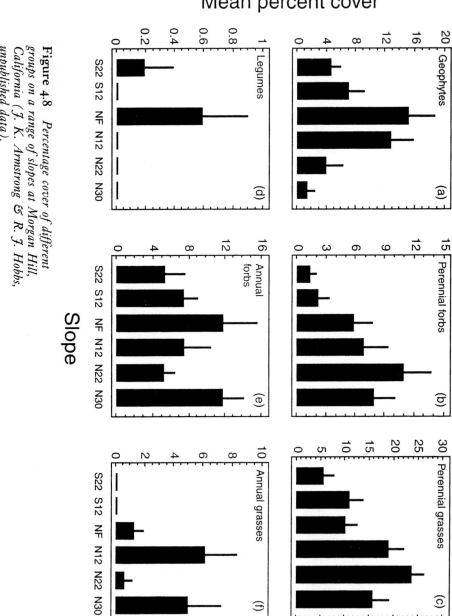

Figure 4.8 *Percentage cover of different groups on a range of slopes at Morgan Hill, California (J. K. Armstrong & R. J. Hobbs, unpublished data).*

Mooney 1991). In particular, the introduced annual grasses increase in abundance following exceptionally wet years, but decline in drier years. Increase following wet years was also related to gopher disturbance, however. There may therefore be some parallels between spatial microclimatic variations and temporal variations in rainfall. Similar switches between forb and grass dominance with rainfall amounts and distributions have been demonstrated by Pitt & Heady (1978) for non-serpentine grasslands.

It should be noted, however, that analyses of functional group responses to microclimatic and rainfall variation obscure considerable variation within groups (Fig. 4.9) (Hobbs & Mooney 1991; J. K. Armstrong & R. J. Hobbs, unpublished data.) Figure 4.9 illustrates the non-uniform response of the two commonest forbs and perennial grasses in the serpentine grassland to slope differences. Hobbs & Mooney (1991) also found that the two forbs *Lasthenia californica* and *Plantago erecta* responded in opposite ways to rainfall variation: *Lasthenia* declined with declining rainfall, whereas *Plantago* increased. Clearly, these two species, which are placed in the same functional group, show quite disparate responses, which may be related to their reproductive strategies (Fig. 4.10). The functional group approach thus cannot capture the full dynamics in this case, unless more detailed groupings are derived. It could be questioned, however, whether the shift between two dominant forb species is important in a functional sense; the community is still forb-dominated and no switch in structure or apparent function is obvious.

Another problem becomes apparent when the data in Figure 4.9 are compared with data collected by McNaughton (1968). He found that the perennial grass *Stipa pulchra* followed the same trend as that in Fig. 4.9, but that the annual grass *Bromus mollis* increased in abundance from north to south exposures; i.e. showed the opposite trend to that found in Fig. 4.9. The reason for this discrepancy is not immediately obvious, but this serves as a warning against generalizing from individual datasets.

Grasslands in a landscape mosaic
Annual grasslands in California frequently exist within a mosaic of other vegetation types, including shrublands and woodlands. Although some reports indicate that this mosaic is relatively stable spatially and temporally (Biswell 1974; Davis & Mooney 1985), there are also cases where the spatial mosaic has changed (Williams *et al.* 1987). Davis & Mooney (1985) suggested that shrub invasion of grassland was prevented by depletion of surface water by grassland species, and Da Silva & Bartolome (1984) demonstrated shrub mortality in the presence of grasses. Grassland annuals also effectively compete with oak (*Quercus douglasii*) seedlings (Gordon *et al.*

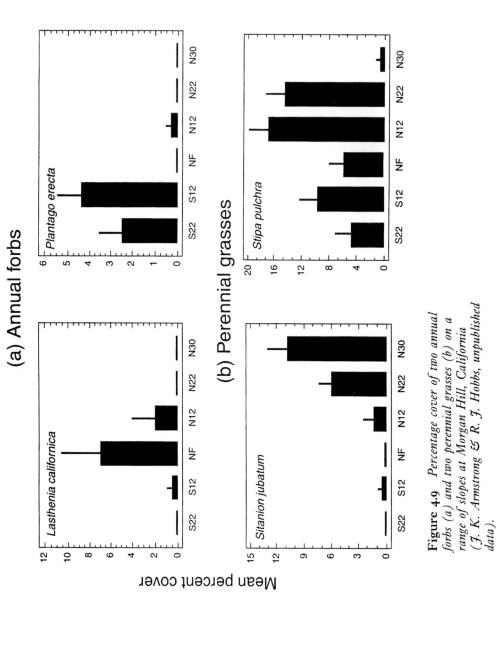

Figure 4.9 *Percentage cover of two annual forbs (a) and two perennial grasses (b) on a range of slopes at Morgan Hill, California (J. K. Armstrong & R. J. Hobbs, unpublished data).*

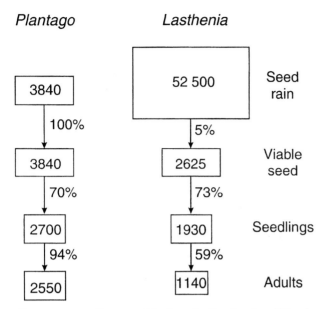

Figure 4.10   *Demographic data for* Lasthenia californica *and* Plantago erecta *at Jasper Ridge, California, 1983–4 (data from Hobbs & Mooney 1985).*

1989; Gordon & Rice 1993). Williams *et al.* (1987) and Williams & Hobbs (1989) demonstrated how shrub invasion of grassland was linked with episodic above-normal rainfall, and that increased water availability increased the likelihood of shrub establishment. Subsequent to invasion, the grassland species quickly disappeared (Hobbs & Mooney 1986), indicating that the transformation to shrubland represented a change to a new stable (or metastable) state.

A recent study by George *et al.* (1992) has built on the approach of Westoby *et al.* (1989) to examine the California grassland in terms of non-equilibrium dynamics, using state and transition models as discussed elsewhere in this volume (Chapters 2, 5 and 13). For the non-serpentine grassland, George *et al.* (1992) recognized a number of alternative states in which the grassland could be, and the factors needed to force transition from one state to another. They recognized three interrelated states where the grassland was dominated by annuals of varying type, one state dominated by perennial grass, a shrub-dominated state and two states dominated by trees (Fig. 4.11). Examinations of the factors thought to influence transitions between states illustrates the importance of rainfall amounts and distribution in determining some of the transitions, but the importance of disturbance (especially fire) and grazing management is also emphasized. The importance of this in the present context is the interaction between disturbance, management and climate. Climate *per se* determines only some

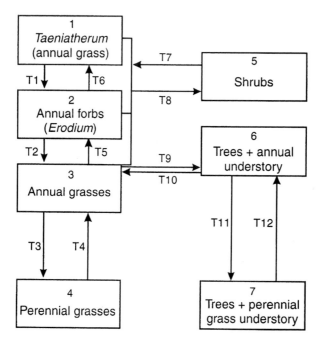

Figure 4.11   *State and transition model for California grasslands, with factors thought to induce transitions between states (modified from George et al. 1992).*

transitions; management and disturbance are implicated in many others, especially those involving large transitions from one life form (or functional group) to another. This effectively lessens the predictability of the vegetation composition at any given site.

The importance of distinguishing between transient dynamics and true shifts in functional groups is again highlighted. George *et al.* (1992) proposed that the degree of difficulty involved in switching from one state to another varied, so that, for instance, it was relatively easy to switch between states 1, 2, and 3, but more difficult to go up from 3 to 4 (Fig. 4.12). In the present context, I have interpreted this as the difference between transitions among members of the same (or similar) functional

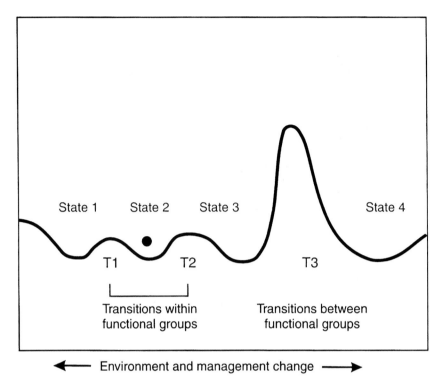

State 1   State 2   State 3   State 4

T1   T2   T3

Transitions within   Transitions between
functional groups   functional groups

◄─── Environment and management change ───►

Figure 4.12 *Degree of difficulty in moving from one state to another in California grassland (states as in Fig. 4.11) (modified from George* et al. *1992).*

groups and transitions between different functional groups. Using the examples given previously, the changes in dominance of herbaceous components of the grassland are more easily produced than the change from grassland to shrub dominance. If this is the case, this could have some predictive value in determining the response to different types of environmental change. Determination of the likelihood of transitions between different states and the events or combination of factors needed to force a transition would be an important step in predicting the likely impact of climatic and land-use changes.

Conclusions

The above discussion has indicated that *a priori* groupings of plants can be used to capture the main features of the response of different vegetation types to a variety of disturbances and climatic events. There is some evidence to suggest that the set of groups recognized in a particular vegetation will respond in a similar way to different types of impact. Hence one might conclude that such groupings based on current levels of understanding of functional responses can be useful in assessing the likely impacts of future environmental changes. Such an approach may not capture the complete dynamics of community or ecosystem change, but the importance of this will be determined by the question being asked and the level of precision required.

Most of the material discussed here dealt with changes within one particular vegetation type, although the possibility of changes between types was also discussed for the Californian grasslands. This example highlighted the notion that the distribution of vegetation types within a given area may be determined largely by the effects of past disturbances, past and present management practices, particular climate events, and chance combinations of these factors. Thus, although regional climate may delimit the broad types of vegetation likely in an area, the actual composition of the landscape mosaic will be determined by the factors listed above (Shugart, Chapter 2, this volume). Any attempt at predicting vegetation response to future environmental change at anything other than a very broad level must therefore take this into account.

Clearly, some components of the landscape mosaic may be relatively stable because they are strongly influenced by edaphic or topographic variation. However, within a given climate or edaphic zone, a number of alternative (meta)stable states may be possible. Switches between these states may be predictable if we know enough about responses to various disturbances, and so on. State and transition models such as that illustrated in Figure 4.11 may therefore be very useful in predicting changes in landscape mosaics in the face of environmental change. To do this successfully, such models need to include all possible states and transitions, including those that are considered unlikely or impossible under current conditions. For instance, in Figure 4.11, although the transition from grassland to woodland (transition 9) currently has a low probability (as inferred from its lack of occurrence), changed climatic and/or management conditions in the future could result in a higher transition probability and hence an increase in the abundance of woodland in the landscape mosaic.

The use of state and transition models that work on *a priori* functional groupings appears to have considerable potential as a relatively simple way to predict vegetation response to environmental change, based on existing knowledge. Clearly, both models and groupings can be improved as more information becomes available, but we can at least start now. Given the scale of the problems to be tackled and the short time-frame in which we have to come up with answers, this approach needs to be considered seriously.

Acknowledgements

I thank SCOPE for financial assistance to attend the workshop in Charlottesville, the workshop organizers and graduate students from the University of Virginia for assistance during my stay in Charlottesville, and Hal Mooney, William Bond, Mark Westoby and other workshop participants for discussions on various components of the paper.

References

Angelstam, P. & Arnold, G. W. (1993) Contrasting roles of remnants in old and newly impacted landscapes: lessons for ecosystem reconstruction. In Saunders, D. A., Hobbs, R. J. & Ehrlich, P. R. (eds), *Nature Conservation 3: Reconstruction of Fragmented Ecosystems, Global and Regional Perspectives*, pp. 109–25. Chipping Norton, NSW: Surrey Beatty and Sons.

Biswell, H. H. (1974) Effects of fire on chaparral. In Kozlowski, T. T. & Ahlgren, C. E. (eds), *Fire and Ecosystems*, pp. 321–64. New York: Academic Press.

Botkin, D. B., Estes, J. E., MacDonald, R. M. & Wilson, M. V. (1984) Studying the earth's vegetation from space. *Bioscience*, 34, 508–14.

Chiariello, N. R. (1989) Phenology of California grasslands. In Huenneke, L. F. & Mooney, H. (eds), *Grassland Structure and Function: California Annual Grassland*, pp. 47–58. Dordrecht: Kluwer.

Clark, J. S. (1991) Ecosystem sensitivity to climate change and complex responses. In Nyman, R. L. (ed.), *Climate Change and Life on Earth*, pp. 65–98. New York: Routledge, Chapman and Hall.

Da Silva, P. G. & Bartolome, J. W. (1984) Interaction between a shrub, *Baccharis pilularis* ssp. *consanguinea* (Asteraceae) and an annual grass, *Bromus mollis* (Poaceae) in coastal California. *Madrono*, 31, 93–101.

Davis, M. B. (1981) Quaternary history and the stability of forest communities. In West, D. C., Shugart, H. H. & Botkin, D. B. (eds), *Forest Succession*, pp. 132–53. New York: Springer.

Davis, M. B. (1986) Climatic instability, time lags and community disequilibrium. In Diamond, J. & Case, T. J. (eds), *Ecology*, pp. 269–84. New York: Harper and Row.

Davis, M. B. (1988) Ecological systems and dynamics. In *Toward an Understanding of Global Change: Initial Priorities for U. S. Contributions to the International Geosphere-Biosphere Programme*, pp. 69–106. (Ed. by the National Research Council.) Washington, D.C.: National Academy Press.

Davis, M. B. (1989) Retrospective studies. In Likens, G. E. (ed.), *Long-Term Studies in Ecology*, pp. 71–89. New York: Springer-Verlag.

Davis, M. B. & Botkin, D. B. (1985) Sensitivity of cool-temperate forests and their fossil pollen record to rapid temperature change. *Quaternary Research*, 23, 327–40.

Davis, S. D. & Mooney, H. A. (1985) Comparative water relations of adjacent California shrub and grassland communities. *Oecologia (Berlin)*, 66, 522–9.

Delcourt, H. R. & Delcourt, P. A. (1991) *Quaternary Ecology: A Paleoecological Perspective*. New York: Chapman and Hall.

George, M. R., Brown, J. R. & Clawson, W. J. (1992) Application of non-equilibrium ecology to management of Mediterranean grasslands. *Journal of Range Management*, 45, 436–40.

Gimingham, C. H. (1972) *Ecology of Heathlands*. London: Chapman and Hall.

Gimingham, C. H. (1975) *An Introduction to Heathland Ecology*. Edinburgh: Oliver and Boyd.

Gordon, D. R. & Rice, K. J. (1993) Competitive effects of grassland annuals on soil water and blue oak (*Quercus douglasii*) seedlings. *Ecology*, **74**, 68–82.

Gordon, D. R., Welker, J. M., Menke, J. W. & Rice, K. J. (1989) Competition for water between annual plants and blue oak (*Quercus douglasii*) seedlings. *Oecologia (Berlin)*, **79**, 533–41.

Graetz, R. D. (1988) Global change and terrestrial vegetation. In *Global Change. Proceedings of the Elizabeth and Frederick White Research Conference*, pp. 126–9. Canberra: Australian Academy of Science.

Graetz, R. D., Walker, B. H. & Walker, P. A. (1988) The consequences of climatic change for seventy percent of Australia. In Pearman, G. I. (ed.), *Greenhouse, Planning for Climatic Change*, pp. 399–420. Melbourne: CSIRO.

Green, B. (1993) The recognition and implementation of landscape management objectives for agriculture in the UK. In Saunders, D. A., Hobbs, R. J. & Ehrlich, P. R. (eds) *Nature Conservation 3: Reconstruction of Fragmented Ecosystems, Global and Regional Perspectives*, pp. 259–66. Chipping Norton, NSW: Surrey Beatty and Sons.

Groves, R. H. & Hobbs, R. J. (1992) Patterns of plant functional responses and landscape heterogeneity. In Hobbs, R. J. (ed.), *Biodiversity of Mediterranean Ecosystems in Australia*, pp. 47–60. Chipping Norton, NSW: Surrey Beatty and Sons.

Henderson-Sellers, A. & Blong, R. (1989) *The Greenhouse Effect: Living in a Warmer Australia*. Kensington, NSW: NSW University Press.

Hnatiuk, R. J. & Hopkins, A. J. M. (1980) Western Australian species-rich kwongan (sclerophyllous shrubland) affected by drought. *Australian Journal of Botany*, **28**, 573–85.

Hobbie, S. E., Jensen, D. B. & Chapin, F. S. III (1993) Resource supply and disturbance as controls over present and future plant diversity. In Schultze, E. D. & Mooney, H. A. (eds), *Biodiversity and Ecosystem Function*, pp. 385–408. Heidelberg: Springer-Verlag.

Hobbs, R. J. (1990) Remote sensing of spatial and temporal dynamics of vegetation. In Hobbs, R. J. & Mooney, H. A. (eds) *Sensing of Biosphere Functioning*, pp. 203–19. New York: Springer-Verlag.

Hobbs, R. J. (1992) Function of biodiversity in mediterranean ecosystems in Australia: definitions and background. In Hobbs, R. J. (ed.), *Biodiversity of Mediterranean Ecosystems in Australia*, pp. 1–25. Chipping Norton, NSW: Surrey Beatty and Sons.

Hobbs, R. J. & Atkins, L. (1991) Interactions between annuals and woody perennials in a Western Australian wheatbelt reserve. *Journal of Vegetation Science*, **2**, 643–54.

Hobbs, R. J., Gulmon, S. L., Hobbs, R. J. & Mooney, H. A. (1988) Effects of fertiliser addition and subsequent gopher disturbance on a serpentine annual grassland community. *Oecologia (Berlin)*, **75**, 291–5.

Hobbs, R. J. & Hopkins, A. J. M. (1990) From frontier to fragments: European impact on Australia's vegetation. *Proceedings of the Ecological Society of Australia*, **16**, 93–114.

Hobbs, R. J. & Hopkins, A. J. M. (1991) The role of conservation corridors in a changing climate. In Saunders, D. A. & Hobbs, R. J. (eds), *Nature Conservation 2: The Role of Corridors*, pp. 281–90. Chipping Norton, NSW: Surrey Beatty and Sons.

Hobbs, R. J. & Mooney, H. A. (1985) Community and population dynamics of serpentine grassland annuals in relation to gopher disturbance. *Oecologia (Berlin)*, **67**, 342–51.

Hobbs, R. J. & Mooney, H. A. (1986) Community changes following shrub invasion of grassland. *Oecologia (Berlin)* **70**, 508–13.

Hobbs, R. J. & Mooney, H. A. (1991) Effects of rainfall variability and gopher disturbance on serpentine annual grassland dynamics. *Ecology*, **72**, 59–68.

Hopper, S. D. (1992) Patterns of plant diversity at the population and species level in south-west Australian mediterranean ecosystems. In Hobbs, R. J. (ed.), *Biodiversity of Mediterranean Ecosystems in*

Australia, pp. 27–46. Chipping Norton, NSW: Surrey Beatty and Sons.

Huenneke, L. F., Hamburg, S. P., Koide, R., Mooney, H. A. & Vitousek, P. M. (1990) Effects of soil resources on plant invasion and community structure in Californian serpentine grassland. *Ecology*, **71**, 478–91.

Huenneke, L. F. & Mooney, H. A. (eds) (1989) *Grassland Structure and Function: California Annual Grassland*. Dordrecht: Kluwer.

Huntley, B. & Webb, T. III (1989) Migration: species' response to climatic variations caused by changes in the earth's orbit. *Journal of Biogeography*, **16**, 5–19.

Koide, R. T. & Mooney, H. A. (1987) Re-vegetation of serpentine substrates: response to phosphate application. *Environmental Management*, **11**, 563–7.

Körner, C. (1993) Scaling from species to vegetation: the usefulness of functional groups. In Schultze, E. D. & Mooney, H. A. (eds), *Biodiversity and Ecosystem Function*, pp. 117–40. Heidelberg: Springer-Verlag.

Lamont, B. B. (1982) Mechanisms for enhancing nutrient uptake in plants, with particular reference to Mediterranean South Africa and Western Australia. *Botanical Review*, **48**, 597–689.

Lamont, B. B. (1992) Functional interactions within plants – the contribution of keystone and other species to biological diversity. In Hobbs, R. J. (ed.), *Biodiversity of Mediterranean Ecosystems in Australia*, pp. 95–127. Chipping Norton, NSW: Surrey Beatty and Sons.

Le Houérou, H. N. (1981) Impact of man and his animals on Mediterranean vegetation. In de Castri, F., Goodall, D. W. & Specht, R. I. (eds), *Mediterranean-type Shrublands*, pp. 479–521. Amsterdam: Elsevier.

Leishman, M. R., Hughes, L., French, K., Armstrong, D. & Westoby, M. (1992) Seed and seedling biology in relation to modeling vegetation dynamics under global climatic change. *Australian Journal of Botany*, **40**, 599–613.

Leishman, M. R. & Westoby, M. (1992) Classifying plants on the basis of associations of individualistic traits: evidence from Australian semi-arid woodlands. *Journal of Ecology*, **80**, 417–24.

McNaughton, S. J. (1968) Structure and function in California annual grasslands. *Ecology*, **49**, 962–72.

Mooney, H. A., Hobbs, R. J., Gorham, J. & Williams, K. (1986) Biomass accumulation and resource utilization in co-occurring grassland annuals. *Oecologia (Berlin)*, **70**, 555–8.

Noble, I. R. & Slatyer, R. O. (1980) The use of vital attributes to predict successional changes in plant communities subject to recurrent disturbances. *Vegetatio*, **43**, 5–21.

Pearman, G. I. (ed.) (1988) *Greenhouse. Planning for Climate Change*. Melbourne: CSIRO.

Pennington, W. (1969) *The History of British Vegetation*. London: English Universities Press.

Pitt, M. D. & Heady, H. F. (1978) Responses of annual vegetation to temperature and rainfall patterns in northern California. *Ecology*, **59**, 336–50.

Saunders, D. A. & Hobbs, R. J. (1992) Impact on biodiversity of changes in land-use and climate. In Hobbs, R. J. (ed.), *Biodiversity of Mediterranean Ecosystems in Australia*, pp. 61–75. Chipping Norton, NSW: Surrey Beatty and Sons.

Scougall, S. A. (1991) *Edge effects in fenced and non-fenced remnants on Jam-York Gum woodlands in the Western Australian wheatbelt*. B. Appl. Sci. thesis, Curtin University of Technology, Perth.

Scougall, S. A., Majer, J. D. & Hobbs, R. J. (1993) Edge effects in grazed and ungrazed Western Australian wheatbelt remnants in relation to ecosystem reconstruction. In Saunders, D. A., Hobbs, R. J. & Ehrlich, P. R. (eds), *Nature Conservation 3: Reconstruction of Fragmented Ecosystems, Global and Regional Perspectives*, pp. 163–78. Chipping Norton, NSW: Surrey Beatty and Sons.

Solbrig, O. T. (1993) Plant traits and adaptive strategies: their role in ecosystem function. In Schultze, E. D. & Mooney, H. A. (eds),

Biodiversity and Ecosystem Function, pp. 97–116. Heidelberg: Springer-Verlag.

Thirgood, J. V. (1981) *Man and the Mediterranean Forest: a History of Resource Depletion*. London: Academic Press.

Tilman, D. (1988) *Plant Strategies and the Dynamics and Structure of Plant Communities*. Princeton, New Jersey: Princeton University Press.

Turner, B. L. II, Clark, W. C., Kates, R. W., Mathews, J. T., Richards, J. R. & Mayer, W. (eds) (1991) *The Earth as Transformed by Human Action*. New York: Cambridge University Press.

US Committee on Global Change (1990) *Research Strategies for U. S. Global Change Research Program*. Washington, D.C.: National Academy Press.

Vernet, J.-L. & Thiebault, S. (1987) An approach to northwestern Mediterranean recent prehistoric vegetation and ecologic implications. *Journal of Biogeography*, 14, 117–27.

Walker, B. H. (1992) Biodiversity and ecological redundancy. *Conservation Biology*, 6, 18–23.

Weiss, S. & Murphy, D. (1993) Climatic considerations in reserve design and ecological restoration. In Saunders, D. A., Hobbs, R. J. & Ehrlich, P. R. (eds). *Nature Conservation 3: Reconstruction of Fragmented Ecosystems, Global and Regional Perspectives*, pp. 89–107. Chipping Norton, NSW: Surrey Beatty and Sons.

Weiss, S. B., Murphy, D. D. & White, R. R. (1988) Sun, slope and butterflies: topographic determinants of habitat quality for *Euphydryas editha*. *Ecology*, 69, 1486–96.

Westoby, M., Walker, B. H. & Noy-Meir, I. (1989) Opportunistic management for rangelands not at equilibrium. *Journal of Rangeland Management*, 42, 266–74.

Williams, K. & Hobbs, R. J. (1989) Control of shrub establishment by spring soil water availability in an annual grassland. *Oecologia (Berlin)*, 81, 62–6.

Williams, K., Hobbs, R. J. & Hamburg, S. (1987) Invasion of annual grassland in Northern California by *Baccharis piluris* ssp. *consanguinea*. *Oecologia (Berlin)*, 72, 461–5.

Wills, R. T. (1994) The ecological impact of *Phytophthora cinnamomi* in the Stirling Range National Park. *Australian Journal of Ecology*, 77, 127–31.

Wolman, M. G. & Fournier, F. G. A. (eds) (1987) *Land Transformation in Agriculture (SCOPE 32)*. New York: SCOPE.

Woodward, F. I. (1992) Predicting plant responses to global environmental change. *New Phytologist*, 122, 239–51.

Wyman, R. I., Steadman, D. W., Sullivan, M. E. & Walters-Wyman, M. F. (1991) Now what do we do? In Wyman, R. I. (ed.), *Global Climate Change and Life on Earth*, pp. 252–63. New York: Routledge, Chapman and Hall.

⑤ Functional types in non-equilibrium ecosystems

B. H. Walker

Introduction

The notion of using plant functional types (PFTs) in global change analyses stems from the need to develop general models of vegetation composition as a function of climate, CO_2 and land use, and then to use these models to predict vegetation changes. In order to relate vegetation to climate there is an assumption, to a greater or lesser extent, of equilibrium between them. At a very coarse scale this is largely true. The Holdridge Classification is in effect a model of vegetation functional types (VFTs). The IIASA Biome I model (Prentice et al. 1992) is a PFT model based on dominant type only, and therefore effectively equates to a VFT model.

At coarse, biome scales, therefore, the use of PFTs or VFTs is appropriate and useful (e.g. Woodward 1987). At patch and landscape scales, however, and especially in areas of high climatic variability, it poses a problem. In widely differing biomes (cf. Westoby et al. 1989 for savannas and Van Cleve & Viereck 1981 for boreal forest) species composition generally does not exhibit a tight correlation with mean climate. It varies over time and may exhibit more than one metastable state, and in each state the vegetation exhibits a wide range in both the kinds and the proportions of species. A change in state is mostly dictated by major episodic events, or combinations of particular conditions, and between these events the changes reflect fluctuations in response to variations in climate.

The description by Austin & Williams (1988), of the changes in vegetation of what is now a forest area of New South Wales, Australia, known as the Pillaga Scrub, illustrates how one site can support a range of quite different vegetation types with no change in mean climate. Over the past 150 years the vegetation of this area has changed from an open, grassy savanna with scattered *Eucalyptus* trees and a few cypress pines (*Callitris*) to a dense, even-aged stand of cypress pine, to a structured forest of cypress pine with some *Eucalyptus*. The changes were associated with a few major events involving such phenomena as the introduction of sheep, cessation of fire, invasion by rabbits, the introduction of myxomatosis, and major droughts. Modelling vegetation composition and predicting its change in response to global changes calls for recognition of these dynamics.

Attributes of non-equilibrium systems

I begin with an assertion that there are no special attributes of non-equilibrium systems and there is no dichotomy between those ecosystems that are in equilibrium and those that are not. Rather, there is a continuum, either in the degree to which the composition of ecosystems is out of phase with that which the current environment favours, or in the propensity for ecosystems to persist in more than one (meta-)stable state under the same environment. At the low end of this continuum the non-equilibrium effects can be ignored; at the high end they cannot. With this in mind let us proceed by examining some of the issues that are of interest in ecosystems far from equilibrium.

The issue of redundancy

Apparent species redundancy is observable in most ecosystems, at least with regard to the major ecosystem processes. Examples of two or more plant species performing the same role in regard to one process (such as primary productivity), but different roles in another (such as herbivory), are common. The special significance of some species may only become apparent under particular circumstances (exceptional droughts, intense fire, etc.). In the context of global change analyses, there is a particular distinction between functional types relating to the role plants perform in ecosystem processes, and their functionally different responses to environmental shifts. As discussed later, these two classifications may result in quite different groupings of species. The species that constitute a single functional type with respect to, say, water-use efficiency may respond quite differently to an increase in minimum temperature. In the absence of a complete understanding, this apparent species redundancy is best regarded as a system adaptation against a highly variable and unpredictable environment. In many cases it is likely to be an essential component of, and in fact may well be used as a measure of, the system's ability to continue functioning when stressed or disturbed. It therefore provides for constancy in ecosystem function, in that species substitution minimizes (or lessens) variation in primary productivity, nutrient cycling, etc. For example, McNaughton (1985) showed that those grass communities in the Serengeti that displayed the highest variability in proportional species composition over time were most constant in terms of production. One mechanism for this is the existence of phenological suites of species, e.g. early, mid- and late-season growers, as shown by Silva (1987) for Venezuelan grasslands. The relative performance (and therefore abundance) of each phenological 'guild' in any year is determined by the timing of the rainfall.

In order to achieve a useful PFT classification there is a need for better definition and understanding of the critical ecosystem processes and environmental factors that determine the structure and function of an ecosystem. One way of approaching the problem in systems that are generally not in equilibrium is to ask what the necessary functional attributes of such systems are, the functions that distinguish them from systems that *are* in equilibrium. All ecosystems fix nitrogen, decompose organic matter, transpire, and so on. Are there particular functions or processes associated only with systems that are not in equilibrium and what are the properties of such systems? If a non-equilibrium system were brought to equilibrium, what changes would occur?

As a first approximation, if the environment of such a system were held constant, species best adapted to particular conditions that no longer occur (albeit rarely), and which are inferior competitors or are unable to reproduce under the prevailing condition, would disappear. In systems prone to alternative states, the analogy would be loss of dominant species or suites of species that characterized the properties of the alternative phases or states.

There are no obvious special ecosystem-level functions that occur only in systems far from equilibrium, but these systems do share one predominant property: they all have unstable mixtures of species that collectively afford a higher diversity of adaptive features than occurs in very stable systems. In terms of the old adage about eggs and baskets, non-equilibrium systems tend to have each of their functionally different eggs replicated in several different kinds of basket.

If we regard substitutability as an ecosystem property particularly associated with non-equilibrium systems, can we measure it (or the lack of it) and then use such a measure to test whether it is associated with differences in functional behaviour of ecosystems over time? What is the nature of these different responses and, though some would be peculiar to particular ecosystems, can we identify any general forms?

At a longer timescale, some systems may not be in equilibrium with the present climate owing to a lag in the response of vegetation to a changed environment. An example is the slow migration of tree species in the northeastern USA, from their glacial refugia into regions where they are now climatically adapted (Brubaker 1981). The consequence of this is an apparent (displaced) optimum environment for the species concerned. Also, any correlative models would give incorrect predictions of potential distributions under future climate scenarios. Use of plant functional attributes (rather than species) will overcome this bias to some extent, but for such systems there is no completely satisfactory solution.

Most commonly, perhaps, non-equilibrium systems occur, on a much shorter timescale, because of differences in the rates at which climate varies and plants can respond. Under a very variable climate the lag in demographic processes means that, especially for long-lived species, species composition never gets into an equilibrium with climate except by chance, for brief periods when the vegetation and climate trajectories coincide (see later discussion of tree : grass composition in savannas).

Approaches to analysing functional types

Appropriate models of vegetation change

For ecosystems that can exist in different states, an appropriate model is one that identifies, for the region in question, the following.

1. The major determinants of vegetation structure and composition (e.g. plant available moisture, temperature, nutrients, fire, herbivory), which particular attributes or characteristic features of these determinants are important (for example, for temperature is it the occurrence of frost or very high temperatures?), and how they affect plants.
2. The major landscape units.
3. For each landscape unit the number and nature of vegetation states, and the extreme events or combinations and sequences of conditions which result in either mortality or successful establishment of various dominant and/or characteristic kinds of plants.

(Note: For global change purposes differences in vegetation composition are recognized as constituting different states if their feedback effects are significantly different. It is a user-defined criterion, and it may be that in some cases change is more continuous than stepped, and the continuum needs to be divided up for convenience.)

In attempting to establish a classification of PFTs for such a model, the set of plant species that needs to be considered is the full complement, including all possible states in all landscape units. Non-equilibrium systems that exhibit alternative states therefore call for a two-stage approach to the use of PFTs: first, the development of the full array of PFTs that accounts for all possible states, and then the grouping of PFTs into (overlapping) sets that constitute the different states, or VFTs. This is perhaps the only special consideration that is required in the application of PFTs to non-equilibrium systems.

Classifying function: the species approach

The classification of functions in vegetation can be considered at various levels; the task becomes unmanageable unless the complexity is somehow reduced. This is best achieved, as described earlier, by distinguishing between the response to global changes and the contribution to, or feedback to, such changes.

Feedback functions

For feedback effects the scale is such that we want an integrated measure over a large area. In other words we want VFTs rather than PFTs. VFTs are of course the sum of their PFTs, but the important parameters for feedback to climate (albedo, evapotranspiration (ET) and turbulent transfer (Z_0)) and atmospheric composition operate at coarse scales. The characteristics of vegetation that are important in this feedback are relatively few (Table 5.1). Out of a total of some 15 ecosystem attributes that together determine the feedbacks, there are about 10 that relate directly to vegetation, and which determine albedo, ET and Z_0.

For feedback modelling, therefore, it would seem more appropriate to go straight from functional attributes to VFTs without the intervening step of having to define PFTs. However, such an approach will not allow for prediction of how the VFT will change in response to a change in environment.

Response functions

In contrast to VFT feedback, determining the functional types that reflect the different responses of plants to global change is much more complex. The problem is one of where to cut off: distinguishing between differences in plant functional response that are important and have significant consequences for ecosystem function, and those that reflect only individual species' traits of an idiosyncratic nature. Furthermore, determining PFTs for response to global change is region-specific, and there may well be no general global PFT classification.

The derived PFTs must relate to global change phenomena, primarily responses to CO_2, temperature (maximum, minimum and mean), available soil moisture, available nutrients and secondary effects such as fire and herbivory (Table 5.2).

The PFTs must also be considered in terms of both the mature (growth and survival) phase and the reproductive (dispersal and establishment) phase. Leishman & Westoby (1992), for example, found that there was little overlap between groups of species classified into functional types according to their reproductive characteristics or their growth characteristics.

Table 5.1 *Plant functional attributes involved in the feedback effects of terrestrial ecosystems on climate and atmospheric composition*

Feedback effect	Feedback processes	Ecosystem		Functional attributes
		properties	processes	
1. Surface energy exchange	reflectance	albedo	FPAR	LAI
				deciduousness
				soil colour
				plant cover
				litter cover/mass
				texture (?)
	latent heat	ET	soil evaporation	G_{max}
				WUE
				LAI
			transpiration	deciduousness
				rooting depth
			turbulent transfer (Z_0)	height
				spatial pattern
2. Water	ET	(as above)		
3. Momentum transfer	drag	structure (Z_0) (as above)		
4. Concentration of radiatively active compounds	net trace gas emission	net CO_2	NPP allocation, decomposition	LAI
				biomass/structure
				leaf C : N
				litter mass
				soil OM
		net CH_4	adsorption, anaerobic decomposition	soil texture
				soil water
				soil surface
				ruminant biomass
				fire
		NO_x, N_2O	N fixation mineralization	?
				C : N

Table 5.2 *Plant functional type attributes involved in the response of plants to environmental changes*

Environmental change	Plant response	Plant attribute
1. Atmospheric CO_2	photosynthetic pathway	C_3/C_4
	water-use efficiency	C_3/C_4
2. Maximum temperature	heat resistance	threshold effect
	water access	root depth
	albedo	colour
3. Minimum temperature	frost resistance	threshold effect
4. Mean temperatures	growth/competition	optimum temperature
5. Water		
total	growth/competition	size
seasonality	growth phenology	phenology
depth	root depth distribution	growth/competition
frequency of droughts	drought resistance	threshold
6. Fire	resistance	?
7. Available nutrients	growth, competition	optimal level
		threshold effects

Grouping these functional attributes for response to environmental change, for both mature and reproductive phases, can potentially lead to an unwieldy number of PFT combinations. Fortunately, the likely combinations are considerably constrained (woody annuals 10 m high do not exist). As an illustration of how the two kinds of functional grouping might be combined, Table 5.3 presents a tentative PFT classification for a generalized tropical savanna. It is not surprising that more than half of the plant functional attributes (PFAs) are determinants of both the feedback and the response functions. Only one of the PFTs (plant structure, or physiognomy, classed as tree, shrub or herbaceous) occurs in easily recognizable, discrete classes, and even in this case there is a considerable range within each. In all of the others there is a continuum (e.g. shallow to deeply rooted) and the usefulness of the classification will depend on how well the classes are associated with significant differences in function.

A classification using attributes of all the species requires that it be based on readily observable features. In practical terms it is simply not possible to characterize all plants in all areas in terms of, for example, the osmotic potential of their cell sap or their DNA mass. Morphological surrogates will have to be used as far as possible if a direct classification of plants is

Table 5.3 *Possible PFTs for a 'generic' savanna*

PFT	Feedback	Response
1. evergreen ... deciduous	albedo, ET	survival
I 4	(FPAR)	(drought avoidance)
2. deep ... shallow roots	ET	competition
I 4		survival
3. water-use efficiency	ET	competition
high ... low		
I 3		
4. CO_2 uptake C4/C3	CO_2, flux, ET	competition
5. drought resistance; H ... L	—	competition
I 3		survival
6. tree/shrub/herb	ET, Z_0	fire, soil water
7. longevity; annual ... centuries	—	survival
I 5		
8. heat resistance; H ... L	—	establishment, survival
9. cold resistance; H ... L	—	establishment, survival
10. fire resistance; H ... L	—	establishment, survival
11. N-fixation: H ... L	N flux	competition, NPP
12. leaf/litter quality (C : N)	CO_2 flux, pools;	herbivory
high ... low	N flux, pools	

attempted. Gillison (1981, 1988), for example, has developed such a scheme for tropical rainforests and imputes a number of physiological attributes on the basis of associated morphological traits. His system has had some success in terms of associating such plant attributes to environmental variables, and has also been shown to predict the distribution of selected fauna better than a vegetation classification based on species composition (A. Gillison, personal communication). It has still to be tested as a system for predicting shifts in vegetation in response to changes in climate.

This distinction between the functional attributes and the observable characteristics has been further developed by Smith *et al.* (1993), who describe them as *intensive* functional-type characteristics and *extensive* descriptor-based functional types. They list ten intensive characteristics (leaf longevity, threshold temperature for growth, etc.) needed to incorporate all the parameters of Woodward's (1993) global photosynthesis and growth model. They then develop nine extensive functional types that would be needed to capture the intensive characters (physiognomy, desiccation features, life-span, seed dispersal, photosynthetic pathway, etc.). Further refinement and development of this approach, to make it applicable beyond just coping with Woodward's global model, would be a profitable line of research.

Classifying function: the environmental approach

We can approach the derivation of a PFT or VFT classification in two ways. One is to begin with the plants, as described in the preceding sections. The reverse procedure is to begin with the environment, by attempting to determine the minimum, sufficient set of climatic, soil-based and disturbance features that will determine which kinds of plant can persist. Such a scheme can be used to classify the PFT 'niches' of a region (let's avoid the semantic argument about 'no such thing as a vacant niche'), and it then needs to be determined which of, or whether any of, the plants in the region fit those requirements (analogous to the 'vital attributes' approach of Noble & Slatyer 1980). A complete, global analysis of such PFT niches (environmental combinations) would be analogous to the periodic classification of the elements. The difference is that the PFT classification would be based on arbitrary subdivisions and it is a matter of utility whether we recognize the plant equivalents of gold and silver as being different, or merely both precious metals.

As a start, the major determinants of plant species distribution and performance, and the ways in which they are important, are as follows.

Plant available moisture (PAM)

the total, annual amount
seasonality (e.g. winter vs. summer)
interannual variation
the extremes of variation: the frequencies of major droughts and very wet years

Temperature (T)

maximum
minimum
optimum (for maximum growth)

The first two are important for plant survival, the third for competitive success. The effects of the three can be expressed in different ways; for example, in tropical savannas two important aspects are the occurrence or non-occurrence of frosts and the radiation load a plant has to endure in the absence of transpirational cooling (i.e. heat stress).

Available nutrients (AN)

There is a continuum from highly fertile (eutrophic) soils to very infertile (dystrophic) soils. Plants have evolved particular mechanisms for adapting to all segments of this continuum, including (at the low end) complex mycorrhizal associations and rhizosphere interactions. Plants generally respond most to changes in soil inorganic nitrogen, but because it is dependent on atmospheric replenishment and is strongly influenced by ecosystem management, a better, inherent measure of AN is available P and the sum of exchangeable cations.

Fire (F)

frequency
seasonal timing
intensity

Collectively, the three combine to give a distinct fire regime.

Herbivory (H)

The two important aspects are the general levels of intensity (usually high in eutrophic and low in dystrophic systems), and their variation over time, especially in the case of insect herbivory which is characterized (e.g. in the case of lepidopteran larvae and locusts) by massive, episodic defoliation events.

In a stable, equilibrium world, the environmental space could be reduced to three simple axes, the means of PAM, AN and T. The other aspects described above would either be non-existent or captured by these three measures. For non-equilibrium systems, however (and for virtually all systems that have significant environmental variability) it is necessary to consider the interactive effects of the determinants as they vary over time. A particular year's weather on a particular site, with a particular fire condition and level of herbivory, will favour a particular PFT.

As an illustration, on the sandy soil site of the South African Savanna Ecosystem Project in the northern Transvaal, a year with 800 mm of rain (above average) and a fire in mid-winter will favour mature individuals of the dominant tree (*Burkea africana*), which is relatively deeply rooted (with 800 mm much of the water will have gone through to the deeper layers) and is fire-resistant. If, however, this year of 800 mm followed a year of 400 mm, although the trees would be strongly favoured the net effect would be different, because the 400 mm year would have caused death of trees and parts of trees, and the remaining amount of tree biomass would be insufficient to use all the extra water. *Burkea africana* cannot grow fast enough to take advantage of a sudden large increase in available water. The response rate of grass, however, is much faster and so grass would be favoured and would increase more than the trees, even though there would be good growth of the latter and an increase in branch density and individuals (which, for trees, means an increase in potential leaf area index). The net effect of all this is that the woody : grass ratio in savannas decreases with increasing interannual variation in rainfall. If the mean annual rainfall fell in exactly the same way every year, tree biomass would increase to the level that the available water could support, and virtually exclude grasses.

This example illustrates an essential consideration in the selection of PFTs in unstable environments: it is the frequency distributions of rainfall, fire and herbivore conditions, together with the *rate* at which these conditions vary, that determine the proportional composition of PFTs (in the example above, of trees and grasses). One important functional attribute of all PFTs, therefore, is the response time in regard to change in the dominant environmental determinant(s).

These effects are not precise and are continuous, and the question for a PFT analysis, therefore, concerns how much of a change in the frequency distribution and its pattern over time is needed to induce a significant change in the proportions of PFTs. In particular, in regard to global change, how is this influenced by superimposing a trend on the change in frequency distribution?

A particular aspect of response to global change is the enhancement of the impact of invasive alien species and other potential 'weeds'. Such changes are likely to manifest themselves as a sudden and broad-scale increase in a particular species. A current example is the rapid dominance by the introduced tree *Acacia nilotica* in the previously treeless Mitchell grass plains of Australia. The invasive species problem is a special case of PFT analysis. Recognising such 'sleepers' will involve identifying threshold conditions that will allow such a species to take over.

A logical way to approach the problem is via analysis of shifts in the PAM– AN–T–F–H environmental space, leading to prediction of the PFT(s) that will be favoured. In the *A. nilotica* example, such an analysis of present conditions predicts the potential presence (if not dominance) of a tree species. The reason that the Mitchell grass plains have until now remained treeless is the absence in the Australian flora of a tree that can cope with the alternate wetting and drying of self-mulching, cracking clay soils. *A. nilotica*, introduced from Pakistan early this century, is such a species.

As a final comment to this section, the threshold, hysteresis effects, and other non-linear responses that are a feature of non-equilibrium systems, combine to give rise to changes that are associated with particular conditions or combinations of conditions. It is therefore not enough to know only the beginning and final environmental conditions of an area in an attempt to predict the change in its vegetation; it is also necessary to know the sequence and magnitudes of the intervening conditions.

Conclusion

I conclude with the initial assertion that there are no special attributes of non-equilibrium systems that warrant a dichotomy between two types. Some attributes, however, become increasingly important the further a system is from a stable, equilibrium state. In terms of the PFTs themselves, there are none that are peculiar to non-equilibrium systems. The same PFTs may occur in both. The combination of PFTs, however, may well be different, and collectively the combination in a non-equilibrium system imparts a set of properties that enable it to persist under a long-term disturbance regime.

The main consideration in using PFTs in such non-equilibrium systems is to distinguish between response to the mean climate and response to the particular climatic and other environmental conditions, and sequences of such conditions, that induce changes in state. Without any change in the mean, the particular conditions can induce differences in vegetation that may well result in significant differences in terms of feedback to climate, or value to managers.

Under present climate regimes, PFTs can be classified in three ways: (1) a relatively small set of PFTs that reflect the responses to mean climatic conditions; (2) a larger set, including the first but with additional FTs that reflect differences in response to particular climatic events or combinations of conditions; and (3) an overlapping but different set classified in terms of their feedback effects on climate. Although this third set has some congruence with the second, there is no necessary relationship and use of either of the first two in a coupled global climate–dynamic global vegetation model will require a translation scheme into the third. For non-equilibrium ecosystems the appropriate PFT classification will be of the second and third kinds, including the translation scheme.

The extra dimension of environmental variation in non-equilibrium systems requires that the PFT classification include all possible states of vegetation, and cannot be based on a single assessment at one point in time. In the development of this full set of potential PFTs, particular attention must be paid to regional disturbance regimes (rare and extreme climate events, fire, herbivory and disease) as well as to the determinants of mean conditions. This leads to the need for defining the environment of a site (region) in terms of the frequency distribution of the 'conditions' of its determinants (PAM, T, AN, F, H) and their rate and magnitude of variation.

I suggest that the best way to proceed in the quest for an appropriate global change-related PFT classification is to adopt both approaches, starting with the plants and starting with the environment, and iterating between them.

References

Austin, M. A. & Williams, O. B. (1988) Influence of climate and community composition on the population demography of pasture species in semi-arid Australia. *Vegetatio*, **77**, 43–9.

Brubaker, L. (1981) Long-term forest dynamics. In West, D. C., Shugart, H. H. & Botkin, D. B. (eds), *Forest Succession: Concepts and Application*, pp. 95–106. Berlin: Springer-Verlag.

Gillison, A. N. (1981) Towards a functional vegetation classification. In Gillison, A. N. & Anderson, D. J. (eds), *Vegetation Classification in Australia*, pp. 30–41. Canberra: CSIRO and Australian National University Press.

Gillison, A. N. (1988) *A plant functional proforma for dynamic vegetation studies and natural resource surveys.* Technical Memorandum 88/3, Division of Water Resources, CSIRO, Australia.

Leishman, M. & Westoby, M. (1992) Classifying plants into groups on the basis of associations of individual traits – evidence from Australian semi-arid woodlands. *Journal of Ecology*, **80**, 417–24.

McNaughton, S. J. (1985) Ecology of a grazing ecosystem; the Serengeti. *Ecological Monographs*, **55**, 259–94.

Noble, I. R. & Slatyer, R. O. (1980) The use of vital attributes to predict successional changes in plant communities subject to recurrent disturbance. *Vegetatio*, **43**, 5–21.

Prentice, I. C., Cramer, W., Harrison, S. P.,

Leemans, R., Monserud, R. A. & Solomon, A. M. (1992) A global biome model based on plant physiology and dominance, soil properties and climate. *Journal of Biogeography*, **19**, 118–34.

Silva, J. F. (1987) Responses of savannas to stress and disturbance: Species dynamics. In Walker, B. H. (ed.), *Determinants of Tropical Savannas*, pp. 141–56. IUBS Monograph Series 3. Oxford: IRL Press.

Smith, T. M., Shugart, H. H., Woodward, F. I. & Burton, P. J. (1993) Plant functional types. In Solomon, A. M. & Shugart, H. H. (eds), *Vegetation Dynamics and Global Change*, pp. 272–92. New York: Chapman & Hall.

Van Cleve, K. & Viereck, L. A. (1981) Forest succession in relation to nutrient cycling in the boreal forest of Alaska. In West, D. C., Shugart, H. H. & Botkin, D. B. (eds), *Forest Succession: Concepts and Application*, pp. 185–211. Berlin: Springer-Verlag.

Westoby, M., Walker, B. H. & Noy-Meir, I. (1989) Opportunistic management for rangelands not at equilibrium. *Journal of Range Management*, **42**, 266–74.

Woodward, F. I. (1987) *Climate and Plant Distribution.* Cambridge University Press.

Woodward, F. I. (1993) Leaf responses to the environment and extrapolation to larger scales. In Solomon, A. M. & Shugart, H. H. (eds), *Vegetation Dynamics and Global Change*, pp. 71–100. New York: Chapman & Hall.

6 Categorizing plant species into functional types

M. Westoby and M. Leishman

Introduction

Ecological classification schemes for plant species have a long and rich tradition in plant geography and ecology. The many schemes that have been proposed embody a great deal of knowledge as to how plants with different attributes are distributed geographically and within landscapes. At the same time, surprisingly few studies have applied explicit criteria to decide how many different categories of plants are worth recognizing, where the boundaries among categories should be drawn, or whether a scheme can be assessed as successful or otherwise. If we are to work towards an agreed, worldwide scheme of plant functional types, explicit criteria will be needed.

Plant species vary in a wide range of attributes, but this variation is not random. Some combinations of attributes are more commonly represented than others, among extant species. For example, species adapted for dispersal by vertebrates tend to have relatively large seeds. This might come about because large seed size is only adaptive when the fruit is dispersed by vertebrates, or because habitats that favour large seed size also favour dispersal by vertebrates, or because plant lineages adapted for dispersal by vertebrates also tend to have large seeds. It is important to appreciate that these three types of interpretation can be complementary; it is an error to treat them as mutually exclusive alternatives.

To the extent that plant attributes tend to occur in some combinations and not others, it should be possible to describe the prolific diversity of the plant world in terms of a relatively small set of ways of making an ecological living, or plant functional types (PFTs). Such an ecological, as opposed to taxonomic, description of the diversity of land plants has long been an implicit objective of evolutionary ecology. In recent decades, however, few evolutionary ecologists have set their sights as high as an actual worldwide PFT classification; rather, they have sought to characterize the evolutionary or environmental trade-offs that would underlie such a typology.

Currently, the ambition of building a PFT classification with worldwide applicability is re-emerging as a credible and widely held research objective. Three circumstances are contributing to this re-emergence. First, a sufficient body of ideas and evidence has accumulated with regards to major trade-offs that might underlay a typology, so that the time is ripe to construct classifications and to test them. Second, accessible bodies of comparative data have been compiled, for example by progressive computerizing of herbarium information over the past 15 years or so, and PC-based software for handling databases is now cheap and routine. Over the forthcoming couple of decades this groundwork of data compilation will be coming to fruition; individual researchers will have convenient access to data such as detailed distributions and flowering dates for large numbers of species, where previously it would have cost months of herbarium work to compile such information for a single species. The third circumstance is the emergence of international scientific collaboration to address problems of global change, in the form of the International Geosphere–Biosphere Programme (IGBP). A major objective of the programme on Global Change and Terrestrial Ecosystems (GCTE) within the IGBP is to develop the capacity to project vegetation dynamics under future global change. Since it is hardly credible that vegetation dynamics models can be parameterized individually for all quarter-million species of the terrestrial flora, the IGBP Operational Plan (Steffen *et al.* 1992) includes Task 2.1.1, 'Global Key of Plant Functional Types'.

In this chapter, we first comment briefly on the many ecological classifications that have been proposed for plants, based on correlations among plant traits or between plant traits and environment. We then ask whether the correlation structure among plant traits has natural discontinuities; in other words, whether a limited number of natural groupings can be recognized, as opposed to a continuous spectrum of variation. Surprisingly, this question does not seem to have been addressed previously. The analysis is for a flora of 300 species from semi-arid woodland in Australia. A wide selection of traits was used, subject to the practical limitation that new experimental work was not required in order to obtain them. This enabled us to consider also to what extent a classification devised for one purpose, using one set of traits, would correspond to classifications devised for other purposes using different traits.

Finally, we take some features of the correlation structure found in the database for semi-arid woodlands, and ask whether the same features are found consistently in floras from other environments and different continents.

Brief history of ecological classifications of plant species

Botanists have been classifying plants on the basis of plant form since the
beginnings of botanical studies. An early system based on physiognomy
(external appearance or characteristic form) was developed by Theophrastus
around 300 BC (du Rietz 1931), and general growth forms such as trees,
shrubs, forbs and grasses have long been recognized.

One major objective has been to devise classification systems that capture
the manner in which vegetation varies geographically in response to climate.
Early classification systems (for example von Humboldt 1806; Kerner 1863;
du Rietz 1931; extensively reviewed by Cain 1950 & Barkman 1988) used
morphological and physiognomic attributes including growth form, height,
stem morphology, leaf type, longevity and seasonality. More recent classifications in this line (e.g. Kuchler 1949; Dansereau 1951; Ellenberg &
Mueller-Dombois 1967; Fosberg 1967; Webb 1959, 1968, 1976; Gillison
1981; Johnstone & Lacey 1984; Barkman 1988) are often elaborate, using
numerous plant attributes and incorporating structural features of vegetation
such as spacing, canopy depth and foliage cover. They can produce a large
number of species groups, for example 23 groups in the system of
Ellenberg & Mueller-Dombois (1967), and 88 groups in the system of
Barkman (1988). The best-justified scheme along these lines is by Box
(1981).

The authors mentioned so far have not, on the whole, set up explicit
criteria for choosing some plant attributes rather than others as discriminating between groups of species. Implicitly, their criterion of a successful
classification scheme has usually seemed to be that it captured important
geographical variation in vegetation.

Other authors have constructed classification systems that begin from a
judgment as to what environmental pressures are of interest, and what plant
traits reflect tolerance or response to those pressures. Early examples
include de Candolle (1818), Drude (1890), and Warming (1923); the system
of Raunkiaer (1934), based on the location of perennating buds during the
most difficult season, remains in wide use today. The present approach of
the IGBP lies firmly within this tradition of identifying relevant traits *a
priori*, then dividing species into groups according to those traits. The expectation (Steffen *et al.* 1992) is that PFTs will be 'based on a minimum set of
functional attributes that are considered to be most critical in reliably
predicting the present-day distribution of plants from climatic input
variables'.

Some classifications (seminal were those of Pound & Clements (1898)
and Braun-Blanquet (1932)) have been relatively less interested in the

geographical distributions of plants in response to climate, and more
concerned with effects within landscapes, for example effects of soil fertility
or of succession. Some examples are: Sarmiento & Monasterio (1983) for
tropical savanna plants; Whitmore (1988) for forest tree species; Shantz
(1927), Solbrig *et al.* (1977) and Orshan (1953) for arid zone species;
Dyksterhuis (1949), Dix (1959), Stoddart *et al.* (1975) and Friedel *et al.*
(1988) for rangeland species in response to grazing; Ellenberg (1952) for
pastures in Europe, and Knight (1965) for prairies in North America.

Certainly the most highly developed within-landscape scheme is the
C–S–R triangle (Grime 1974, Grime *et al.* 1987, 1988; Chapter 7, this
volume). Grime's scheme asserts that life-history opportunities for plants
vary along two main gradients. One gradient runs from C situations, where
a fast growth rate can be supported, to S situations where only slow
biomass accumulation is possible. The other gradient runs from R situations, where accumulated biomass is frequently destroyed, to other situations not subject to recurrent disturbance. Grime has done a great deal of
work positioning species of the British herbaceous flora in relation to these
two gradients, and identifying plant attributes that correlate with the strategies.

In summary, ecological classification schemes for plant species have a
long and rich tradition within ecology and plant geography. Unques-
tionably, these classifications summarize very extensive experience as to
which plant attributes are correlated with which, and with different environmental pressures. Up to the present time, however, the tradition has taken
remarkably few steps to move into hypothesis-testing mode. To illustrate
this, we will briefly consider three sorts of hypothesis test that one might
seek to implement.

First, one might use a classification scheme to work from climate variables at a location to an *a priori* prediction of the mixture of plant types
found there. Box (1981) synthesized much existing information to develop a
scheme that could make testable predictions of this kind. Surprisingly,
though, the decade since his book appeared has not seen a flow of tests of
his model, followed by refinements where it failed the test.

Second, principles of parsimony might be applied to the proliferation of
categories evident in some classification schemes. Does the introduction of
each further plant attribute into a key really discriminate a group of plants
that has distinctly different ecological responses?

Third, groupings could be constructed by working from observed
patterns of correlations among attributes, rather than selecting indicator
attributes on the strength of field experience and knowledge of the floras.
Again surprisingly, we know of only two previous attempts (Parsons 1976;

Grime *et al.* 1987) to classify plants into ecological groups by assembling a data table of plant attributes, then applying quantitative clustering methods to it, without prejudging the question of which variables are most important.

It seems evident to us that if a single, worldwide classification is to be constructed, it will be necessary to use explicit and therefore replicable methods for identifying natural clusters of species and of attributes, and to apply principles of parsimony.

Purposes for a classification of plant functional types

It is widely held (e.g. Gitay & Noble and Walker, Chapters 1 and 5, this volume) that PFT classifications will necessarily be purpose-dependent; that is, that different plant attributes will be relevant depending on the purpose of the classification, and further that different groupings of plants will result, depending on the attributes chosen.

In our laboratory up to the present, we have worked on the premise that five main types of research question about differences among plant species need to be considered, for purposes of projecting the future dynamics of vegetation under global change. These are as follows.

1. Which plant species will be most capable of dispersing in long jumps? (The following outline calculation indicates what is meant by 'long' (Leishman *et al.* 1992). The predicted poleward shift in temperature zones is *ca.* 100–500 km by 2050. Thus even for a generation length of 1 year, the mean jump length per generation is 2 km, substantially longer than any measured seed shadows. For a generation length of 50 years, relevant jump lengths could be up to 500 km.)

2. Which plant species will be most capable of establishing as seedlings, especially under preexisting vegetation other than where they currently occur?

3. Which plant species are most capable of growth and competitive persistence as established plants?

4. How do plant species differ with regards to their influence on canopy properties, especially albedo, seasonality of evapotranspiration, and the roughness parameter?

5. How do plant species differ with regards to the effects of elevated CO_2 on their other properties, especially (2) and (3) above?

Of these five, we have deliberately sought to compile information on plant traits that appear relevant to questions (1)–(4). Unfortunately, far too few species have been studied physiologically under elevated CO_2 for it to be possible to include information on question (5) in databases on hundreds of species from particular floras.

It is important to note a particular omission from the list of questions above: there is no direct mention of differences between species with regard to climatic preferences or tolerances, such as frost tolerance, minimum temperature for flowering, etc. In this respect our approach has differed from that adopted by Woodward (1987; Chapter 3, this volume), and implied by the IGBP Operational Plan. According to Woodward's approach, climatic responses would be the primary basis for dividing species into groups, and the resulting functional types would essentially be climate-response types. By contrast, the questions we have focused on are those relevant to how species will behave in patch-scale models of vegetation dynamics, either vital attributes models or models of the gap-dynamics lineage. The effect of the approach we have adopted will be to discriminate functional types that are vegetation-dynamic types, that is, groups of species that would respond similarly in a patch-dynamic model being implemented with a particular disturbance regime. Within each vegetation-dynamic type we envisage that there could be a large number of species, showing a range of tolerances to temperature, rainfall and other macroclimate variables. Thus our current opinion is that a vegetation-dynamic classification would be complementary to a climate-tolerance classification.

Plant functional types in Australian semi-arid woodland

We turn now to summarizing results from a particular study (Leishman & Westoby 1992) in which a species × attributes table was analyzed by using quantitative clustering methods. We will focus on two questions: how do the groupings that emerge relate to groupings recognized by previous authors using more subjective methods; and to what extent would the groupings be different if different groups of traits (corresponding to different purposes for the classification) were used?

Data were compiled for *ca.* 300 species from temperate semi-arid woodland in western New South Wales, Australia. Semi-arid woodlands were chosen because: they are important on a global scale; they include a range of growth forms; the growth form mix is susceptible to being much changed by grazing and fire management; and changes in the growth form mix are likely to have feedback effects to the climate system via influences on albedo, canopy roughness, and the seasonality of evapotranspiration.

We compiled information on 43 attributes describing 8 vegetative, 9 life history, 15 phenology and 11 seed biology features, relevant to questions (1)–(4) above.

Dendrograms were constructed by hierarchical agglomerative polythetic clustering methods. For the dendrogram in Fig. 6.1, all variables were given equal weighting, and a Bray–Curtis distance measure was used. However, the essential features of the dendrogram also emerged when variables were weighted such that each of the four main groups of attributes was equally important, and also when other distance metrics were used. Accordingly we are reasonably confident that the features we draw attention to are not arte-facts of having collected more attributes within some categories than others.

Five main groups emerged (Fig. 6.1). Group 1 was a mixture of forbs and low shrubs, nearly all C3. None were strict annuals, but some were facultative perennials, by which we mean that in many years they would behave as annuals, but given a sufficient sequence of good rains they might survive for more than one year. Group 2 consisted entirely of chenopod shrubs and subshrubs. Group 3 consisted almost entirely of grasses, none of

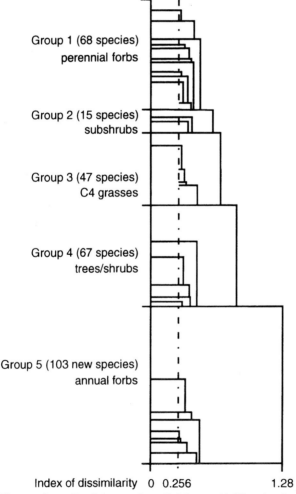

Group 1 (68 species)
perennial forbs

Group 2 (15 species)
subshrubs

Group 3 (47 species)
C4 grasses

Group 4 (67 species)
trees/shrubs

Group 5 (103 new species)
annual forbs

Index of dissimilarity 0 0.256   1.28

Figure 6.1   *Dendrogram from Leishman & Westoby (1992).*

them strictly annual, and most summer-growing and C4; a summer-annual chenopod also fell into this group. Group 4 consisted entirely of trees and shrubs, all but one of them C3. Group 5 consisted of strict annuals and some facultative perennials, including both forbs and grasses. The following main conclusions can be drawn.

Conclusion 1

The boundaries of four of the five main groupings represented discontinuities in patterns of variation among species. Consider group 5 (Fig. 6.1), for example. It was different from all other groups at a dissimilarity level of 1.28, whereas the highest level of dissimilarity within the group was less than 0.5. Similarly for groups 2–4, there was a distinct step up from the highest level of dissimilarity within each group, to the dissimilarity between that group and other groups. In this sense groups 2–5 were 'natural' groupings. Group 1 was not so distinctly different from the other groups.

The distinct step up from within-group to between-group dissimilarities in our results can be appreciated more clearly by contrasting our dendrogram (Fig. 6.1) with the results of Parsons (1976; Fig. 6.2), where no discrete natural groupings appear. Parsons' analysis was for 64 shrub species from Mediterranean scrub in Chile and California, described by 24 structural and functional attributes. His results are consistent with ours in as much as his whole species list would have fallen within one of our major groupings. The analysis by Grime *et al.* (1987) of the Sheffield database proceeded by first extracting a small subset of variables that were 'optimal' in terms of being correlated with a large proportion of other variation, then using this subset to categorize species. Accordingly their analysis does not provide a fair test of the question of whether there are natural discontinuities in the variation between species.

Conclusion 2

There was no evidence for discontinuous variation, or distinct subclusters, within any of the five major groups. Rather, the dendrogram cascaded smoothly from the level of the five major groups down to the level of individual species, reflecting continuous variation among species within each major group. Thus if the species were to be categorized into more than five groups (as most existing classification schemes would suggest), those further groupings should be interpreted as arbitrary subdivisions of continuous variation.

Interesting questions remain about the organization of variation within each of the main groups. Can it be represented as a single continuum, or is the variation less organized than that? If there is a single main axis of variation, of what does it consist? Are the patterns of variation within each group similar to those within other groups? Study of these questions requires data sets having at least 20–30 species within a group.

Conclusion 3

The five major groups are ones that experienced field botanists would always have recognized, and that previous classifications have recognized. In this sense the formal clustering approach has not added anything to knowledge already in existence.

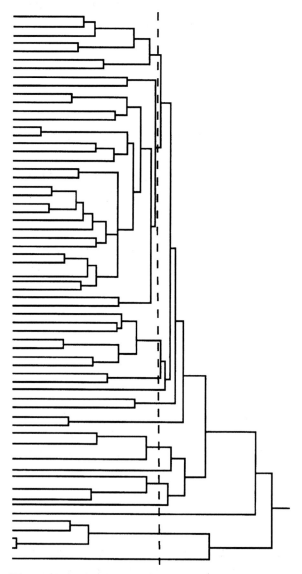

Figure 6.2   *Dendrogram from Parsons (1976).*

On the other hand, it should be noted that the formal clustering approach chose not to recognize a number of distinctions that might have been predicted. Shrubs were not split from trees. Although longer-lived chenopods were thrown into a distinct cluster, shorter-lived chenopod genera were variously placed with perennial forbs or annual forbs. Summer annuals and facultative perennials did not appear as a distinct group; some were placed with winter annuals in group 5, some with perennial forbs in group 1, some with summer C4 grasses in group 3. There was no general split between grasses and dicotyledons; rather, summer C4 grasses were separated as a group but C3 grasses were grouped with annual or perennial dicotyledonous herbs. In these regards the formal clustering approach is not simply recapitulating existing knowledge.

Conclusion 4

The five major groups reflect differences between species with regard to vegetative, life history and phenological attributes (Table 6.1). This outcome indicates that the groups recognized would capture variation relevant to both questions (3) and (4) above, possibly even to question (2), depending whether seedling establishment is controlled mainly by seed attributes or by vegetative growth attributes. To this extent the outcome suggests that classifications for multiple purposes can indeed be successfully constructed.

At the same time, it should be noted that the five major groups do not capture much variation between species with regard to seed and dispersal biology (Table 6.1). The pattern of weak correlation between attributes of adult and of juvenile traits has been found also by Shipley *et al.* (1989) and by Grime *et al.* (1988) (Table 6.2), so on present evidence it appears to be generally applicable.

Correspondingly, a wide range of seed sizes and dispersal modes can be found within several of the five major groups. This suggests that answers to question (1) ('Which plant species will be most capable of dispersing in long jumps?') are not likely to be captured by the five-group classification outlined above. To arrive at functional types that might be effective in discriminating species with respect to question (1) as well as to the other questions, further subdivisions would seem to be necessary.

Consistency between floras

The basis for plant functional types is the underlying pattern of correlations among traits across species. If functional type classifications devised for particular regions or data sets are to be extended and synthesized to build a

Table 6.1 *Attributes found in more than 50% of species in each species group produced from the cluster analysis*

The numbers in parentheses show the actual percentage of species with that attribute for each species group. Only categorical variables are included.

	Group 1	Group 2	Group 3	Group 4	Group 5
Vegetative	forb (75%) soft leaves (82%) shallow roots (90%)	subshrub (93%) succulent leaves (67%) no hairs (60%) spines (60%) shallow roots (100%)	grass (98%) soft leaves (77%) shallow roots (83%) no hairs (51%)	tree/shrub (39%/51%) stiff leaves (52%) no hairs (72%)	forb (93%) soft leaves (86%) shallow roots (100%)
Life history	perennial (82%) resprout ability (85%) fairly drought-resistant (62%) C_3 (87%)	phanerophyte (60%) perennial (73%) medium drought resistance (53%) resprout ability (67%) C_3 (67%) high fire mortality (73%) low palatability (67%) wind flowers (100%)	hemicryptophyte (72%) perennial (70%) resprout ability (79%) C_4 (81%) wind flowers (100%)	phanerophyte (100%) perennial (67%) resprout ability (85%) low palatability (60%) C_3 (100%)	therophyte (100%) annual (83%) low drought resistance (97%) total fire mortality (94%) C_3 (91%) insect flowers (53%)
Phenology	growth spring (54%) flowers spring (82%) flowers summer (66%) seed release spring (51%) seed release summer (71%)	growth spring (60%) growth summer (73%) flowers spring (87%) flowers summer (73%)	growth summer (81%) flowers spring (57%) flowers summer (64%) flowers autumn (70%) seed release summer (68%)	growth with rain (87%) flowers spring (87%) flowers summer (58%) seed release summer (66%)	growth winter (70%) growth spring (81%) flowers spring (93%) seed release spring (72%)
Seed biology		unassisted diaspores (73%)			autumn germination (65%) unassisted diaspores (50%)

Source: Modified from Leishman & Westoby (1992).

Table 6.2 *Percentages of correlations significant at the 5% level within and between groups of adult and juvenile attributes*

	Adult attributes	Juvenile attributes
(a) Adult attributes	44	4
Juvenile attributes	—	43
(b) Adult attributes	36	21
Juvenile attributes	—	33

Sources: (a) Data from Shipley *et al.* (1989) for 20 attributes of 25 species of emergent macrophytes. (b) Data from Grime *et al.* (1988) for 27 attributes of 288 species from the Sheffield region, England. (Table from Leishman *et al.* 1992.)

worldwide scheme, a critical issue is to what extent the correlation patterns among traits are consistent in different floras. It is not a foregone conclusion that they must be consistent. The ancestors of floras on different continents might have had different preadaptations; or the environmental pressures and life-history opportunities might favour different constellations of attributes.

We have some evidence on this question from our research on understanding the evolution of between-species differences in seed size in relation to other traits. Within the Australian semi-arid woodland database, seed size was correlated with a large number of other plant attributes, but most of these can parsimoniously be interpreted as secondary correlations, arising via the correlation with growth form (Leishman & Westoby 1994). Here we consider only the relationship of seed size with growth form and with dispersal mode, concentrating on the consistency of these relationships across different floras. Besides the western NSW data, we have used data for central Australian arid woodland (Jurado *et al.* 1991), for temperate sclerophyll and rainforest vegetation near Sydney, Australia (Westoby *et al.* 1990), for the Indiana Dunes near the Great Lakes in USA (Mazer 1989), and for vegetation near Sheffield in the UK (Grime *et al.* 1988).

Seeds tend to be larger in woody than in herbaceous plants, and the nature of the pattern is consistent in different floras (Fig. 6.3). Similarly the relationship with dispersal mode is consistent (Fig. 6.4). Seeds larger than about 100 mg tend to be adapted for dispersal by vertebrates; seeds smaller than about 0.03 mg tend to be unassisted; but most seeds are in the middle range of size, and in this middle range all dispersal modes are adopted (Hughes *et al.* 1994). In most of the floras, growth form and dispersal mode together were capable of accounting for about 35% of the variance in $\log_{10}$ seed mass (Table 6.3). The percentage accounted for was lower in the

Table 6.3 *Summary of percentage of variation in log seed mass that can be accounted for by reference to the categorical variables growth form and dispersal mode (GLM analysis)*

Significant at $p < 0.001$ unless indicated.

Flora	Growth form	Dispersal mode	Both dispersal mode and growth form
Western NSW	26	18	34
Central Australia	36	12	39
Sydney	24	32	43
Indiana Dunes	20	45	47
Sheffield	9	16	21

Sheffield flora, for reasons we do not yet understand. There was partial overlap between the portions of the variance explained by growth form and dispersal mode, but they had significant explanatory power independent of each other in all the floras except that of central Australia (where dispersal mode could not explain significant further variance after growth form; Table 6.3).

In summary, comparisons between floras so far are consistent with the idea that similar patterns of correlation among traits recur, with regard to both the forms of the relationships and their strength.

Discussion

The clustering analysis reported in this chapter has arrived at a categorization that has limitations of several kinds. First, it does not capture dispersal attributes. This limitation has already been discussed above. It arises because dispersal biology appears to be only weakly correlated with the biology of vegetative growth.

Second, it is not known how well the categorization reflects the impact of elevated CO_2 on other plant traits. This is because we took the approach of compiling data for hundreds of species, at the expense of obtaining data on physiological attributes beyond a simple C3/C4/CAM attribution. The database from the Sheffield Integrated Screening Program (Grime *et al.*, Chapter 7, this volume) is unique in including physiological information for substantial numbers of species. In our view, the best way forward is not to undertake detailed descriptive work in enriched CO_2 growth chambers for thousands of species. Rather, people doing physiological studies should be encouraged to choose species with a view to hypothesis-testing the

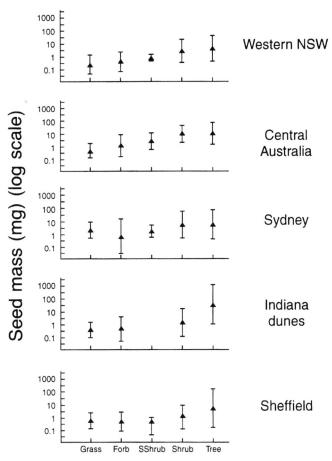

Figure 6.3 *Mean log seed mass (±s.d.) in relation to growth form, in five different floras.*

proposition that physiological attributes, including effects of enhanced CO_2, are consonant with PFT groupings defined on attributes of plant structure and phenology. For example, a physiological study might be designed by selecting three species at random from each of three PFT groups, with the hypothesis being that variation in CO_2 response between groups is much greater than variation within groups.

Third, the categorization derived is at a rather broad level of definition, and major groups recognized in traditional classifications were again recognized by this study using more plant attributes and formal multivariate methods. This should not be surprising, since the classifications were devised by able botanists with much field experience. The prospects are that further research will consolidate and systematize our knowledge of patterns already recognized, more than discover wholly new patterns. Some may regard this as an unsatisfactory prospect, because at present the

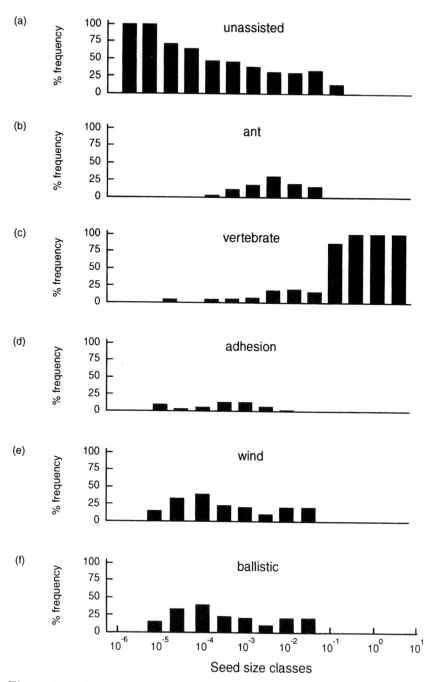

Figure 6.4   *Percentage of all species using different dispersal modes, in relation to log seed mass; compiled from the same five floras as in Fig. 6.3. References in the text; details in Hughes* et al. *(1994).*

incentive system in ecology strongly favours reporting one's results as showing something different from previous research by others. In our view this feature of the ecological research scene is unfortunate. The advance of knowledge will be best served if the research ethos can be shifted, such that where results confirm and generalize items of knowledge, these are equally likely to be reported and publicized.

Finally, the most important limitation of the categorization presented is of course that it has been developed only for a particular region. In our view, it should be quite feasible to broaden the categorization incrementally to a wider range of vegetation types and to floras on other continents. We envisage that studies like the one we have described would be repeated for other large sets of species, using similar clustering algorithms, and variables not requiring quantitative experimentation. Where species overlapped between two databases, it would be possible to confirm that they classified into the same PFT consistently, irrespective of the set of other species being different. This process would naturally find that further PFTs were required beyond those described here, as coverage was extended into different vegetation types.

In summary, our results so far give grounds for optimism that PFT classifications can be constructed that (1) differentiate plants for purposes of several different questions at the same time, and (2) cover more than one region or continent. In order for agreement to be achieved among workers from different regions, it will be important to apply explicit procedures (such as quantitative clustering algorithms) to species × attribute data sets. Over time we expect that different research groups will develop data sets having many attributes in common, and will apply each other's clustering methods to them; in this way it will become apparent which plant functional types are robust and which are consequences of choosing particular plant attributes or clustering procedures.

Acknowledgments

This work is supported by the Australian Research Council, and by an Australian Postgraduate Research Award to M. Leishman. It has benefited from very useful discussions with many of the participants in the workshop at Charlottesville in January 1993. This chapter is Contribution No. 205 from the Research Unit for Biodiversity and Bioresources, Macquarie University.

References

Barkman, J. J. (1988) New systems of plant growth forms and phenological plant types. In Werger. M. J. A. *et al.* (eds), *Plant Form and Vegetation Structure. Adaptation, plasticity and relation to herbivory*, pp. 9–44. The Hague: SPB Academic Publishing.

Box, E. O. (1981) *Macroclimate and Plant Forms*. The Hague: W. Junk.

Braun-Blanquet, J. (1932) *Plant Sociology: the Study of Plant Communities*. (Translated by G. D. Fuller & H. S. Conard.) New York: Hafner Publishing Company.

Cain, S. A. (1950) Lifeforms and phytoclimate. *Botanical Review*, **16**, 1–32.

Candolle, A. P. de (1818) *Regni Vegetabilis Systema Naturale*. Vol. 1. Parisiis.

Dansereau, P. (1951) Description and recording of vegetation upon a structural basis. *Ecology*, **32**, 172–229.

Dix, R. L. (1959) The influence of grazing on the thin-soil prairies of Wisconsin. *Ecology*, **40**, 36–49.

Drude, O. (1890) *Handbuch der Pflanzengeographie*. Bibliothek geographischer Handbucher. Stuttgart: F. Ratzel.

Dyksterhuis, E. J. (1949) Condition and management of rangeland, based on quantitative ecology. *Journal of Rangeland Management*, **2**, 104–15.

Ellenberg, H. (1952) *Landwirtschaftliche Pflanzensociologie. II. Wiesen und Weiden und ihre standortliche Bewertung*. Stuttgart: Eugen Ulmer.

Ellenberg, H. & Mueller-Dombois, D. (1967) Tentative physiognomic-ecological classification of plant formations of the earth. *Bericht über das Geobotanische Forschunginstitut Rubel in Zurich*, **37**, 21–55.

Fosberg, F. R. (1967) Classification of vegetation for general purpose. In Peterken, G. F. (ed.), *IBP Handbook No.4. Guide to the checksheet for IBP areas*, pp. 73–120. Oxford: Blackwell Scientific Publications.

Friedel, M. H., Bastin, G. N., Griffen, G. F. (1988) Range assessment & monitoring in arid lands: the derivation of functional groups to simplify data. *Journal of Environmental Management*, **27**, 85–97.

Gillison, A. N. (1981) Towards a functional vegetation classification. In Gillison, A. N. & Anderson, D. J. (eds), *Vegetation Classification in Australia*, pp. 30–41. Canberra: ANU Press.

Grime, J. P. (1974) Vegetation classification by reference to strategies. *Nature*, **250**, 26–31.

Grime, J. P., Hodgson, J. G. & Hunt, R. (1988) *Comparative Plant Ecology. A functional approach to common British species*. London: Unwin Hyman.

Grime, J. P., Hunt, R. & Krzanowski, W. J. (1987) Evolutionary physiological ecology of plants. In Calow, P. (ed), *Evolutionary Physiological Ecology*, pp. 105–25. Cambridge University Press.

Hughes, L., Dunlop, M., French, K., Leishman, M., Rice, B., Rodgerson, L. & Westoby, M. (1994) Predicting dispersal spectra: a minimal set of hypotheses based on plant attributes. *Journal of Ecology*, **82**, 933–50.

Humboldt, A. von (1806) *Ideen zu einer Physiognomik der Gewachse*. Tübingen.

Johnstone, R. D. & Lacey, C. J. (1984) A proposal for the classification of tree-dominated vegetation in Australia. *Australian Journal of Botany*, **32**, 529–49.

Jurado, E., Westoby, M. & Nelson, D. J. (1991) Diaspore weight, dispersal, growth form and perenniality of central Australian plants. *Journal of Ecology*, **79**, 811–30.

Kerner, von Marilaun A. (1863) *Das Pflanzenleben der Donaulander*. Innsbruck.

Knight, D. H. (1965) Gradient analysis of Wisconsin prairie vegetation on the basis of plant structure and function. *Ecology*, **46**, 744–7.

Kuchler, A. W. (1949) A physiognomic classification of vegetation. *Annals of the Association of American Geographers*, **39**, 201–10.

Leishman, M., Hughes, L., French, K., Armstrong, D. & Westoby, M. (1992) Seed and seedling biology in relation to modelling vegetation dynamics under global climate change. *Australian Journal of Botany*, **40**, 599–613.

Leishman, M. R. & Westoby, M. (1992) Classifying plants into groups on the basis of associations of individual traits – evidence from Australian semi-arid woodlands. *Journal of Ecology*, **80**, 417–24.

Leishman, M. R. & Westoby, M. (1994) Hypotheses on seed size: tests using the semi-arid flora of western New South Wales, Australia. *American Naturalist*, **143**, 890–906.

Mazer, S. J. (1989) Ecological, taxonomic and life-history correlated of seed mass among Indiana Dune angiosperms. *Ecological Monographs*, **59**, 153–75.

Orshan, G. (1953) Note on the application of Raunkiaer's system of life forms in arid regions. *Palestine Journal of Botany, Jerusalem*, **6**, 120–2.

Parsons, D. J. (1976) Vegetation structure in the Mediterranean scrub communities of California and Chile. *Journal of Ecology*, **64**, 435–47.

Pound, R. & Clements, F. E. (1898) *The Phytogeography of Nebraska. I.* Lincoln, Nebraska.

Raunkiaer, C. (1934) *The Life Forms of Plants and Statistical Plant Geography*. Oxford: Clarendon Press.

Rietz, G. E. du (1931) Lifeforms of terrestrial flowering plants. *Acta Phytogeographica Suecica*, **3**, 1–95.

Sarmiento, G. & Monasterio, M. (1983) Life forms and phenology. In Bourliere, F. (ed.), *Ecosystems of the World 13: Tropical Savannas*, pp. 79–108. Amsterdam: Elsevier Scientific Publishing Company.

Shantz, H. L. (1927) Drought resistance and soil moisture. *Ecology*, **8**, 145–57.

Shipley, B., Keddy, P. A., Moore, D. R. J. & Lemkt, K. (1989) Regeneration and establishment strategies of emergent macrophytes. *Journal of Ecology*, **77**, 1093–100.

Solbrig, O. T., Barbour, M. A., Cross, J., Goldstein, G., Lowe, C. H., Morello, J. & Yang, T. W. (1977) The strategies and community patterns of desert plants. In Orians, G. H. & Solbrig, O. T. (eds), *Convergent Evolution in Warm Deserts*, pp.

67–106. Stroudsberg, Pennsylvania: Dowden, Hutchinson & Ross.

Steffen, W. L., Walker, B. H., Ingram, J. S. & Koch, G. W. (eds) (1992) *Global Change and Terrestrial Ecosystems: The Operational Plan*. Stockholm: International Geosphere–Biosphere Program, International Council of Scientific Unions.

Stoddart, L. A., Smith, A. D. & Box, T. W. (1975) *Range Management*. 3rd edn. New York: McGraw Hill.

Warming, E. (1923) Okologiens grundformer. Udkast til en systematisk ordning. *Kongelige Danske Videnskabernes Selskabs Skrifter, Naturvidenskabelig og Mathematisk Afdeling* **8(4)**, 121–87.

Webb, L. J. (1959) Physiognomic classification of Australian rain forests. *Journal of Ecology*, **47**, 551–70.

Webb, L. J. (1968) Environmental relationships of the structural types of Australian rainforest vegetation. *Ecology*, **47**, 296–311.

Webb, L. J. (1976) A general classification of Australian Rainforests. *Australian Plants*, **9**, 349–63.

Westoby, M., Rice, B. & Howell, J. (1990) Seed size and plant stature as factors in dispersal spectra. *Ecology*, **71**, 1307–15.

Whitmore, T. C. (1988) The influence of tree population dynamics on forest species competition. In Davy, A. J., Hutchings, M. J. & Watkinson, A. R. (eds), *Plant Population Ecology*, pp. 271–292. Oxford: Blackwell Scientific Publications.

Woodward, F. I. (1987). *Climate and Plant Distribution*. Cambridge University Press.

⑦ Functional types: testing the concept in Northern England

J. P. Grime, J. G. Hodgson, R. Hunt, K. Thompson, G. A. F. Hendry, B. D. Campbell, A. Jalili, S. H. Hillier, S. Diaz and M. J. W. Burke

Introduction

After a long gestation period (MacLeod 1894; Ramenskii 1938; Hutchinson 1959; MacArthur & Wilson 1967; Pianka 1970; Greenslade 1972, 1983; Grime 1974, 1977; Southwood 1977; Whittaker & Goodman 1979; Noble & Slatyer 1979; Pugh 1980) the notion of functional types has entered the planning processes of both climate-change studies and investigations of declining biodiversity with recent and remarkable speed. In the small community of scientists dedicated to this approach but hitherto operating under different colours ('strategies', 'syndromes', 'vital attributes' and 'sets of traits') this is likely to be seen as a vindication of the view that community and ecosystem analysis will not progress until it finds a way of bypassing the infinite variety of species and populations and establishing a coherent predictive framework based on a relatively small number of universal functional types of plants and animals. However, as the concept of functional types is developed and applied, it will be necessary to justify carefully the rationale behind work at this broad scale; past experience suggests (Grime 1989) that biologists accustomed to working at a finer scale and familiar with the genetically labile nature of populations will not readily accept the virtues of this broad-brush approach. With this problem in mind, our objective here is not restricted to that of identifying functional types; an equally important purpose is to consider the measures required to establish the rigour and to test the usefulness of this approach.

This chapter draws upon the results of a programme of field and laboratory research relating to herbaceous plants in an area of 3000 km^2 surrounding Sheffield in northern England (Grime et al. 1988). In the first part we use these studies to review briefly various sets of functional types which have been recognized in Britain. The second part reports progress in a current research programme, the Integrated Screening Programme (ISP), which attempts a formal search for functional types. The final part illustrates the way in which vegetation monitoring and experimental manipulations are being used to test the predictive value of ISP outputs.

Some general principles

Classifications of functional types differ with respect to the geographical and taxonomic scale at which they are applied and the criteria they use.

Geographical and taxonomic scale

In studies of climate change and biodiversity it is often useful to recognize a continuous hierarchy or nesting of functional types. At the largest scale are systems of classification (e.g. r- and K-selected organisms (MacArthur & Wilson 1967)) which include all plants and animals and attempt to separate them into a small number of basic types. Here, generality is sought at the expense of precision, and criteria are restricted by the need to work with attributes common to all organisms. At the other extreme, it is possible to define functional types extremely narrowly by restricting attention to one functional aspect of one taxon in one habitat (e.g. temperature responses of spore germination in the bryophyte flora of arable fields). In order to draw maximum benefit from the use of functional types, it is usually an advantage to draw inferences from classifications applied at several geographical and taxonomic scales.

Criteria

Some very different criteria have been employed to define functional types. By measuring demographic patterns, population biologists have often sought to define functional types (e.g. Deevey 1947; Whittaker & Goodman 1979) and there is no doubt that knowledge relating to the spatial and temporal distribution of reproduction, mortality, dispersal and dormancy can contribute vital information to a functional typology. Some population biologists (e.g. Harper 1982) have argued until recently that demography, particularly when allied to studies of plant morphology, provides an adequate basis for interpretation and prediction of vegetation processes. However, this school of thought has been notably silent on the topic of climate change where the need for physiological information is indisputable.

This chapter works on the assumption that it is now generally agreed that functional types must be defined by reference to both demographic criteria *and* those features of life-history, physiology and biochemistry that determine the responsiveness of plants to soils, land-use and climatic factors. Rather than a return to the 'old conflicts' of population biologists and physiologists, what is now required is a precise definition of when and where to use both kinds of criteria in various appropriate combinations. Here, a very welcome development is the recent attempt by Silvertown *et al.* (1993) to assemble and compare a large number of demographic studies conducted on a wide range of ecologically contrasted plant species. This attempt at synthesis brings nearer not only the prospect of classification by demography but also the opportunity to assess the extent to which types of demography are a predictable consequence of life-history and physiological specialization.

Functional types in the British flora

Primary functional types: established phase

A comprehensive functional classification of common British vascular plants using a mix of demographic and physiological information is attempted in the C–S–R system, the uses and limitations of which have been discussed elsewhere (Grime 1988; Grime *et al.* 1988). The C–S–R system relies heavily upon features related to life span, growth rates, resource capture and retention, and the timing of and commitment to reproduction. In consequence, C–S–R is a useful predictor of sensitivity to changes in land use, particularly those mediated by vegetation destruction, eutrophication and dereliction. It is also helpful in predicting rates of expansion and decline in plant populations, under changing soil fertility and in the recovery following disruptive effects of climate extremes on vegetation. A particularly interesting example of such an analysis is contained in the study of resistance and resilience of vegetation exposed to the severe drought in Czechoslovakia in 1976 (Leps *et al.* 1982).

Most British plants can now be classified into C–S–R categories and accessed by computer systems designed for retrospective analysis of vegetation change (FIBS (Hodgson 1991)) or prediction (TRISTAR (Hunt *et al.* 1991b)).

Primary functional types: regenerative phase

From many different sources (e.g. Stebbins 1971; Gill 1978; Grime 1979) there is evidence of recurrent patterns of specialization in the regenerative phase of plant life histories. For British vascular plants, these patterns have been summarized as five primary regenerative strategies (Table 7.1), which differ in attributes such as propagule size, number, dispersal and dormancy. Three of the primary regenerative functional types (S, B_s and W in Table 7.1) are of special significance in relation to both the immediate and long-term responses of vegetation to climate change in that:

Table 7.1 *Five regenerative strategies of widespread occurrence in terrestrial vegetation (from Grime 1989)*

Strategy		Functional characteristics	Conditions under which strategy appears to enjoy a selective advantage
Vegetative expansion	(V)	New shoots vegetative in origin and remaining attached to parent plant until well-established	Productive or unproductive habitats subject to low intensities of disturbance
Seasonal regeneration	(S)	Independent offspring (seeds or vegetative propagules) produced in a single cohort	Habitats subjected to seasonally predictable disturbance by climate or biotic factors
Persistent seed or spore bank	(B₁)	Viable but dormant seeds or spores present throughout the year, some persisting more than 12 months	Habitats subjected to temporally unpredictable disturbance
Numerous widely dispersed seeds or spores	(W)	Offspring numerous and exceedingly buoyant in air, widely dispersed and often of limited persistence	Habitats subjected to spatially unpredictable disturbance or relatively inaccessible (cliffs, walls, tree trunks, etc)
Persistent juveniles	(Bₛ)	Offspring derived from an independent propagule but seedling or sporeling capable of long-term persistence in a juvenile state	Unproductive habitats subjected to low intensities of disturbance

■ *S* predicts the capacity for exploitation of seasonal events such as drought and frost damage;

■ B_s predicts (1) population persistence in circumstances unfavourable to the established plant and (2) potential for dispersal by soil transfer;

■ *W* predicts capacity for long-distance dispersal and 'island-hopping'.

A database for 500 common British plant species already exists for S, B_s and W (Grime *et al.* 1988); K. Thompson and J. Bakker are currently extending the coverage for B_s and W to include the published evidence for the flowering plants of western Europe. Recently, an attempt has been made to test the predictive value of regenerative strategies at a landscape scale (Grime & Hillier 1992). A preliminary attempt to incorporate regenerative strategies in vegetation processes is made in the predictive rule-base TRISTAR and in retrospective analyses of vegetation change by FIBS. In their utilization of propagule number, size, dispersal and dormancy, both FIBS and TRISTAR show clear affinities with the frameworks developed by Salisbury (1942), Noble & Slatyer (1979) and van der Valk (1981).

Phenological functional types

Temperature, daylength and moisture supply change substantially over the growing season in Britain and in consequence many plant communities contain populations that grow and flower at different times and show profound differences in physiology. Some useful predictions of community responses to climate change or seasonal impacts of vegetation management can be based upon plant morphology (Raunkiaer 1934) or quantitative studies of shoot phenology (Al-Mufti *et al.* 1977). There is an urgent need, however, to replace these correlative and essentially non-mechanistic procedures with a system based on functional criteria; this will require an expansion in studies of developmental responses to daylength and temperature and physiological tolerance of drought and frost.

Hydrophytes, mesophytes and xerophytes

From field data collected in western and central Europe, it is already possible to recognize strong correlations between the distribution of species and water supply (e.g. Ellenberg 1978). For this information to be translated into a functional understanding and secure basis for prediction of responses to climate change, it is essential to know certain basic features of the water relations of each species. This is because radically different mechanisms of regulating water relations often coexist within communities. In order to differentiate between these mechanisms it is necessary to refer to features such as desiccation tolerance, root penetration and plasticity, xylem structure and stomatal index. Standardized sources of information on these features are rarely available for more than a handful of species.

Calcicoles and calcifuges

In Britain, as elsewhere, there are latitudinal and altitudinal trends in which gradients in precipitation : evaporation ratio dictate close parallels between the temperature and moisture supplies of plant habitats and the vertical movement of cations. Hence, the concentration of calcicoles in southern Britain and preponderance of calcifuges in the north is not simply a function of geology. In the cool, high-rainfall areas of the north and west, many of the soils, including those on calcareous substrata, are surface-leached and support calcifuge vegetation. This has two important implications for predictions of the impact of climate change in Britain: (1) calcicoles and calcifuges need to be characterized with respect to differences in mineral nutrition and potential sensitivity to reduction or reversal of the downward movement of cations under future climates; (2) calcicoles and calcifuges may be expected to differ consistently in rooting habit and moisture relations. Such differences may be more important than those in mineral nutrition in devising predictions of response to climate change.

Other functional types

The preceding paragraphs identify some of the major criteria that have guided the definition of functional types in the British flora. Others can be suggested and may need to be acted upon. Several relate to life-form, phylogeny, nutrition and mutualistic associations (e.g. orchids, legumes, graminoids and ericoids). In practice the definition of functional types and the choice of species for study will be affected by additional considerations. Where impacts of climate change are the main object of study, it is likely that investigations will include the following.

1. Plants of contrasted geographical distribution and with sharply defined climatic boundaries in Britain
2. Plants that are dominant components of major elements in the landscape and are important economically and aesthetically
3. Plants for which there are existing studies of relevance to climate change
4. Food plants of herbivores that are the subject of climate change investigations
5. Plants in danger of extinction by climate change
6. Aliens likely to invade and expand in the event of climate change.

Of these additional criteria for selection, (2) is most likely to cause complications. Many of the species that fall into this category, e.g. *Calluna vulgaris, Pteridium aquilinum, Agrostis capillaris, Festuca rubra* and *Arrhenatherum elatius*, are widespread in Britain and may be genetically variable in attributes that determine responses to specific climate factors. There is an urgent need to assess the extent to which genetic variation widens the functional amplitude and geographical range of these important species.

The Integrated Screening Programme: a formal search for functional types

Origins

As explained earlier in this chapter, the existence of recurring and predictable functional types of plants within the British flora has been suspected for many years, mainly on the basis of field observations and correlative studies on habitats, plant distribution and phenology. In order to elevate the search for functional types from natural history to science, formal procedures and tests are required. These usually involve the collection of standardized information on species attributes measured under controlled conditions. In order to allow reliable inference, such screening operations must involve substantial numbers of species (Grime 1965; Clutton-Brock & Harvey 1979; Keddy 1992) and must beware of the confounding effects of phylogeny (Stebbins 1971; Hodgson & Mackey 1986; Givnish 1987; Harvey & Pagel 1992). Tests may involve single attributes by examining their variation in relation to habitat and ecology (e.g. Grime & Hunt 1975) but greater rigour occurs where many attributes are screened and tests can be performed to confirm (or not) the recurrence of traits in the sets predicted to occur in particular functional types.

In preliminary attempts to recognize plant functional types (Grime *et al.* 1987, 1988), multivariate clustering analyses were conducted on large data sets of species attributes, assembled from existing published sources for common British herbaceous plants. With the notable exception of the regenerative strategies *W* and *S* (Table 7.1), these analyses failed to reveal clear patterns of ecological specialization. In retrospect, this outcome was inevitable. The databases were heavily biased towards morphological attributes which, although universally available as a byproduct of taxonomy, are of uncertain relevance to ecological prediction. The Integrated Screening Programme (ISP), initiated at UCPE in 1987, is a conscious effort to provide a more balanced and useful data set for an objective (Baconian) search for functional types.

Structure of the ISP

The ISP applies standardized experimental procedures and measurements for the ecological, physiological and biochemical characterization of a plant species, population or cultivar. Procedures have been devised to cover many important aspects of the established and regenerative phase. A majority of the screening operations are conducted in Sheffield but some parts of the ISP involve collaborators at other laboratories. A laboratory manual is available (Hendry & Grime 1993) describing each procedure and illustrating its use in ecological prediction. Some tests examine basic features of anatomy, morphology, physiology and biochemistry at different stages of development (seed, seedling, established plant); others measure survivorship and growth responses under particular environmental stresses. Further sets of procedures are designed with specific relevance to impacts of land use and climate change (temperature, moisture supply, CO_2 concentration), and efforts have been made to measure attributes relevant to competitive interactions and the potential to dominate plant communities. In the initial phase of development, the ISP has been applied to 43 species (one seed source per species) of which two are crop plants and the remainder are native British herbs from a wide range of habitats. In 1993, the ISP will be expanded to accommodate a range of native trees and shrubs.

The majority of the procedures in the ISP are conducted over a relatively short period of time, under standard conditions, on material of known and consistent genetic origin. It is recognized, however, that some important attributes are not yet amenable to this approach and for the present we must rely upon data collected from the field or reported in the literature. Examples in this category are seed persistence in the soil and the nature and extent of mycorrhizal infections. However, extreme caution must be applied in using these sources of data, particularly where they allow contamination of the database by information of weaker genetic definition. As we explain in relation to seed persistence later in this chapter (p. 141) our ultimate objective is to replace these 'softer' sources with objective data.

Results

The formal search for functional types using the clustering techniques of Grime *et al.* (1987) must await the completion, late in 1993, of all screening procedures on the first 43 species. Already, however, patterns of specialization can be recognized in the data obtained for individual parts of the ISP, and these have been discussed by Hendry & Grime (1993). Certain of these patterns appear to be particularly relevant to prediction of vegetation responses to changing land use and climate; they provide the subject matter of the next section of this chapter.

Testing the predictive value of ISP outputs

Protocol

Figure 7.1 summarizes two complementary sets of procedures which have been used to test the predictive value of ISP outputs. The first compares ISP predictions of the responses of populations or communities of plants against data collected by direct monitoring of events in natural conditions. The second matches ISP predictions against the results of manipulative experiments in which controlled impacts of ecological factors are applied to natural vegetation or to plant communities synthesized in microcosms or mesocosms. As indicated in the scheme in Fig. 7.1, the logical pathway associated with the use of monitored data is the same as that involving manipulation experiments. In both cases discrepancies between prediction and reality drive further cycles, each consisting of ISP screening, refinement of prediction and tests of prediction.

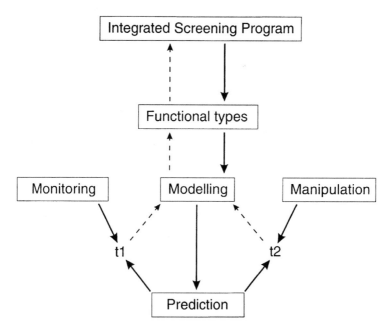

Figure 7.1 *Protocol for testing the predictive value of ISP outputs. Discrepancies revealed in the tests at t1 and t2 initiate further modelling cycles, each of which may necessitate refinement of the functional types or even additional screening.*

Examples

1. Competitive ability

Several ISP procedures measure attributes that are suspected to determine competitive ability and status within communities. Two provide assays of the capacity of the leaf canopy and root system to exploit standardized resource patchiness, simulating aspects of the conditions experienced in a perennial community (Campbell et al. 1991). Both techniques create patchiness in resource supply without using partitions or barriers, which could impede growth between patches. In both root and shoot assays, measurement is made of the partitioning of dry matter allocation between depleted and undepleted sectors imposed after an initial growth period in uniform, productive conditions.

In order to test the predictive value of data obtained by using the two procedures, a comparison, following the protocol in the right-hand side of Fig. 7.1, was made between ISP outputs for eight species and the status achieved in a conventional competition experiment in which the eight species were grown together in an equiproportional mixture under productive glasshouse conditions for sixteen weeks. The results (Fig. 7.2) reveal a consistent relationship between the increment of dry matter to the undepleted sectors in both assays and the capacity for dominance in the competition experiment. Covariance between roots and shoots in their responses to resource patchiness is also apparent. This relationship is maintained despite the consistent tendency for the scale of leaf canopy adjustment to exceed that of the root system. This, of course, arises from the freedom of movement of leaves in air and the encasement of roots in soil.

These results suggest that the status of plants in perennial herbaceous communities of high productivity may be predictable from specific, measurable, plant attributes. They also tend to confirm the strong interdependence of competitive abilities for light and mineral nutrients (Donald 1958). The significance of these results is considered in relation to the current debate about root–shoot trade-offs and primary functional types on pp. 144–5.

2. Root penetration and drought avoidance

Among several ISP procedures designed to quantify attributes related to moisture relations, one is specifically concerned with the capacity of the root system to penetrate vertically during seedling establishment. A detailed account of the experimental method used, together with results for a range of species, is provided by Reader et al. (1992). One of the most informative outputs of this screening procedure is the mean maximum root penetration ($n = 6$), measured as the position of the deepest root tip six weeks after

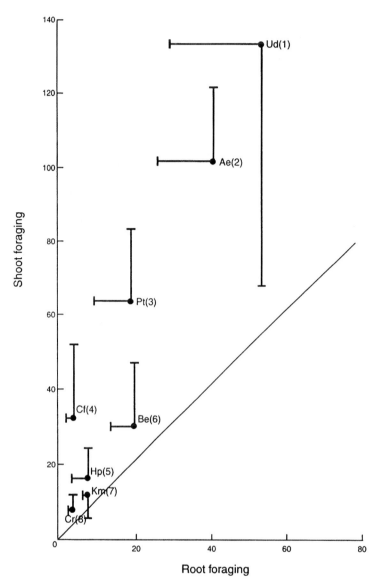

Figure 7.2   *An examination of the relationship between root and shoot responses to resource heterogeneity in eight herbaceous species of contrasted ecology. A description of the methods used to expose the plants to resource patchiness is provided in Campbell et al. (1991). Scales of foraging by the roots and shoots in the foraging assays are expressed at the respective increments of biomass (mg) to two undepleted quadrants, which in both assays constitute 50% of the available volume. The vertical and horizontal bars incicate 95% confidence limits. The numbers in parentheses refer to the species ranking in a conventional competition experiment in which all eight species were grown together in an equi-proportional mixture on fertile soil for 16 weeks.*

Key to species: Ae, Arrhenatherum elatius; *Be,* Bromus erectus; *Cf,* Cerastium fontanum; *Cr,* Campanula rotundifolia; *Hp,* Hypericum perforatum; *Km,* Koeleria macrantha; *Pt,* Poa trivialis; *Ud,* Urtica dioica.

commencing radicle growth down a vertical tube (1 m in length) of moist, nutrient-sufficient sand. Using the protocol of the left-hand side of Figure 7.1, the predictive value of this output has been tested for a group of fifteen coexisting species by comparing root penetration achieved in the ISP test against the success of seedling establishment in natural cohorts monitored in permanent quadrats situated on a droughted shallow soil over fissured limestone in north Derbyshire (Hillier 1984). The results (Fig. 7.3) indicate greatest mortality in the small-seeded, shallow-rooted seedlings of *Briza media* and *Origanum vulgare* and much greater survival in large-seeded species with deep roots, such as *Centaurea scabiosa* and *Plantago lanceolata*. However, the relationship is only weakly defined and, as we might expect from local observations based on a single cohort of naturally established seedlings, the analysis is complicated by the small number of seedlings observed for some species and by the presence of other mortality factors such as seed and seedling predation.

In the second phase of the investigation (Jalili 1991), following the alternative protocol of Fig. 7.1, the estimates of root penetration from the

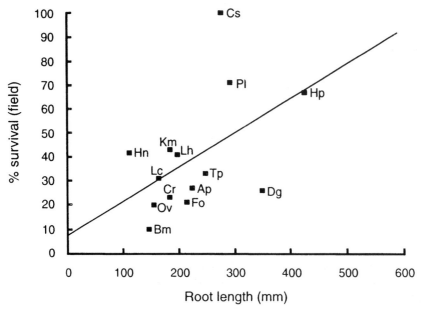

Figure 7.3 *Seedling survival at the Miller's Dale site as a function of root penetration measured in the Integrated Screening Programme; r=0.53, p<0.05.*
    Key to species: Ae, Arrhenatherum elatius; *Ap,* Avenula pratensis; *Bm,* Briza media; *Cf,* Cerastium fontanum; *Cr,* Campanula rotundifolia; *Cs,* Centurea scabiosa; *Dg,* Dactylis glomerata; *Fo,* Festuca ovina; *Hn,* Helianthemum nummularium; *Hp,* Hieracium pilosella; *Km,* Koeleria macrantha; *Lc,* Lotus corniculatus; *Lh,* Leontodon hispidus; *Ov,* Origanum vulgare; *Pl,* Plantago lanceolata; *Ra,* Rumex acetosa; *Tp,* Thymus praecox.

ISP were compared with survivorship of seedlings subjected to controlled drought in experimental plant communities. These were synthesized from seed in experimental microcosms, which provided local, randomly located opportunities for root penetration into continuously moist subsoil. The data reveal an even more consistent relationship between seedling survivorship and rooting habit (Fig. 7.4) and suggest that data from microcosms can provide a valuable supplement to those obtained by direct observation of natural systems.

3. Growth responses to elevated carbon dioxide

Screening of responses to elevated CO_2 in the ISP is conducted under non-limiting conditions of mineral nutrient supply and temperature, in natural summer daylight, in the comparatively high-radiation environment of Littlehampton on the south coast of England. Growth of seedlings for 6–8 weeks at each of four levels of CO_2 (ambient, 500, 650 and 800 ppm) has allowed curve-fitting procedures to be used to characterize the response of each species. Experimental methods and results for 40 native C_3 species are described by Hunt *et al.* (1991a, 1993). From these data it is evident that, under the conditions of the screening experiment, there were large species-specific differences in response to elevated CO_2, varying from zero in some ephemerals and slow-growing perennials to very large increases in dry

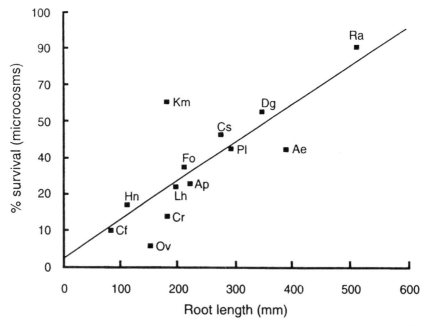

Figure 7.4   *Survival in droughted experimental communities as a function of root penetration measured in the Integrated Screening Programme; r=0.81, p<0.001. Key to species as in Fig. 7.3.*

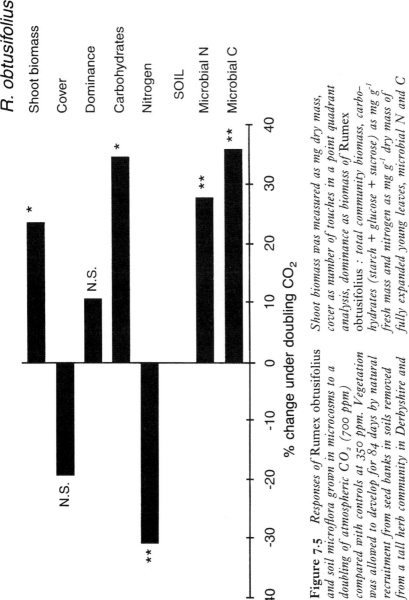

Figure 7·5 *Responses of* Rumex obtusifolius *and soil microflora grown in microcosms to a doubling of atmospheric CO_2 (700 ppm) compared with controls at 350 ppm. Vegetation was allowed to develop for 84 days by natural recruitment from seed banks in soils removed from a tall herb community in Derbyshire and placed in microcosms (six replicates per treatment) in cabinets without nutrient addition.*

Shoot biomass was measured as mg dry mass, cover as number of touches in a point quadrant analysis, dominance as biomass of Rumex obtusifolius : *total community biomass, carbohydrates (starch + glucose + sucrose) as mg g^{-1} fresh mass and nitrogen as mg g^{-1} dry mass of fully expanded young leaves, microbial N and C as mg g^{-1} dry soil; N.S., non-significant; * p<0.05; ** p<0.01 (ANOVA).*

matter production in some species. Among the latter, potentially large, fast-growing, clonal perennials of productive habitats (e.g. *Chamaenerion angusti-folium* and *Urtica dioica*) are conspicuous, prompting the hypothesis that responsiveness to elevated CO_2 may be dependent on the existence of strong carbon sinks. Here it is interesting to note that fast-growing clonal herbs are prominent among the life forms identified as currently expanding in abund-ance in the British flora (Grime *et al.* 1988). In previous attempts to explain this phenomenon (e.g. Hodgson 1989), emphasis has been placed on changes in land use and eutrophication by mineral fertilizers and atmospheric depo-sition of nitrogen. The ISP data suggest that a stimulatory effect of elevating CO_2 could be contributing to the observed shift in functional types.

The ISP screening is conducted on plants grown for a limited period as isolated individuals under non-limiting conditions. As many workers in this field have recognized (Woodward 1992; Korner 1993), it is vital that any hypotheses arising from differential responses observed in such experiments are subjected to tests in more natural conditions. Again following the protocol in the right-hand side of Fig. 7.1, hypotheses have been tested by examining the effect of elevated CO_2 on the structure of early secondary-successional, herbaceous communities. These were allowed to regenerate from natural seedbanks on disturbed, fertile soils transferred to laboratory microcosms. The results of these manipulative experiments with natural communities (Díaz *et al.* 1993) are strongly at variance with the ISP predic-tions in that promotory effects of CO_2 were not confirmed in potentially responsive species. It is particularly interesting to note that under a doub-ling of CO_2 concentration the early-successional, fast-growing and poten-tially large *Rumex obtusifolius*, although attaining a higher shoot biomass, did not increase its relative abundance in the community; it also showed marked symptoms of leaf stunting and had reduced foliar nitrogen and carbohydrate accumulation (Fig. 7.5). These patterns were associated with a stimulation of microbial sequestration of carbon and nitrogen in the soil. Leaf nitrogen content of plants grown at elevated CO_2 did not increase when additional mineral nutrients were provided in a second experiment on the same system (Fig. 7.6), suggesting that factors other than the absolute fertility of the soil determine these responses. Although these results do not invalidate the patterns of response evident in the ISP data, they are a salu-tary reminder of the complexities that distinguish the screening laboratory from a multispecies ecosystem. They suggest that sink strength is not the only trait accounting for CO_2 responsiveness. Other characteristics, such as species' nutritional requirements and the nature of their interaction with the microbial flora, may be important in predicting responses to high CO_2 at the community level.

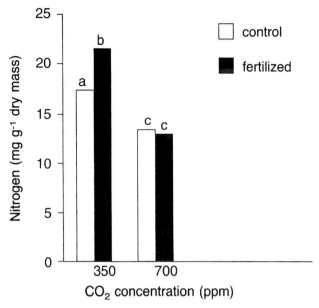

Figure 7.6 *Effects of atmospheric doubling of CO_2 concentration (ppm) and fertilizer addition on foliar N content of* Rumex obtusifolius, *grown in microcosms for 60 days. Deionized water (control) and full-strength Rorison solution (fertilized) were added throughout the experiment as 100 ml per microcosm every 4 days. Bars designated by the same letter are not different at* $p < 0.05$ *(ANOVA).*

4. Nuclear DNA amounts

From large-scale screening operations (e.g. Stebbins 1956; Bennett & Smith 1976; Levin & Funderburg 1979) and various other investigations (Hartsema 1961; Bennett 1971, 1976; Grime & Mowforth 1982; Grime 1983; Grime et al. 1985), it has been established that there is more than a thousand-fold variation in nuclear DNA amount in vascular plants and that differences in DNA amount in cool temperate regions such as the British Isles coincide with differences in the timing of shoot growth (Fig. 7.7). A mechanism interpreting variation in DNA amount, cell size and the length of the cell cycle (the three attributes are inextricably linked) as a consequence of climatic selection has been proposed (Grime & Mowforth 1982) but will not be reviewed here. In the context of the ISP and the search for functional types in the British flora it is pertinent to focus on one specific prediction concerning the potential impact of global warming. An essential precursor to this prediction is the hypothesis that the delayed phenology of many British plants (see Fig. 7.7) is imposed by the greater sensitivity of small-celled, low-DNA species to the inhibitory effects of low temperature on cell division. If this hypothesis is correct, we may expect that a rise in temperature would confer an advantage on low-DNA plants by

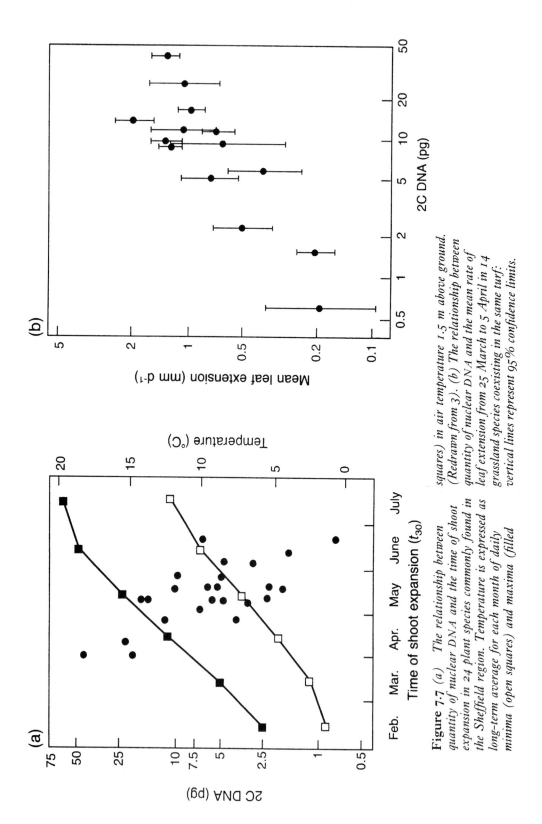

Figure 7.7 (*a*) *The relationship between quantity of nuclear DNA and the time of shoot expansion in 24 plant species commonly found in the Sheffield region. Temperature is expressed as long-term average for each month of daily minima (open squares) and maxima (filled squares) in air temperature 1.5 m above ground. (Redrawn from 3). (b) The relationship between quantity of nuclear DNA and the mean rate of leaf extension from 25 March to 5 April in 14 grassland species coexisting in the same turf: vertical lines represent 95% confidence limits.*

differentially lengthening their growing seasons (Fig. 7.8). Recently, this prediction has been tested (Willis *et al.* 1993) by reference to the results of a 37-year monitoring study conducted on roadside vegetation at Bibury in Gloucestershire. The results of this analysis confirm that plants of low DNA amount are promoted by warm winters.

5. Seed size

In common with several earlier large-scale comparative studies (Salisbury 1942; Baker 1972; Grime *et al.* 1981; Jurado *et al.* 1991) the ISP provides strong circumstantial evidence that seed size is a crucially important attribute with important implications for both dispersal and seedling establishment. Two hypotheses have sought to explain the adaptive significance of large seeds. The first, based upon the observation that many plants of droughted habitats tend to have large seeds (Baker 1972), proposes that the greater storage reserves of large seeds facilitate rapid root penetration to subsoil moisture; some evidence supporting this interpretation is contained in Figs. 7.3 and 7.4. The second hypothesis (Salisbury 1942) is almost independent of the first in the sense that it applies to a quite different circumstance in which there is a preponderance of large-seeded species. This occurs among the trees, shrubs and herbs of vegetation types providing a mature, closed cover. Here, Salisbury suggests that the greater resources of large seeds confer the potential for seedling persistence and establishment in conditions where resource supplies are limited by the presence of established vegetation. Although the advantage of large seeds in escaping from shade has been demonstrated under laboratory conditions (Grime & Jeffrey 1965), there have been few comparative tests of the ability of large-seeded species to establish from seed in closed vegetation (Thompson & Baster 1992).

Following the protocol of Figure 7.1 (right-hand side), a large-scale experiment has been initiated in 2.0 m × 2.0 m field plots situated in ancient, continuously moist, derelict sheep pasture at Harpur Hill in north Derbyshire (Burke 1993). Each plot has been sown with a high density of seeds of native herbaceous species chosen to represent a wide range in seed size and ecology. Of the 67 species sown into the plots, the majority (54) are not present in the established vegetation. This has allowed an annual census to be conducted to record the establishment and persistence of seedlings originating from a single, known inoculum. A further refinement in the experimental design has been the maintenance within each plot of orthogonally crossed, five-step gradients in soil fertility and vegetation disturbance (Fig. 7.9).These gradients were applied prior to seed sowing and are reinforced annually. The advantage of the 5 × 5 matrix in each plot

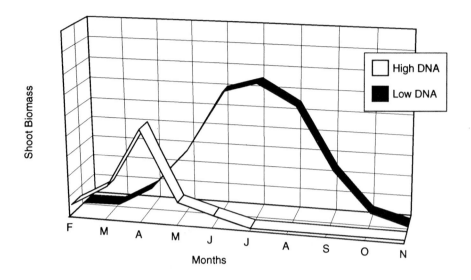

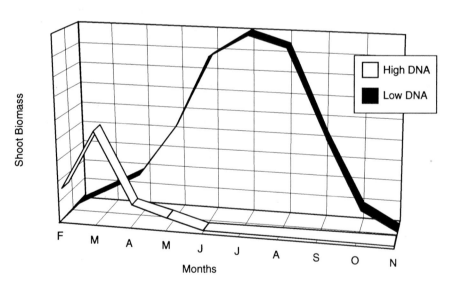

Figure 7.8 *Diagram illustrating the predicted differential effect of elevated winter and spring temperatures on the shoot biomass of two plants of contrasted DNA content. In the determinate phenology of the vernal geophyte (high DNA), shoot expansion occurs earlier but is terminated by the limited supply of preformed tissue. Although still relatively delayed by its dependence upon warm temperatures and current cell division, the indeterminate summer green herb (low DNA) experiences a lengthening of the growing season and is predicted to show an immediate increase in shoot biomass.*

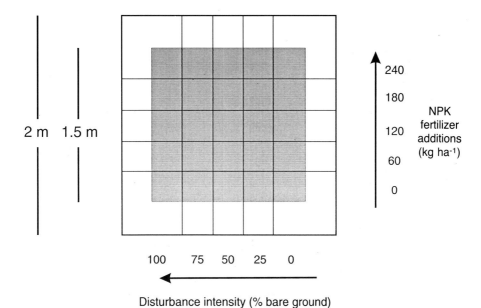

Figure 7.9 *The layout of the productivity – disturbance matrix.*

is that it presents the colonizing species with an extremely diverse matrix of
regenerative opportunities. This allows the capacity for invasion of the established vegetation to be assessed by reference to the performance of seedlings
of the same species across a wide range of conditions. Study of the establishment of species in productive, undisturbed turf over the duration of the
experiment (1991 and 1992) allows identification of those sown species that
are most successful at establishing in conditions of intense competition from
established vegetation. Equally important, the matrix approach allows recognition of species in which regeneration fails throughout the plot and is the
result of factors (e.g. seed predation, frost) unrelated to impacts of the established vegetation.

   Fig. 7.10 provides an illustration of data obtained from this experiment.
Persistence of colonizing seedlings and population expansion in the undisturbed turf is clearly higher in large-seeded species.

6. Seed persistence

Seed persistence in the soil is of great importance to both ecology and agriculture, yet data in the literature are scattered and of variable reliability,
while determination of seed content of soil samples is both time- and
labour-intensive. Recently, two aspects of the ISP have contributed to the
goal of predicting seed persistence in soil.

   The first (Thompson *et al.* 1993) involves the recognition that probability
of seed burial depends strongly on seed size and shape. Small, more or less

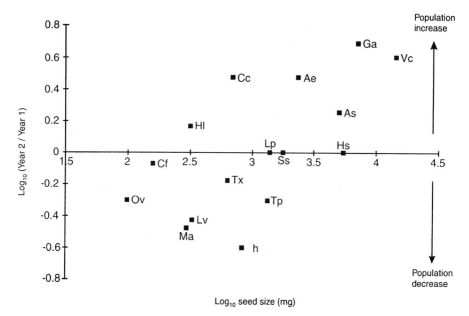

Figure 7.10   *The establishment of sown species in highly productive, undisturbed turf at Harpur Hill. Regenerative success is measured as the relative change in frequency of a species from 1991 to 1992.*

Key to species: Ae, Arrhenatherum elatius; *As*, Anthriscus sylvestris; *Cc*, Cynosurus cristatus; *Cf*, Cerastium fontanum; *Ga*, Galium aparine; *Hs*, Heracleum sphondylium; *Hl*, Holcus lanatus; *Lh*, Leontodon hispidus; *Lv*, Leucanthemum vulgare; *Lp*, Lolium perenne; *Ma*, Myosotis arvensis; *Ov*, Origanum vulgare; *Ss*, Stachys sylvatica; *Tx*, Taraxacum *spp.*; *Tp*, Trifolium pratense; *Vc*, Vicia cracca.

spherical seeds are readily incorporated into a persistent seed bank, whereas large, flattened or attenuated seeds rarely become buried. The measure of 'shape' we have adopted is the variance of the three linear seed dimensions, transformed so that the largest dimension is unity, while seed size is simply mean mass. When these two variables are plotted against each other, it is apparent that nearly all the species with persistent seed banks occupy a compact region near the origin, whereas all species distant from the origin are transient. Overlap between these regions is slight, suggesting that in the limited number of species so far examined the method can discriminate between the two types with fair accuracy.

The second example involves the attempt to discriminate biochemically between persistent and transient seeds. It is commonly assumed that predation and microbial pathogens must play a role in the mortality of buried seeds, but evidence is hard to find. Predators are probably an insignificant source of mortality for small buried seeds, but pathogens, especially fungi, may be much more important. Strong circumstantial evidence that this is so comes from ISP data on seed persistence in the soil and on the quantities of *ortho*-dihydroxyphenols in seeds (Fig. 7.11). Seeds of species with persistent soil seedbanks contain significantly higher concentrations of *ortho*-dihydroxyphenols than seeds of species with transient seedbanks. Other published evidence strongly suggests that phenols are important anti-fungal compounds in seeds, but published data are largely confined to crops and a handful of weeds. The ISP data are the first convincing demonstration of a role for antimicrobial compounds in the long-term persistence of seeds in soil.

Discussion

Current efforts to model the future of terrestrial vegetation at a global or regional scale are limited not only by uncertainties about climate and land use but also by weaknesses in our understanding of the mechanisms that control the structure and dynamics of vegetation. Contemporary ecological journals bear witness to major unresolved debates concerning central issues such as resource competition, successional processes and coexistence mechanisms. Why have these debates remained unresolved for so long?

One answer to this question is that much of the observational and experimental research in plant ecology has retreated into the precise and the particular and, in consequence, attempts at synthesis have been difficult and

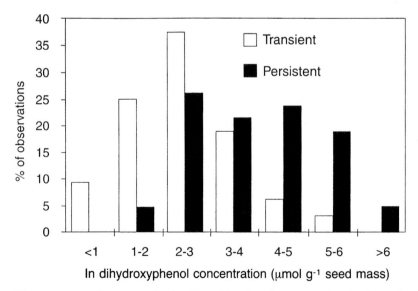

Figure 7.11 *Frequency distribution of In phenol concentration in the seeds of species with transient and persistent seedbanks. Distributions are significantly different at p<0.02 (Mann–Whitney test); n=83 (Hendry, Moss, Thorpe & Thompson, unpublished).*

delayed. Theoretical models of wider amplitude have been developed, but too often these have remained untested for lack of data at a matching scale. If the present search for functional types signals a return to the broader research agendas and databases of E. J. Salisbury, G. L. Stebbins and H. G. Baker, it is much to be welcomed and we draw encouragement from the persistent efforts of research groups such as those led by I. Washitani, J. M. and C. C. Baskin, P. Keddy, P. Jordano, N. C. Garwood, H. Poorter and M. Westoby, all of whom are building, in various parts of the world, standardized data sources broad enough to test the generalizing principles relevant to the definition of functional types of plants.

In this chapter we have attempted to illustrate one approach to the recognition of functional types, and we have explored various ways of testing the results experimentally. The examples chosen reveal several axes of functional specialization but more screening effort and cluster analysis will be required before it is possible to judge the extent to which these patterns are capable of crystallizing into a unified typology. A more distant goal is to compare the spectrum (cf. Raunkiaer 1934; Willson *et al.* 1990) of functional types in local floras in various parts of the world.

The most serious impediment in current efforts to define primary functional types is the continuing argument about the mechanism of resource competition in productive and unproductive vegetation and the importance of evolutionary trade-offs between root and shoot functions. Some ecologists have argued that there is an essential interdependence between competitive abilities above and below ground (Donald 1958; Grime 1977; Chapin 1980; Colasanti & Grime 1993) whereas others (Newman 1973; Huston & Smith 1987; Tilman 1988; Smith *et al.* 1993) have attached central importance to the concept of trade-offs between competitive abilities for above- and below-ground resources. Evidence of a trade-off between root and shoot function has been drawn from experiments (Brouwer 1962a,b; Corré 1983a,b; Hunt & Nicholls 1986) in which partitioning of dry matter between shoots and roots has been modified by mineral nutrient stress and shade. Such data have sometimes encouraged the view that plant evolution under various conditions of soil fertility, shade and moisture limitation will have resulted in predictable differences in root–shoot allocation and an uncoupling of competitive abilities for light, water and mineral nutrients. Recent investigations (Olff *et al.* 1990; Berendse & Elberse 1989; Tilman & Cowan 1989; Gleeson & Tilman 1990; Aerts *et al.* 1991; Campbell *et al.* 1991; Shipley & Peters 1990; Reich *et al.* 1992) do not support this hypothesis. This result is not surprising for three reasons:

1. Many habitats cannot be simply classified with respect to single limiting factors. Although shading, for example, is often a conspicuous

feature near the ground surface of grasslands and woodlands, mineral nutrients also frequently limit plant production at particular sites.

2. Models such as those of Huston & Smith (1987) and Tilman (1988) and experiments such as those of Brouwer (1963) and Hunt & Nicholls (1986) involve circumstances in which resource depletion is imposed uniformly within the aerial or rooting environment. In particular, omission of the depletion zones that surround the root surfaces of fast-growing plants growing on fertile soils (Bhat & Nye 1973) has led to serious underestimation of the expenditure of assimilate required to sustain the continuous process of root growth necessary to escape from the local but expanding zones of nutrient exhaustion, which are an inescapable consequence of the physics of nutrient uptake (Nye & Tinker 1977) from the rhizosphere.

3. There are biochemical limits to the trade-off between root and shoot. Autotrophy involves the assembly of components derived from both parts of the plant. Chemical analyses of plants reveal that the ratio of root- to shoot-derived elements remains relatively constant across a wide range of ecologies. Interdependence is further enforced by the carbon and energy demand of roots and the mineral nutrient demand of leaves. Scope for variation in root : shoot ratio across plant functional types is restricted by the fact that species with the capacity for high rates of photosynthesis and dry matter production have higher concentrations of leaf nitrogen (Sharkey 1985; Field & Mooney 1986) and leaf phosphorus (Band & Grime 1981) and in consequence are dependent upon high rates of nutrient capture by the root system.

From these arguments and from a large body of experimental evidence we conclude that the most important distinction between plant functional types of productive and unproductive vegetation relates not to the relative abundance of above- and below-ground resources but to the different rates at which captured resources are utilized in the two circumstances. In productive vegetation, the survival and status of component plants depend on continuous projection of new leaves and new roots into the undepleted sectors of the aerial and soil environment, processes that involve a swift commitment of captured mineral nutrients and photosynthate into plant structure, high tissue turnover, and rapid resource loss through senescence, herbivory and decomposition. In marked contrast, the dominant functional types of unproductive vegetation are characterized by more conservative mechanisms of resource capture and utilization. These involve uncoupling of resource capture from growth and the development of internal storage pools, many of which are protected by antiherbivore defences. In addition, resource foraging above and below ground tends to be less dynamic than that observed in plants of productive habitats and often involves pulse interception through long-lived surfaces.

Conclusions

As Stearns (1976) recognized, it is relatively easy to devise theories of adaptive specialization in plants and animals but much more difficult to test them. For the practical purposes of predicting future impacts of land use and climate change on vegetation, we recommend that modelling should be informed by close contact with experimental studies. In particular we advocate a protocol involving (1) formal searches for plant functional types by screening and (2) hypothesis testing by field monitoring studies and ecosystem manipulations. For too long, an unresolved debate about the role of competition in plant communities has remained an obstacle to progress in the effort to find a unified theory of primary plant functional types. A resolution of this debate is at hand as the focus of argument shifts from field observation and speculation to mechanistic analysis of resource capture and utilization in specified conditions.

Acknowledgement

The research drawn upon in this paper was supported by the Natural Environment Research Council. Parts of the Integrated Screening Programme reported here involved contributions from J. M. L. Mackey, A. M. Neal, S. R. Band, D. W. Hand, M. A. Hannah and N. Matthews.

References

Aerts, R., Boot, R. G. A. & van der Aart, P. J. M. (1991) The relation between above- and below-ground biomass allocation patterns and competitive ability. *Oecologia*, **87**, 551–9.

Al-Mufti, M. M., Sydes, C. L., Furness, S. B., Grime, J. P. & Band, S. R. (1977) A quantitative analysis of shoot phenology and dominance in herbaceous vegetation. *Journal of Ecology*, **65**, 759–92.

Baker, H. G. (1972) Seed weight in relation to environmental conditions in California. *Ecology*, **53**, 997–1010.

Band, S. R. & Grime, J. P. (1981) Chemical composition of leaves. *Annual Report, 1981, Unit of Comparative Plant Ecology (NERC)*, pp. 6–8. University of Sheffield.

Bennett, M. D. (1971) The duration of meiosis. *Proceedings of the Royal Society of London*, B **178**, 277–99.

Bennett, M. D. (1976) DNA amount, latitude and crop plant distribution. *Environmental and Experimental Botany*, **16**, 93–108.

Bennett, M. D. & Smith, J. P. (1976) Nuclear DNA amounts in angiosperms. *Philosophical Transactions of the Royal Society*, B **274**, 227–74.

Berendse, F. & Elberse, W. T. (1989) Competition and nutrient losses from the plant. In Lambers, H., *et al.* (eds.), *Causes and Consequences of Variation in Growth Rate and Productivity of Higher Plants*, pp. 269–84. The Hague: SPB Academic Publishing.

Bhat, K. K. S. & Nye, P. H. (1973) Diffusion of phosphate to plant roots in soil. 1. Quantitative autoradiography of the depletion zone. *Plant and Soil*, **38**, 161–75.

Brouwer, R. (1962a) Distribution of dry matter in the plant. *Netherlands Journal of Agricultural Science*, **10**, 361–76.

Brouwer, R. (1962b) Nutritive influences on the distribution of dry matter in the plant. *Netherlands Journal of Agricultural Science*, **10**, 399–408.

Brouwer, R. (1963) Some aspects of the equilibrium between overground and underground plant parts. *Jaarboek Instituut voor Biologisch en Scheikundig Onderzock van Landbouwgewassen*, pp. 31–9.

Burke, M. J. W. (1993) *Plant community responses to fertilisation, disturbance and seed inoculation in limestone grassland.* PhD thesis, University of Sheffield.

Campbell, B. D., Grime, J. P. & Mackey, J. M. L. (1991) A trade-off between scale and precision in resource foraging. *Oecologia*, **87**, 532–8.

Chapin, F. S. (1980) The mineral nutrition of wild plants. *Annual Review of Ecology and Systematics*, **11**, 233–60.

Clutton-Brock, T. H. & Harvey, P. H. (1979) Comparison and adaptation. *Proceedings of the Royal Society of London*, B **205**, 547–65.

Colasanti, R. L. & Grime, J. P. (1993) Resource dynamics and vegetation processes: A deterministic model using two-dimensional cellular automata. *Functional Ecology* **7**, 169–76.

Corré, W. J. (1983a) Growth and morphogenesis of sun and shade plants. I. The influence of light intensity. *Acta Botanica Neerlandica*, **32**, 49–62.

Corré, W. J. (1983b) Growth and morphogenesis of sun and shade plants. III. The combined effects of light intensity and nutrient supply. *Acta Botanica Neerlandica*, **32**, 277–94.

Deevey, E. S. (1947) Life tables for natural populations of animals. *Quarterly Review of Biology*, **22**, 283–314.

Díaz, S., Grime, J. P., Harris, J. & McPherson, E. (1993) Evidence of a feedback mechanism limiting plant response to elevated carbon dioxide. *Nature* **364**, 616–17.

Donald, C. M. (1958) The interaction of competition for light and for nutrients. *Australian Journal of Agricultural Research*, **9**, 421–32.

Ellenberg, H. (1978) *Vegetation Mitteleuropas mit den Alpen in Ökologischer Sicht.* Stuttgart: Ulmer.

Field, C. & Mooney, H. A. (1986) The photosynthesis–nitrogen relationship in world plants. In Givnish, T. V. (ed.), *On the Economy of Plant Form and Function*, pp. 25–55. Cambridge University Press.

Gill, D. E. (1978) On selection at high population density. *Ecology*, **59**, 1289–91.

Givnish, T. J. (1987) Comparative studies of leaf form: assessing the relative roles of selective pressures and phylogenetic constraints. In Rorison, I. H., *et al.* (eds), *Frontiers of Comparative Plant Ecology. New Phytologist*, **106**, (Suppl.), pp. 131–60.

Gleeson, S. K. & Tilman, D. (1990) Allocation and the transient dynamics of succession on poor soils. *Ecology*, **71**, 1144–55.

Greenslade, P. J. M. (1972) Distribution patterns of *Priochirus species* (Coleoptera; Staphylinidae) in the Solomon Islands. *Evolution, Lancaster, Pa.*, **26**, 130–42.

Greenslade, P. J. M. (1983) Adversity selection and the habitat templet. *American Naturalist*, **122**, 352–65.

Grime, J. P. (1965) Comparative experiments as a key to the ecology of flowering plants. *Ecology*, **45**, 513–15.

Grime, J. P. (1974) Vegetation classification by reference to strategies. *Nature*, **250**, 26–31.

Grime, J. P. (1977) Evidence for the existence of three primary strategies in plants and its relevance to ecological and evolutionary theory. *American Naturalist*, **111**, 1169–94.

Grime, J. P. (1979) *Plant Strategies and Vegetation Processes.* Chichester: Wiley.

Grime, J. P. (1983) Prediction of weed and crop response to climate based upon measurements of nuclear DNA content. In *Aspects of Applied Biology. 4: Influence of Environmental Factors on Herbicide Performance and Crop and Weed Biology.* National Vegetable Research Station, Wellesbourne.

Grime, J. P. (1988) The C-S-R model of primary plant strategies – origins, implications and tests. In Gottlieb, L. D. & Jain, S. K. (eds), *Plant Evolutionary Biology*, pp. 371–93. London: Chapman and Hall.

Grime, J. P. (1989) The stress debate: symptom of impending synthesis? In Calow, P. (ed.), Evolution, Ecology and Environmental Stress. *Biological Journal of the Linnean Society*, **37**, 3–17.

Grime, J. P. & Hillier, S. H. (1992) The contribution of seedling regeneration to the structure and dynamics of plant communities and larger units of landscape. In Fenner, M. (ed.) *The Ecology of Regeneration in Plant Communities*, pp. 349–64. Wallingford: C. A. B. International.

Grime, J. P., Hodgson, J. G. & Hunt, R. (1988) *Comparative Plant Ecology: A Functional Approach to common British Species.* London: Unwin Hyman.

Grime, J. P. & Hunt, R. (1975) Relative growth rate; its range and adaptive significance in a local flora. *Journal of Ecology*, **63**, 393–422.

Grime, J. P., Hunt, R. & Krzanowski, W. J. (1987) Evolutionary physiological ecology of plants. In Calow, P. (ed.), *Evolutionary Physiological Ecology*, pp. 105–26. Cambridge University Press.

Grime, J. P. & Jeffrey, D. W. (1965) Seedling establishment in vertical gradients of sunlight. *Journal of Ecology*, **53**, 621–42.

Grime, J. P., Mason, G., Curtis, A. V., Rodman, J., Band, S. R., Mowforth, M. A., Neal, A. M. & Shaw, S. C. (1981) A comparative study of germination characteristics in a local flora. *Journal of Ecology*, **69**, 1017–59.

Grime, J. P. & Mowforth, M. A. (1982) Variation in genome size – an ecological interpretation. *Nature*, **299**, 151–3.

Grime, J. P., Shacklock, J. M. L. & Band, S. R. (1985) Nuclear DNA contents, shoot phenology and species coexistence in a limestone grassland community. *New Phytologist*, **100**, 435–44.

Harper, J. L. (1982) After description. In Newman, E. I. (ed.), *The Plant Community as a Working Mechanism* (Special Publi-

cation No. 1, The British Ecological Society), pp. 11–25. London: Blackwell.

Hartsema, A. M. (1961) Influence of temperature on flower formation and flowering of bulbous and tuberous plants. In Ruhland, W. (ed.), *Handbuch der Pflanzenphysiologie. 16. Ansenfaktoren in Wachstum und Entwicklung*, pp. 123–67. Berlin: Springer.

Harvey, P. H. & Pagel, M. D. (1992) *The Comparative Method in Evolutionary Biology*. Oxford University Press.

Hendry, G. A. F. & Grime, J. P. (1993) *Comparative Plant Ecology – A Laboratory Manual*. London: Chapman and Hall.

Hillier, S. H. (1984) *A quantitative study of gap recolonization in two contrasted limestone grasslands*. Ph.D. thesis, University of Sheffield.

Hodgson, J. G. (1989) What is happening to the British flora? An investigation of commonness and rarity. *Plants Today*, **2**, 26–32.

Hodgson, J. G. (1991) The use of ecological theory and autecological data-sets in studies of endangered plant and animal species and communities. *Pirineos*, **138**, 3–28.

Hodgson, J. G. & Mackey, J. M. L. (1986) The ecological specialization of dicotyledonous families within a local flora: some factors constraining optimization of seed size and their possible evolutionary significance. *New Phytologist*, **104**, 479–515.

Hunt, R., Hand, D. W., Hannah, M. A. & Neal, A. M. (1991a) Response to CO_2 enrichment in 27 herbaceous species. *Functional Ecology*, **5**, 410–21.

Hunt, R., Hand, D. W. Hannah, M. A. & Neal, A. M. (1993) Further responses to CO_2 enrichment in British herbaceous species. *Functional Ecology*, **7**, 661–8.

Hunt, R., Middleton, D. A. J., Grime, J. P. & Hodgson, J. G. (1991b) TRISTAR: an expert system for vegetation processes. *Expert Systems*, **8**, 219–26.

Hunt, R. & Nicholls, A. O. (1986) Stress and the course control of root-shoot partitioning in herbaceous plants. *Oikos*, **47**, 149–58.

Huston, M. A. & Smith, T. M. (1987) Plant

succession, life history and competition. *American Naturalist*, **130**, 168–98.

Hutchinson, G. E. (1959) Homage to Santa Rosalia or why are there so many kinds of animals? *American Naturalist*, **93**, 145–59.

Jalili, A. (1991) *An investigation of the role of drought and mineral nutrients in the structure and dynamics of a calcareous grassland community*. PhD thesis, University of Sheffield.

Jurado, E., Westoby, M. & Nelson, D. (1991) Diaspore weight, dispersal, growth form and perenniality of central Australian plants. *Journal of Ecology*, **79**, 811–28.

Keddy, P. (1992) A pragmatic approach to functional ecology. *Functional Ecology*, **6**, 621–6.

Korner, C. (1993) CO_2 fertilization: The great uncertainty in future vegetation development. In Solomon, A. M. & Shugart, H. H. (eds), *Vegetation Dynamics and Global Change*, pp. 53–70. London: Chapman and Hall.

Leps, J., Osbornova-Kosinova, J. & Rejmanek, M. (1982) Community stability, complexity and species life-history strategies. *Vegetatio*, **50**, 53–63.

Levin, D. A. & Funderburg, S. W. (1979) Genome size in angiosperms: temperate versus tropical species. *American Naturalist*, **114**, 784–95.

MacArthur, R. H. & Wilson, E. O. (1967) *The Theory of Island Biogeography*. Princeton: Princeton University Press.

MacLeod, J. (1894) Over de bevruchting der bloemen in het Kempisch gedeelte van Vlaanderen. *Deel II. Bot. Jaarboek Dodonaea*, **6**, 119–511.

Newman, E. I. (1973) Competition and diversity in herbaceous vegetation. *Nature*, **243**, 244–310.

Noble, I. R. & Slatyer, R. O. (1979) The use of vital attributes to predict successional changes in plant communities subject to recurrent disturbances. *Vegetatio*, **43**, 5–21.

Nye, P. H. & Tinker, P. B. (1977) *Solute Movement in the Soil–Root System*. Oxford: Blackwell Scientific Publications.

Olff, H., van Andel, J. & Bakker, J. P. (1990) Biomass and shoot/root allocation of five

species from a grassland succession series at different combinations of light and nutrient supply. *Functional Ecology*, **4**, 193–200.

Pianka, E. R. (1970) On r- and K-selection. *American Naturalist*, **104**, 592–7.

Pugh, G. J. F. (1980) Strategies in fungal ecology. *Transactions of the British Mycorrhizal Society*, **75**, 1–14.

Ramenskii, L. G. (1938) *Introduction to the geobotanical study of complex vegetations*. Moscow: Selkzgiz.

Raunkiaer, C. (1934) *The Life Forms of Plants and Statistical Plant Geography; being the collected papers of C. Raunkiaer, translated into English by H. G. Carter, A. G. Tansley and Miss Fansboll*. Oxford: Clarendon Press.

Reader, R. J., Jalili, A., Grime, J. P., Spencer, R. & Matthews, N. (1992) A comparative study of plasticity in seedling rooting depth in drying soil. *Journal of Ecology*, **81**, 543–50.

Reich, P. B., Walters, M. B. & Ellsworth, D. S. (1992) Leaf life-span in relation to leaf, plant and stand characteristics among diverse ecosystems. *Ecological Monographs*, **62**, 365–92.

Salisbury, E. J. (1942) *The Reproductive Capacity of Plants*. London: George Bell.

Sharkey, T. D. (1985) Photosynthesis in intact leaves of C3 plants: Physics, physiology and rate limitations. *Botanical Review*, **51**, 53.

Shipley, B & Peters, R. H. (1990) A test of the Tilman model of plant strategies: Relative growth rate and biomass partitioning. *American Naturalist*, **136**, 139–53.

Silvertown, J., Franco, M., Pisanty, I. & Mendoza, A. (1993) Comparative plant demography: relative importance of life-cycle components to the finite rate of increase in woody and herbaceous perennials. *Journal of Ecology*, **81**, 465–76.

Smith, T. M., Shugart, H. H., Woodward, F. I. & Burton, P. J. (1993) Plant functional types. In Solomon, A. M. & Shugart, H. H. (eds), *Vegetation Dynamics and Global Change*, pp. 272–92. London: Chapman and Hall.

Southwood, T. R. E. (1977) Habitat, the

templet for ecological strategies? *Journal of Animal Ecology*, **46**, 337–65.

Stearns, S. C. (1976) Life-history tactics; a review of the ideas. *Quarterly Review of Biology*, **51**, 3–47.

Stebbins, G. L. (1956) Cytogenetics and evolution of the grass family. *American Journal of Botany*, **43**, 890–905.

Stebbins, G. L. (1971) *Chromosomal Evolution in Higher Plants*. London: Edward Arnold.

Thompson, K., Band, S. R. & Hodgson, J. G. (1993) Seed size and shape predict persistence in soil. *Functional Ecology*, **7**, 236–41.

Thompson, K. & Baster, K. (1992) Establishment from seed of selected Umbelliferae in unmanaged grassland. *Functional Ecology*, **6**, 346–52.

Tilman, D. (1988) *Plant Strategies and the Structure and Dynamics of Plant Communities*. Princeton: Princeton University Press.

Tilman, D. & Cowan, M. L. (1989) Growth of old-field herbs on a nitrogen gradient. *Functional Ecology*, **3**, 425–38.

Valk, A. G. van der (1981) Succession in wetlands: a Gleasonian approach. *Ecology*, **62**, 688–96.

Whittaker, R. H. & Goodman, D. (1979) Classifying species according to their demographic strategy. 1: Population fluctuations and environmental heterogeneity. *American Naturalist*, **113**, 185–200.

Willis, A. J., Hunt, R., Grime, J. P., Dunnett, N. P., Sutton, F., Band, S. & Neal, A. M. (1993) *A thirty-five year study of vegetation and climate in road verges at Bibury, Gloucestershire*. The NERC Unit of Comparative Plant Ecology, Terrestrial Ecology Research on global warming: Phase 2. Contract report to Nuclear Electric plc.

Willson, M. F., Rice, B. L. & Westoby, M. (1990) Seed dispersal spectra: a comparison of temperate plant communities. *Journal of Vegetation Science*, **1**, 547–62.

Woodward, F. I. (1992) Predicting plant responses to global environmental change. *New Phytologist*, **122**, 239–51.

Part three

8 Plant functional types and ecosystem change in arctic tundras

G. R. Shaver, A. E. Giblin, K. J. Nadelhoffer and E. B. Rastetter

Introduction

The arctic region includes a remarkably diverse array of terrestrial ecosystems, with dramatic differences in the kinds of plants that dominate the vegetation over large areas. The dominant plant of an arctic ecosystem might be woody, graminoid or herbaceous; it might be evergreen or deciduous; it might be erect, prostrate or caespitose; it might even be a moss or a lichen (Wielgolaski 1975; Bliss 1981; Bliss & Matveyeva 1992). Each of these distinctively different plant types dominates over hundreds of thousands of square kilometres within the Arctic. On a local basis, similarly dramatic variation in vegetation of arctic ecosystems occurs over short topographic gradients (Billings & Mooney 1968; Bliss 1977; Miller 1982; Shaver *et al.* 1991).

Because the variation in the dominant plant types among arctic ecosystems is so striking and is clearly related to climate, microclimate and topography, many ecologists have investigated the functional differences among arctic plant types in attempts to understand the mechanisms controlling their distribution. The concept of plant functional types has proven to be useful in these efforts, and reasonably accurate predictions of the functional characteristics of plants that should be found in a given location in the Arctic are now possible, as are predictions of change in the relative abundance of different plant functional types in response to disturbance or climate change. The ability to make such predictions rests on several decades of research on carbon and nutrient uptake kinetics, water relations, growth rates, allocation patterns and life histories of arctic plants (reviewed by Bliss 1962; Billings & Mooney 1968; Savile 1972; Lewis & Callaghan 1976; Bliss *et al.* 1981; Chapin & Shaver 1985a; Chapin *et al.* 1992a; · Reynolds & Tenhunen 1995).

On the other hand, our understanding of the importance of these plant functional characteristics to overall ecosystem function in the Arctic is much less well developed (Chapin *et al.* 1992b). This is because the focus of most past research on arctic plants has been on understanding the rela-

tive fitness or competitive superiority of individual ecotypes, species or
species groups, either within the Arctic or in comparison to more temperate
regions (e.g. Mooney & Billings 1961). Although small differences in physiology or morphology can often explain the patterns of species abundance
occurring in the field, the implications of these small differences in plant-
level functional characteristics for large differences in ecosystem-level characteristics, such as net carbon storage or long-term productivity, are often
unclear. Simulation models have been used occasionally in the past to
predict the implications of a given species composition for primary
production in a given climate, but the predictions are generally useful only
in the short term (Reynolds & Leadley 1992).

The first aim of this chapter is to review what is known about the
ecosystem-level implications of variation in the functional characteristics of
arctic plants. More specifically, 'How important is the species composition
or dominant plant type of arctic vegetation to the accumulation and turnover of whole-ecosystem carbon and nutrient stocks?' The answer to this
question is still unclear and depends greatly on the timescale of interest and
the specific characteristics under investigation. This review then leads to a
second important question: 'How useful to predictions of ecosystem
response to disturbance are classifications of plant functional types based on
small differences in relative fitness among species?' Here, the answer is
more clear-cut and it is possible to say that existing, commonly used classifications of arctic plant functional types are useful in predicting constraints
on ecosystem-level response. It is clear that new experimental and analytical
approaches are needed, however, to clarify the controlling mechanisms.

Comparisons of whole vegetation among undisturbed ecosystems

Variation in ecosystem-level characteristics such as primary production,
total biomass, and soil organic matter and element stocks is clearly related
to variation in the dominant plant types of arctic vegetation. On local,
regional and global scales the most productive arctic ecosystems are
invariably those in which the vegetation is dominated by woody, erect,
deciduous shrubs (e.g. Fig. 8.1, Tables 8.1 and 8.2). The most productive
arctic ecosystems also have a high total vegetation mass, but they generally
do not accumulate a thick, peaty organic mat at the surface of the soil. Less
productive ecosystems are typically dominated by sedges, by erect or prostrate evergreen shrubs, and by herbs, mosses or lichens. These less
productive ecosystems often accumulate large amounts of soil organic

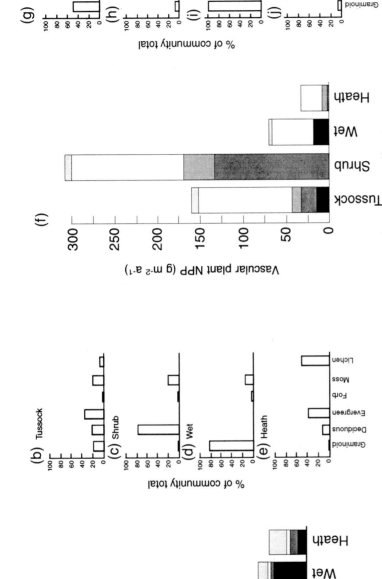

Figure 8.1 Vegetation biomass and annual vascular plant production in four contrasting ecosystem types near Toolik Lake, Alaska (compiled from data in Shaver & Chapin 1991). The four ecosystems include a moist tussock tundra ('tussock'), a deciduous shrub tundra ('shrub'), a wet sedge tundra ('sedge'), and an evergreen shrub–lichen heath ('heath'). The first panel (a) shows whole community biomass at each site. Solid black portions of each bar indicate below-ground stem and rhizome mass, hatched portions indicate above-ground vascular stem mass, clear portions indicate leaf mass, and dotted portions indicate live moss and lichen mass. Panels (b)–(e) show the percentage of the community total biomass at each site that is accounted for by graminoids, deciduous shrubs, evergreen shrubs, forbs, mosses, and lichens.

Panel (f) shows annual vascular plant production at each site (not including root production). Solid black portions of each bar indicate below-ground stem and rhizome production, lower hatched portions indicate secondary wood production in stems, upper hatched portions indicate apical stem (twig) production, clear portions indicate leaf production, and vertically striped portions indicate inflorescence production. Panels (g)–(j) show the percentage of the community total production at each site that is accounted for by graminoids, deciduous shrubs, evergreen shrubs, forbs, mosses, and lichens.

matter and can have higher total stocks of C and other elements than the most productive ecosystems.

Overall, primary productivity varies by three orders of magnitude from the herb- and cryptogram-dominated polar deserts of the High Arctic to the deciduous shrub tundras of the Low Arctic (Table 8.1). Although the productivity of a given arctic ecosystem type also varies with climate and latitude (as in the near-doubling of productivity of wet sedge tundra from the High Arctic to the Low Arctic), variation in productivity among ecosystem types is greater than variation within ecosystem types. Local variation is somewhat less dramatic, but still great; for example, along a single toposequence in northern Alaska, productivity varies four-fold from wet sedge tundra to a deciduous shrub–lupin zone that is protected by winter snow cover (Table 8.2). Regional variation in productivity can be even greater: about 10-fold among the major ecosystem types near Toolik Lake in northern Alaska (Table 8.3) and 40-fold on the Truelove Lowland, Devon Island, Canada (Bliss 1977).

Productivity gradients in arctic ecosystems are also related to functional differences in the dominant plants along those gradients. Perhaps the most consistent observation is that the species and plant types that dominate on productive sites in the Arctic are those with the highest capacity for photosynthesis and for soil nutrient uptake when nutrient resource availability is high (Johnson & Tieszen 1976; Chapin & Shaver 1985a; Kielland & Chapin 1992). These species also have high growth rates (at least in disturbed or resource-rich sites), high element concentrations in leaves, and high rates of leaf, twig and root turnover, suggesting that high uptake capacity is linked to high carbon and element requirements in support of new growth. Common examples include *Salix*, *Betula* and *Vaccinium* species among the deciduous shrubs. The same pattern shows up, however, in other plant types when comparisons are made among ecotypes within the same species, or among species within the same genus growing on sites with different resource availability. For example, within the common sedge genera *Carex* and *Eriophorum*, populations and species from high-resource sites consistently have higher carbon and nutrient uptake capacity, higher potential growth rate, and apparently higher carbon and nutrient requirements for growth than do populations and species from low-resource sites (Chapin & Oechel 1983; McGraw & Chapin 1989; Fetcher & Shaver 1990).

With this background of research on individual species it would be logical to predict that the most productive arctic ecosystems should also have vegetation with higher rates of carbon and nutrient turnover, higher overall nutrient concentrations, and greater dependence on nutrient uptake as opposed to internal recycling in support of primary production.

Table 8.1 *Dominant plant types, productivity and organic matter content of major arctic ecosystems*

	Low Arctic					High Arctic		
	Tall shrub	Low shrub	Tussock–dwarf shrub	Wet sedge	Semidesert	Wet sedge	Semidesert	Polar desert
Dominant plants	Deciduous shrubs	Deciduous and evergreen shrubs	Sedges and dwarf shrubs	Sedges, mosses	Dwarf shrubs, cryptogams, herbs	Sedges, mosses	Dwarf shrubs, cryptogams, herbs	Dwarf shrubs, Herbs, cryptogams
NPP $(g\ m^{-2})$[†]	1000*	375*	225*	220	45	140	35	1
Vegetation mass $(g\ m^{-2})$[†]	5800*	3100*	7400*	4560	1470	2360	1155	24
Soil organic matter $(g\ m^{-2})$	400*	1000*	1000*	38750	2540	2100	1030	22
Area $(\times 10^3\ km^2)$	174	1282	922	880	358	132	1005	847
Total organic matter $(10^{15}\ g)$	1.08	5.26	7.74	38.11	1.44	3.08	2.20	0.04
Total C $(10^{15}\ g)$	0.47	2.32	3.41	17.13	0.65	1.36	0.98	0.02

* Reported by Bliss & Matveyeva (1992) as 'estimates' for data from specific Alaskan sites, see Tables 8.2 and 8.3.

[†] Sum of above- and below-ground production.

Source: Compiled from Bliss & Matveyeva (1992).

Table 8.2 *Dominant plant types, productivity, and organic matter and element content of six ecosystem types along a single toposequence near the Sagavanirktok River, Alaska*

	Tussock tundra	Hilltop heath	Hillslope shrub–lupin	Footslope Equisetum	Wet sedge tundra	Riverside willow
Dominant plants	Sedges, dwarf shrubs	Dwarf shrubs, lupins	Deciduous and dwarf shrubs, lupins	Equisetum, sedges	Sedges	Deciduous shrubs
NPP $(g\ m^{-2})$*	118	170	263	99	82	175
Vegetation mass $(g\ m^{-2})$*	462	971	928	522	161	833
Peat mass $(g\ m^{-2})^{+}$	102 000	8140	41 400	29 400	49 000	7000
Peat C $(g\ m^{-2})^{+}$	21 633	3085	10 151	9937	19 747	1365
Peat N $(g\ m^{-2})^{+}$	1242	112	443	470	1293	85.4
Peat P $(g\ m^{-2})^{+}$	110	9.0	38.5	41.6	31.0	6.3
Peat thickness $(cm)^{+}$	30	3.7	15	21	40	3.5

* Includes above-ground parts and below-ground stems and rhizomes but not roots.
+ Includes thawed portions of upper organic horizons only: mineral soils beneath hilltop heath and riverside willow also contain significant organic matter and element content but depth of annual soil thaw is highly variable.

Sources: Soils data compiled and recalculated from Nadelhoffer *et al.* (1991) and Giblin *et al.* (1991). Production and biomass data are from G. R. Shaver *et al.* (unpublished), collected using the methods of Shaver & Chapin (1991).

Table 8.3 *Intersite variation in biomass and NPP of arctic vegetation is matched by a nearly equal variation in N and P content of biomass and NPP, with much smaller intersite variation in overall N and P concentrations*

For the four ecosystems described in Fig. 8.1, the data below indicate the range of variation among sites as the quotient of the largest value among the four sites for the variable indicated, divided by the smallest value. Nitrogen is the principal limiting element except perhaps in the wet sedge vegetation, and the greater variability in P concentration relative to N concentration is probably due to the fact that P is non-limiting.

Variable	Variation
Total biomass	9.1×
N mass in total biomass	8.0×
P mass in total biomass	8.1×
Per cent N in biomass	1.3×
Per cent P in biomass	3.2×
NPP	9.5×
N mass in NPP	10.5×
P mass in NPP	9.2×
Per cent N in NPP	1.2×
Per cent P in NPP	2.3×

Source: Original data in Shaver & Chapin (1991).

However, comparisons among arctic ecosystems at the level of the whole vegetation do not support such predictions (Shaver & Chapin 1991; Shaver *et al.* 1992). For example, primary production and plant biomass show a nearly constant relationship among the ten sites described in Fig. 8.1 and Table 8.2, indicating relatively constant biomass turnover (Fig. 8.2; $r^2=0.83$, $p<0.01$). In fact, the trend indicated by linear regression of production vs. biomass in Fig. 8.2 is towards higher production : biomass ratios (i.e. shorter turnover times) in the *least* productive sites, although the y-intercept is not significantly different from zero. Wielgolaski *et al.* (1981) obtained similar results in a comparison of total production vs. total biomass at 19 high-latitude sites. Regression of production vs. biomass in their study yielded a slope of about 0.2, compared with 0.16 in Fig. 8.3 (compare with Fig. 6.1b in Wielgolaski *et al.* (1981)). They also found relatively low product : biomass ratios at most sites with high production and biomass.

Whole-vegetation element concentrations and element concentrations in the current year's production also vary only slightly among arctic ecosystems (Table 8.3), with the most productive ecosystems actually having low to intermediate overall element concentrations (Shaver *et al.* 1992). This suggests that element requirements for biomass accumulation do not differ greatly among ecosystems despite dramatic differences in the dominant plant types of the vegetation. (Note that although the data in Tables 8.2 and 8.3 and Fig. 8.1 and 8.2 do not account for root mass, Shaver & Chapin (1991) have shown that inclusion of estimates of root mass does not produce the expected pattern, either.)

Why is there so little variation in the overall rate of biomass turnover (Fig. 8.2) and overall nutrient requirements (Table 8.3) among whole arctic vegetation types when the species that compose the vegetation differ strongly in their organ turnover patterns and nutrient use? The reason is that previous research on functional characteristics of arctic plants has focused on the most actively growing plant parts with the highest turnover rates and the highest element requirements, especially leaves and young stems. When only these parts are compared, large differences in biomass turnover and nutrient use are apparent (e.g. Fig. 6.1a in Wielgolaski *et al.* 1981). However, when whole vegetation types are compared it becomes clear that the carbon and nutrient budgets in arctic vegetation are

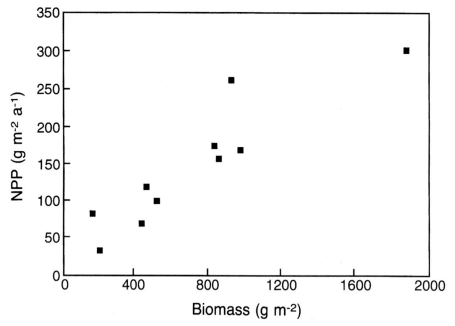

Figure 8.2 *Net primary production vs. biomass for the whole vegetation at ten sites in arctic Alaska (data from Table 8.2 and Fig. 8.1). The correlation coefficient (r²) for these data is 0.83 (p<0.01); linear regression yields the formula NPP = 0.16 × biomass + 33.35.*

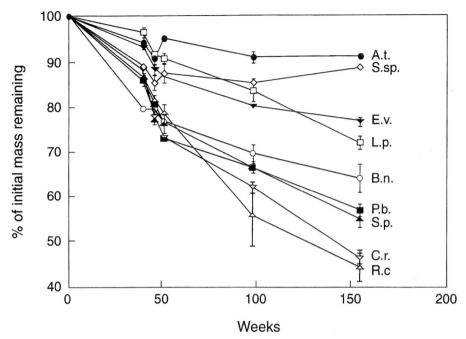

Figure 8.3 *Three-year pattern of mass loss of leaf litter from six common vascular plant species, and of whole, dead plants of two moss and one lichen species, placed in litter bags in moist tussock tundra at Toolik Lake, Alaska. A.t., Aulacomnium turgidum; S.sp., Sphagnum species; E.v., Eriophorum vaginatum; L.p., Ledum palustre; B.n., Betula nana; P.b., Polygonum bistorta; S.p., Salix pulchra; C.r., Cetraria richardsonii; R.c., Rubus cham-aemorus. (G. R. Shaver, unpublished data).*

dominated by a very large stem or rhizome mass, much of which is below ground (Shaver & Chapin 1991). Stems and rhizomes are the largest portion of the standing crop of biomass in most sites, and the relative importance of stems *increases* with both biomass and productivity (Fig. 8.1). Thus, as productivity increases the increasing abundance of species with high leaf turnover, high growth rate and high leaf nutrient requirement is countered by an increased proportional allocation to low-turnover, slow-growing stems with a low nutrient requirement. The net result is similar production : biomass ratios and nutrient requirements for the whole vegetation at all sites.

The main conclusion to be drawn from this comparison of relatively undisturbed tundra ecosystems is that, although there is clear evidence for systematic and predictable change in the functional characteristics of the dominant plant types along gradients of productivity in the Arctic, these changes appear to have little net effect on key indicators of ecosystem function at the level of the whole vegetation, such as production : biomass relationships, overall nutrient use and overall carbon and nutrient turnover.

However, the higher carbon and nutrient uptake capacity and more rapid growth potential of deciduous shrubs may not be realized in the relatively stable, undisturbed ecosystems compared in Fig. 8.1 and 8.2 and Tables 8.2 and 8.3. In disturbed ecosystems or under conditions of climatic change the differences in functional characteristics of different plant types may result in very different *responses* by each vegetation. There is also the possibility that functional differences among plant types that have been studied in previous work are themselves less important than their consequences for other ecosystem processes such as decomposition and element cycling in soils. These alternative effects of the change in plant types along productivity gradients in the Arctic are discussed below.

Feedbacks through decomposition

The chemical quality of plant litter is an important determinant of its decomposability and thus the rate at which mineral nutrients are recycled from litter into plant-available forms (Swift *et al.* 1979; Aber & Melillo 1982; McClaugherty *et al.* 1985). Litter chemical quality also varies widely among plants and is known to be related to functional characteristics such as photosynthetic capacity, leaf nutrient concentration and leaf longevity (Reich *et al.* 1992). For these reasons a number of researchers have suggested that ecosystems dominated by different plant functional types should also differ in their rates and patterns of soil element cycling (e.g. Berendse & Elberse 1990; Pastor & Naiman 1992).

In tundra ecosystems, a link between the dominant plant functional types in the vegetation and element turnover in the soil would be of particular significance because of the consistently strong nutrient limitation to productivity of tundras, and because recycling through decomposition is by far the most important source of nutrients for plant uptake (Jonasson 1983; Shaver *et al.* 1992). Evidence from litter bag experiments (e.g. Fig. 8.3) indicates that the plant functional types that dominate in the most productive sites within the Arctic (i.e. deciduous shrubs) produce some of the most rapidly decomposable litter, resulting in more rapid element turnover in the soil and probably contributing to the maintenance of high productivity. Similarly, in a laboratory experiment, Nadelhoffer *et al.* (1991) incubated soils from the six sites described in Table 8.2 and found that the soils with the highest respiration and N mineralization rates were those formed beneath the vegetation that was most strongly dominated by deciduous shrubs (Fig. 8.4). At least for some northern heathland ecosystems (Berendse & Elberse 1990; Berendse & Jonasson 1992) and in boreal forests

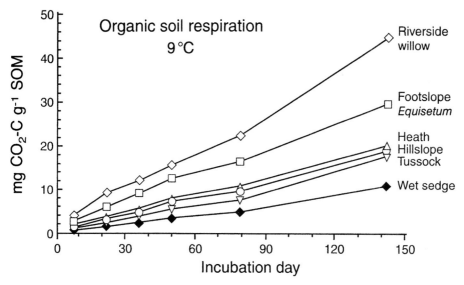

Figure 8.4 *Cumulative respiratory CO_2 losses by six tundra soils incubated aerobically at 9 °C for 150 days (Nadelhoffer et al. 1991). These soils were taken from the sites described in Table 8.2.*

(Pastor & Naiman 1992), this correlation between species composition, litter chemistry, and decomposition is an important factor driving ecosystem response to disturbance by nutrient deposition and by herbivory.

On the other hand, Jonasson (1983) showed in a comparison of three tundras in northern Sweden that rapid decomposition of deciduous leaf litter may be countered by much slower decomposition of the relatively large amounts of woody stem litter produced by deciduous shrubs. The net result was that changes among sites in the overall turnover rate of soil organic matter were better correlated with soil moisture than with any changes in vegetation composition. Jonasson's conclusions were thus consistent with those of Shaver & Chapin (1991) described above for production–biomass relationships and whole-vegetation nutrient use. That is, the distinct differences in ecosystem processes related to leaves (e.g. photosynthesis or leaf decomposition) that are associated with different plant functional types tend to disappear when the whole vegetation (especially including above- and below-ground woody stems) is considered.

Additional work is needed to establish clearly the importance or lack of importance of a feedback relationship between generalized plant functional types and decomposition processes in arctic ecosystems. However, it is clear that *species-level* differences can be important in ecosystem structure and element cycling in northern areas. For example, Johnson and co-workers (Johnson *et al.* 1990; Johnson & Damman 1991) have recently shown that the existence of hummocks and hollows in Swedish bogs is largely explained by differences in decomposability of different *Sphagnum* species growing in the two microsites (Moore 1991). This species-level difference, with the more decomposable *Sphagnum* growing in hollows and the less decomposable species forming the hummocks, is a major factor determining microenvironmental variation and peat accumulation in these bogs.

Responses to experimental manipulation, disturbance and other changes

The importance of plant functional types to overall ecosystem function in the Arctic is probably greatest in ecosystems undergoing rapid change due to disturbance or experimental manipulation. The reason for this is that the rate of change in ecosystem processes such as primary production is strongly dependent on the rate at which plants can change the number, size and metabolism of their resource-acquiring organs and the capacity of the sinks to which resources are allocated. These features are known to vary consistently among the major arctic plant functional types, although there are also important individualistic, species-level differences within functional types (Chapin & Shaver 1985b). Under conditions of rapid change it is often possible to predict which species or functional types should be most responsive, particularly when the disturbance or experimental treatment results in an increase in availability of previously limiting resources. The basis of such predictions is the same understanding of comparative physiology and demography that has been so useful in the past in explaining species distribution patterns and relative fitness in different environments.

The sequence of responses of Alaskan moist tussock tundra to disturbance by tracked vehicles or to fertilizer treatment provides a good example of how plant functional types regulate change in ecosystem function in the Arctic (Chapin & Shaver 1981; Shaver & Chapin 1986; Walker *et al.* 1987). Often, the initial response to either of these events is a dramatic increase in the productivity and abundance of grasses and sedges relative to evergreen or deciduous shrubs (Fig. 8.5). These graminoids respond first because they are initially most able to take advantage of the increased nutrient availability that results from fertilizer addition or from the warmer, often better-drained soils in vehicle tracks. One reason for the greater overall responsiveness of graminoids is their more flexible morphology and vegetative demography. Many arctic graminoids, particularly the grasses, are not constrained by the need to form new buds in order to add new leaves; the size of graminoid leaves can also be increased dramatically, and rapid tillering is possible. A second reason is that graminoids dominate the buried

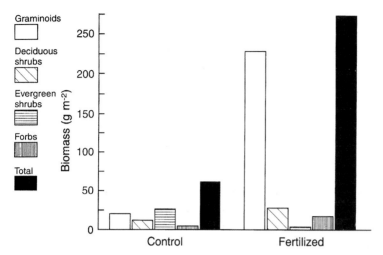

Figure 8.5 *Above-ground biomass in control and fertilized plots in moist tussock tundra at Imnavait Creek, Alaska. Data were collected during the middle of the growing season (late July), following a single fertilizer treatment in June of the previous year. Original data in Shaver & Chapin (1986).*

seed pool of many arctic ecosystems. If living vegetation is partly or largely killed by disturbance, graminoids are among the first plant types to establish on the disturbed site (Chester & Shaver 1982; Gartner *et al.* 1983, 1986). A third reason is that graminoids are less apt to be killed by physical disturbance than shrubby species. Within 5–10 years, however, deciduous shrubs typically become dominant in disturbed or fertilized moist tussock tundra, unless the site becomes less well drained following the disturbance. The reason for this is that deciduous shrubs have the highest potential resource uptake rates and growth rates under conditions of high resource availability (Chapin & Shaver 1985a): it just takes them a year or two longer than the graminoids to reorganize their allocation patterns and to add new sinks for the increased resource supply. The rate at which deciduous species (such as *Betula nana*) can establish dominance also appears to be a function of temperature in some conditions (Chapin *et al.* 1995).

This sequence of changes in productivity and species composition of moist tussock tundra following disturbance or fertilizer addition suggests that the rate of ecosystem-level change in response to environmental change should be predictable, at least in part, from the initial species composition of the vegetation. If the vegetation initially includes species or functional types that are capable of rapid response (grasses in the first year or two, deciduous shrubs in the longer term), then the ecosystem should come into equilibrium with the new conditions more rapidly than if the vegetation is dominated by less responsive species (evergreen shrubs). However, as long as the new level of resource supply in the disturbed or fertilized ecosystem remains constant over time, a more slowly responding evergreen community might eventually reach the same equilibrium productivity and biomass as the communities dominated by other functional types because the equilibrium resource-use efficiencies of tundra ecosystems are so similar (Table 8.3).

Finally, it is important to recognize that when resource availability is increased in moist tussock tundra, species of all functional types respond initially with increased growth (Shaver & Chapin 1980; Chapin & Shaver 1985b). Changes in species and functional type composition of the vegetation are thus determined more by differences in the initial rate and eventual extent of their responses than as a result of some species responding negatively or not at all (Rastetter & Shaver 1994). At the ecosystem level, changes in productivity and plant biomass with increased resources are determined not only by species-level carbon and nutrient uptake kinetics but also by species-level constraints on flexibility in morphology and allocation patterns, and on vegetative and sexual demography. Eventually, the less responsive species are eliminated from the vegetation (or reduced in relative abundance) by competition, not because of any directly detrimental effect of increased resource availability.

Conclusions

Perhaps the best way to summarize the results of this review is with reference to a diagram in which the effects of arctic plant functional types on ecosystem characteristics in relatively stable conditions are distinguished from their effects on ecosystem change. In this diagram (Fig. 8.6) the state of some important ecosystem characteristic (such as net primary production or soil N turnover rate) is shown on the vertical axis, with time on the horizontal axis. Assume that initially this ecosystem state is near equilibrium with the environment and that the environment then undergoes a stepwise change (such as increased nutrient deposition or increased temperature). The ecosystem state changes in response to the change in environment and eventually reaches a new equilibrium. The effects of plant functional type on ecosystem state are indicated in Figure 8.6 as upper and lower bounds on ecosystem state as it changes over time. These bounds are narrow at the beginning and end of the illustration, when the ecosystem is near equilibrium with the environment, and widest during the period of adjustment to environmental change. The bounds are also narrower in the first equilibrium state (before the environment changes) than in the second equilibrium state, reflecting the possibility that effects of plant functional

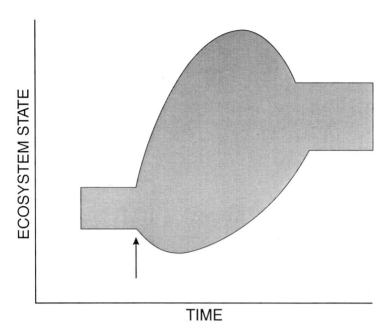

Figure 8.6 *Summary diagram showing the effects of plant functional types on ecosystem state under equilibrium conditions and on changes in ecosystem state over time in response to environmental change. Initially, the ecosystem is assumed to be in equilibrium with the environment, with the upper and lower bounds on ecosystem state being determined by plant functional type composition of the vegetation. A change in the environment is introduced at the time indicated by the arrow, resulting in a change in ecosystem state that is again bounded by effects due to plant functional types. When the ecosystem reaches a new equilibrium with the new environment, the bounds on ecosystem state are wider than initially because of possible effects of plant functional type on cumulative resource accumulation or loss during the period of adjustment to the environmental change.*

type on cumulative resource accumulation or loss during the period of adjustment might influence the final equilibrium state. (As discussed below, it is also possible that some types of environmental change might lead to narrower bounds on ecosystem state following disturbance, rather than the wider bounds shown here.) Not indicated in Figure 8.6 is the possibility that plant functional types might affect the length of time it takes to go from the first equilibrium to the second.

Now we can address the questions raised in the Introduction: first, 'How important is the species composition or dominant plant type of arctic vegetation to the accumulation and turnover of whole-ecosystem carbon and nutrient stocks?' Given our present state of knowledge, the best answer we can give is that *arctic plant species or functional types are usually more important to ecosystem characteristics during periods of change than during periods of relative environmental stability*, as shown in Fig. 8.6. For example, because equilibrium productivity of tundras is strongly nutrient-limited and overall nutrient concentrations of tundra vegetation differ only slightly, when nutrient supply is normalized both production and biomass turnover appear to be relatively independent of the composition of stable vegetation (Fig. 8.2, Table 8.3). This is not true in rapidly changing vegetation, when different functional types change in abundance at very different rates (Fig. 8.5) and probably have very different nutrient-use efficiencies.

Additional research is needed, however, to quantify more narrowly the controls over the vague bounds on ecosystem state shown in Fig. 8.6. In stable ecosystems, there are still important uncertainties about the effects of plant functional type on accumulation and turnover of carbon and nutrients in soils (e.g. Fig. 8.4 vs. Jonasson 1983). In changing ecosystems, more work is needed in particular to quantify variation in the morphological, demographic, and allocation-related constraints on the ability of different functional types to change their productivity, biomass, and resource-use efficiencies (Callaghan 1988; Heal *et al.* 1989; McGraw & Fetcher 1992). More work is also needed on the relationship between the trajectory of ecosystem response to environmental change and the eventual, equilibrium ecosystem state. The plant functional types that are present and dominant during the period of response might significantly affect the final equilibrium state, as shown in Fig. 8.6, but this relationship has not been studied in previous work in the Arctic.

The answer to the second question, 'How useful to predictions of ecosystem response to disturbance are classifications of plant functional types based on small differences in relative fitness among species?', is some-what more clear-cut. The widely used categories of deciduous vs. evergreen shrubs, graminoids, forbs, mosses and lichens (Chapin & Shaver 1985a) are in fact quite useful in predicting relative responsiveness to disturbance (e.g. Fig. 8.5). The known differences in plant functions (such as carbon and nutrient uptake kinetics) among these plant types are probably also impor-tant in determining their responsiveness to change. However, this logic is also tautological, because in part the classification itself is derived from observations of plant distribution and abundance patterns in the field. More importantly, it is not clear that existing knowledge of variation in processes such as photosynthesis and nutrient uptake is sufficient to explain the observed differences in response to disturbance of deciduous, evergreen and graminoid plants. Additional work is needed to evaluate the contributions of other traits that may be strongly correlated with differences in photosyn-thesis and nutrient uptake, including morphology, allocation and demography.

Finally, it is appropriate to ask, 'How much of what is known about the role of plant functional types in the Arctic can be used in explaining or predicting the responses of other ecosystems?' Here the conceptual model outlined in Fig. 8.6 may prove useful as a heuristic device for defining clearer comparisons among major global ecosystem types. Based on the evidence presented above, several of the characteristics related to productivity of undisturbed arctic ecosystems appear to be relatively unaffected by the presence or absence of a particular, dominant plant functional type. However, productivity of virtually all undisturbed arctic ecosystems is strongly nutrient-limited, with nutrient supply to the plants coming mainly from internal recycling through decomposition rather than external inputs such as N fixation or atmospheric deposition (Shaver *et al.* 1992). Under strong nutrient limitation it simply may not be possible to achieve much variation in overall nutrient-use efficiency and carbon and nutrient turnover among plant functional types. In other ecosystems (outside the Arctic) where nutrients are not as limiting, the effects of plant type on equilibrium nutrient-use efficiency and element turnover may be more important; this would be reflected in wider bounds on equilibrium state in Fig. 8.6 for less nutrient-limited ecosystems and narrower bounds on equilibrium state for more strongly nutrient-limited ecosystems such as arctic tundras. If the disturbance indicated by the arrow in Figure 8.6 leads to a reduction, rather than an increase, in nutrient availability to the vegetation, the bounds on the final equilibrium state might be narrower than in the initial equilibrium state. In disturbed or rapidly changing ecosystems, where nutrient availability is often high, plant functional types may have a large effect on ecosystem state in both arctic and non-arctic ecosystems (Fig. 8.6).

A second way that the role of plant functional types in arctic ecosystems may differ from their role in less extreme ecosystems is in the relative importance of the partitioning of resource uptake among different plant types. Although there is some evidence that arctic plant types do obtain their nutrients from different locations in the soil (Shaver & Billings 1975; Shaver & Cutler 1979) and perhaps at different times of year (Shaver & Kummerow 1992) or in different chemical forms (G. R. Shaver *et al.*, unpublished data), the general expectation is that partitioning of resource uptake among species within a vegetation should be greater in more temperate environments than in extreme, resource-poor environments like the Arctic. If different plant types obtain an essential resource (such as nitrogen) from different locations in the soil, at different times of the year, or in different chemical forms, then the presence or absence of a given plant type may have important effects on total resource uptake by the vegetation (e.g. McKane *et al.* 1990). The degree of overlap in the partitioning of resource uptake among plant types should thus affect the bounds on equilibrium state in Fig. 8.6, such that where there is a great deal of overlap the bounds should be narrower than where each plant type obtains an essential resource in different places, times or chemical forms.

Clearly, there are many gaps in our present understanding of the importance of plant functional types to terrestrial biogeochemistry and its changes in response to climate change and human-caused disturbances. Simple conceptual models like Figure 8.6 are only the first steps in developing such understanding; different models might be more appropriate to understanding and predicting different ecosystem characteristics. The main purpose of such models is to clarify comparisons and focus interpretations of existing knowledge, and to develop hypotheses for future research. The Arctic, because it includes a diverse array of ecosystem types dominated by clearly different plant functional types, may make an excellent model region for the initial development and testing of such hypotheses.

References

Aber, J. D. & Melillo, J. M. (1982) Nitrogen immobilization in decaying hardwood leaf litter as a function of initial nitrogen and lignin content. *Canadian Journal of Botany*, **60**, 2263–9.

Berendse, F. & Elberse, W. T. (1990) Competition and nutrient availability in heathland and grassland ecosystems. In Grace, J. B. & Tilman, D. (eds), *Perspectives on Plant Competition*, pp. 93–116. San Diego: Academic Press.

Berendse, F. & Jonasson, S. (1992) Nutrient use and nutrient cycling in northern ecosystems. In Chapin, F. S. III *et al.* (eds), *Arctic Ecosystems in a Changing Climate: An Ecophysiological Perspective*, pp. 337–56. New York: Academic Press.

Billings, W. D. & Mooney, H. A. (1968) The ecology of arctic and alpine plants. *Biological Reviews*, **43**, 481–530.

Bliss, L. C. (1962) Adaptations of arctic and alpine plants to environmental conditions. *Arctic*, **15**, 117–44.

Bliss, L. C. (ed.) (1977) *Truelove Lowland, Devon Island, Canada: A High Arctic Ecosystem*. Edmonton: University of Alberta Press.

Bliss, L. C. (1981) North American and Scandinavian tundras and polar deserts. In Bliss, L. C., Heal, O. W. & Moore, J. J. (eds), *Tundra Ecosystems: A Comparative Analysis*, pp. 8–24. Cambridge University Press.

Bliss, L. C., Heal, O. W. & Moore, J. J. (eds) (1981) *Tundra Ecosystems: A Comparative Analysis*. Cambridge University Press.

Bliss, L. C. & Matveyeva, N. V. (1992) Circumpolar arctic vegetation. In Chapin, F. S., III *et al.* (eds), *Arctic Ecosystems in a Changing Climate: An Ecophysiological Perspective*, pp. 59–89. New York: Academic Press.

Callaghan, T. V. (1988) Physiological and demographic implications of modular construction in cold environments. In Davy, A. J., Hutchings, M. J. & Watkinson, A. R. (eds), *Plant Population Ecology: The 28th Symposium of the British Ecological Society*, pp. 111–35. Oxford: Blackwell Scientific Publications.

Chapin, F. S. III, Jefferies, R., Reynolds, J., Shaver, G. & Svoboda, J. (eds) (1992a) *Arctic Ecosystems in a Changing Climate: An Ecophysiological Perspective*. New York: Academic Press.

Chapin, F. S. III, Jefferies, R., Reynolds, J., Shaver, G. & Svoboda, J. (1992b) Arctic plant physiological ecology: A challenge for the future. In Chapin, F. S. III *et al.* (eds), *Arctic Ecosystems in a Changing Climate: An Ecophysiological Perspective*, pp. 441–52. New York: Academic Press.

Chapin, F. S. III & Oechel, W. C. (1983) Photosynthesis, respiration, and phosphate absorption by *Carex aquatilis* ecotypes along latitudinal and local environmental gradients. *Ecology*, 64, 743–51.

Chapin, F. S. III & Shaver, G. R. (1981) Changes in soil properties and vegetation following disturbances of Alaskan arctic tundra. *Journal of Applied Ecology*, 18, 605–17.

Chapin, F. S. III & Shaver, G. R. (1985a) Arctic. In Chabot, B. & Mooney, H. A. (eds), *Physiological Ecology of North American Plant Communities*, pp. 16–40. London: Chapman and Hall.

Chapin, F. S. III & Shaver, G. R. (1985b) Individualistic growth response of tundra plant species to manipulation of light, temperature, and nutrients in a field experiment. *Ecology*, 66, 564–76.

Chapin, F. S. III, Shaver, G. R., Giblin, A. E., Nadelhoffer, K. J. & Laundre, J. A. (1995) Responses of arctic tundra to experimental and observed changes in climate. *Ecology*, 76, 694–711.

Chester, A. L. & Shaver, G. R. (1982) Seedling dynamics of some cottongrass tussock tundra species during the natural revegetation of small disturbed areas. *Holarctic Ecology*, 5, 207–11.

Fetcher, N. & Shaver, G. R. (1990) Environmental sensitivity of ecotypes as a potential influence on primary productivity. *American Naturalist*, 136, 126–31.

Gartner, B. L., Chapin, F. S. III & Shaver, G. R. (1983) Demographic patterns of seedling establishment and growth of native graminoids in an Alaskan tundra disturbance. *Journal of Applied Ecology*, 20, 965–80.

Gartner, B. L., Chapin, F. S. III & Shaver, G. R. (1986) Reproduction by seed in

Eriophorum vaginatum in Alaskan tussock tundra. *Journal of Ecology*, 74, 1–19.

Giblin, A. E., Nadelhoffer, K. J., Shaver, G. R., Laundre, J. A. & McKerrow, A. J. (1991) Biogeochemical diversity along a riverside toposequence in arctic Alaska. *Ecological Monographs*, 61, 415–36.

Heal, O. W., Callaghan, T. V. & Chapman, K. (1989) Can population and process ecology be combined to understand nutrient cycling? In Clarholm, M. & Bergstrom, L. (eds), *Ecology of Arable Land*, pp. 205–16. Dordrecht: Kluwer Academic Publishers.

Johnson, D. A. & Tieszen, L. L. (1976) Aboveground biomass, leaf growth, and photosynthesis patterns in tundra plant forms in arctic Alaska. *Oecologia*, 24, 159–73.

Johnson, L. C. & Damman, A. W. H. (1991) Species-controlled *Sphagnum* decay on a south Swedish raised bog. *Oikos*, 61, 234–42.

Johnson, L. C., Damman, A. W. H. & Malmer, N. (1990) *Sphagnum* microstructures as an indicator of decay and compaction in peat cores from an ombrotrophic south Swedish peat bog. *Journal of Ecology*, 78, 633–47.

Jonasson, S. (1983) Nutrient content and dynamics in north Swedish shrub tundra. *Holarctic Ecology*, 6, 295–304.

Kielland, K. & Chapin, F. S. III (1992) Nutrient absorption and accumulation in arctic plants. In Chapin, F. S. III *et al.* (eds), *Arctic Ecosystems in a Changing Climate: An Ecophysiological Perspective*, pp. 321–36. New York: Academic Press.

Lewis, M. C. & Callaghan, T. V. (1976) Tundra. In Monteith, J. L. (ed.), *Vegetation and the Atmosphere (Vol. 2)*, pp. 399–433. New York: Academic Press.

McClaugherty, C. A., Pastor, J., Aber, J. D. & Melillo, J. M. (1985) Forest litter decomposition in relation to soil nitrogen dynamics and litter quality. *Ecology*, 66, 266–75.

McGraw, J. B. & Chapin, F. S. III (1989) Competitive ability and adaptation to fertile and infertile soils in two *Eriophorum* species. *Ecology*, 70, 736–49.

McGraw, J. B. & Fetcher, N. (1992) Response of tundra plant populations to climatic change. In Chapin, F. S. III *et al.* (eds), *Arctic Ecosystems in a Changing Climate: An Ecophysiological Perspective*, pp. 359–76. New York: Academic Press.

McKane, R. B., Grigal, D. F. & Russelle, M. P. (1990) Spatiotemporal differences in ^{15}N uptake and the organization of an old-field plant community. *Ecology*, **71**, 1126–32.

Miller, P. C. (1982) Environmental and vegetational variation across a snow accumulation area in montane tundra in central Alaska. *Holarctic Ecology*, **5**, 85–98.

Mooney, H. A. & Billings, W. D. (1961) Comparative physiological ecology of arctic and alpine populations of *Oxyria digyna*. *Ecological Monographs*, **31**, 1–29.

Moore, P. (1991) Ups and downs in peatland. *Nature*, **353**, 299–300.

Nadelhoffer, K. J., Giblin, A. E., Shaver, G. R. & Laundre, J. A. (1991) Effects of temperature and organic matter quality on C, N, and P mineralization in soils from six arctic ecosystems. *Ecology*, **72**, 242–53.

Pastor, J. & Naiman, R. J. (1992) Selective foraging and ecosystem processes in boreal forests. *American Naturalist*, **139**, 690–705.

Rastetter, E. B. & Shaver, G. R. (1994) Functional redundancy and process aggregation: Linking ecosystems to species. In Jones, C. & Lawton, J. H. (eds), *Proceedings of the Fifth Cary Conference, 'Linking Species to Ecosystems'*, pp. 215–23. New York: Chapman & Hall.

Reich, P. B., Walters, M. B. & Ellsworth, D. S. (1992) Leaf life-span in relation to leaf, plant, and stand characteristics among diverse ecosystems. *Ecological Monographs*, **62**, 365–92.

Reynolds, J. F. & Leadley, P. W. (1992) Modeling the response of arctic plants to changing climate. In Chapin, F. S. III *et al.* (eds), *Arctic Ecosystems in a Changing Climate: An Ecophysiological Perspective*, pp. 413–40. New York: Academic Press.

Reynolds, J. F. & Tenhunen, J. (1995) *Landscape Function and Disturbance in Arctic Tundra*. Ecological Studies Series no. 120. Berlin: Springer–Verlag.

Savile, D. B. O. (1972) *Arctic adaptations in plants*. Canada Department of Agriculture Research Branch Monograph No. 6.

Shaver, G. R. & Billings, W. D. (1975) Root production and root turnover in a wet tundra ecosystem, Barrow, Alaska. *Ecology*, **56**, 401–10.

Shaver, G. R., Billings, W. D., Chapin, F. S. III, Giblin, A. E., Nadelhoffer, K. J., Oechel, W. C. & Rastetter, E. B. (1992) Global change and the carbon balance of arctic ecosystems. *BioScience*, **42**, 433–41.

Shaver, G. R. & Chapin, F. S. III (1980) Response to fertilization by various plant growth forms in an Alaskan tundra: nutrient accumulation and growth. *Ecology*, **61**, 662–75.

Shaver, G. R. & Chapin, F. S. III (1986) Effect of fertilizer on production and biomass of tussock tundra, Alaska, U.S.A. *Arctic and Alpine Research*, **18**, 261–8.

Shaver, G. R. & Chapin, F. S. III (1991) Production/biomass relationships and element cycling in contrasting arctic vegetation types. *Ecological Monographs*, **61**, 1–31.

Shaver, G. R. & Cutler, J. C. (1979) The vertical distribution of phytomass in cottongrass tussock tundra. *Arctic and Alpine Research*, **11**, 335–42.

Shaver, G. R. & Kummerow, J. (1992) Phenology, resource allocation, and growth of arctic vascular plants. In Chapin, F. S. III *et al.* (eds), *Arctic Ecosystems in a Changing Climate: An Ecophysiological Perspective*, pp. 193–212. New York: Academic Press.

Shaver, G. R., Nadelhoffer, K. J. & Giblin, A. E. (1991) Biogeochemical diversity and element transport in a heterogeneous landscape, the North Slope of Alaska. In Turner, M. G. & Gardner, R. H. (eds), *Quantitative Methods in Landscape Ecology*, pp. 105–26. New York: Springer-Verlag.

Swift, M. J., Heal, O. W. & Anderson, J. M. (1979) *Decomposition in Terrestrial Ecosystems*. (Studies in Ecology, Vol. 5.) Berkeley: University of California Press.

Walker, D. A., Cate, D., Brown, J. & Racine, C. (1987) *Disturbance and Recovery of Arctic*

Alaskan Tundra Terrain. US Army CRREL, Report 87–11.

Wielgolaski, F. E. (ed.) (1975) *Fennoscandian Tundra Ecosystems. Part 1: Plants and Microorganisms.* New York: Springer-Verlag.

Wielgolaski, F. E., Bliss, L. C., Svoboda, J. & Doyle, G. (1981) Primary production of tundra. In Bliss, L. C., Heal, O. W. & Moore, J. J. (eds), *Tundra Ecosystems: A Comparative Analysis*, pp. 187–226. Cambridge University Press.

9 Functional types for predicting changes in biodiversity: a case study in Cape fynbos

W. J. Bond

Introduction

Centres of biodiversity pose acute problems for extrapolating and predicting responses to environmental changes. The scale of Global Change and Terrestrial Ecosystems predictions requires the collapse of large numbers of species into a handful of manageable functional types. However, the fascination of species-rich systems for many biologists is their complexity, the rich potential for interactions, which makes such systems resistant to simplification (e.g. Janzen 1988). We still have very little understanding of the ecological significance of diversity. If systems are highly integrated with intricate webs of species interactions, diversity may feed on diversity with so many cross-linkages that prediction of global perturbations will be virtually impossible. However, if systems are loose assemblages of functionally equivalent species (showing high redundancy), prediction may be more feasible (Walker 1991; Lawton & Brown 1993). Most systems probably lie somewhere in between because closely related species, by definition, must differ along *some* axis of variation and are therefore not complete functional equivalents.

Biodiversity is still recorded almost exclusively in taxonomic units. The central goal of modern systematics is to show genetic relatedness through depicting phylogenetic history. This may seem far removed from ecologically relevant criteria of functional classifications, but taxonomic units, especially at the genus or family level, can be powerful predictors of functional types (FTs). One need only think of grasses or conifers as growth-form FTs, legumes as nutritional FTs, and the phylogenetic groupings of C3 and C4 photosynthetic pathways.

Taxonomic classifications can be very misleading, however, when unrelated organisms share similar structures through convergent evolution rather than common ancestry. Strong character convergence is noise to a systematist but signal to an ecologist seeking functional interpretations of diversity. This is perhaps the best justification for seeking functional classifications. The problem, for a systematist, is the lack of any logical basis for a 'best' functional classification. A plethora of ecologically based functional taxonomies have been devised using a host of diverse criteria (see Smith *et al.* 1993; Gitay & Noble, Chapter 1 of this volume, for reviews). The purpose of these classifications is seldom well defined but two broad groupings seem apparent. One seeks a 'natural' classification of types based on suites of correlated traits. Examples include Grime's C–S–R strategy system (e.g. Grime 1979, 1988) and the functional types emerging from multivariate analyses of numerous traits (e.g. Leishman & Westoby 1992). These hold the promise of universality and generality for identifying patterns and perhaps predicting responses. The alternatives are special purpose classifications devised for specific objectives. No claims are made for generality or utility of the classification beyond the problem for which it was designed. Noble & Slatyer's (1980) vital attributes system, which predicts the effects of disturbance, is a good example. Chapters 6 and 7 of this volume discuss the utility of the former approach in the context of climate change. Here I explore the use of the latter approach in the diverse Cape flora of the southern tip of Africa.

Functional classifications for climate change

Simply put, the problems that a functional classification needs to address are (1) how will the distribution and diversity of vegetation change in response to climate change; and (2) at what rate? The first question has received much more attention than the second. There are obvious associations between vegetation and climate which suggest climate has a direct effect on plant distribution. The simplest approach to predicting the effects of climate change on plant distribution is to seek contemporary correlations between climate variables and vegetation and extrapolate them into the future. This assumes that vegetation is more or less in equilibrium with climate. The assumption is often wrong, especially since human actions overwhelmingly shape vegetation in large parts of the globe (see Hobbs, Chapter 4 of this volume). More subtly, climate may determine disturbance regimes, which in turn partly determine 'equilibrial' states of the vegetation. Large areas of the earth with seasonally dry climates experience fires. Fires favour grasslands in areas with a climate suitable for trees and can have major effects on the distribution of vegetation in the landscape (Heinselman 1973; Wright & Bailey 1982). Will disturbance regime and vegetation type move together as climate changes? Probably not, since the propagation of fires depends on other factors such as ignition sources, fuel types and loads and the extent of different fuels in the landscape.

More interesting, and more challenging, approaches than the correlative method are now being attempted. There are the beginnings of mechanistic models which predict climatic distribution limits for plant functional types from first principles (Woodward 1987, 1993; Friend 1993). The degree to which such predicted limits coincide with what is observed in nature would give us a good indication of the degree to which climate, operating directly, influences vegetation distribution. There are also the beginnings of mechanistic models for predicting vegetation responses to changing fire regimes (Noble & Slatyer 1980, and below). However, there have been few attempts to model the effects of changing climate on disturbance regime and that in turn on vegetation (but see Clark 1988; Fosberg *et al.* 1992). Such models may prove essential for depicting climate–vegetation interactions in the large parts of the globe where fires regularly burn the vegetation.

Rates of change

Disturbance regime may also be central to predicting rates of vegetation change. Plant lifespans vary from a few weeks to several thousand years (Harper & White 1974; Loehle 1988). Clearly there could be tremendous time-lags in vegetation change if plants persist in the same site for two or three thousand years. What controls rates of change and where can long time-lags be expected in climate-driven vegetation change? Important functional attributes are those relating to *persistence* at a site and *mobility* to colonize new sites. In large parts of the globe, both site persistence and potential mobility are closely related to response to disturbance.

Functional classifications and climate change in a species-rich system: a case study

Here I discuss the way in which climate change may impact biodiversity in the species-rich fynbos biome. The fynbos biome is dominated by evergreen shrublands of the Cape Floristic Region. The region has been distinguished as one of the world's six floristic kingdoms because of its distinct flora (Good 1974) and has the highest species density in the world for equivalent-sized areas (Table 9.1; Bond & Goldblatt 1984; Cowling & Holmes 1992). The only temperate floras that approach the richness of the Cape flora are the ecologically similar heathlands of Australia, especially south-western Australia, although other mediterranean climate regions also have rich floras. The Cape Floral Kingdom is tiny (*ca.* 90 000 km^2, or less than 4% of the area of southern Africa) so the central interest is prediction of the effect of climate change on biodiversity and not the feedback of diversity on climate change.

Table 9.1 *Species richness of the Cape flora compared with other regions*

Region	Area (10^3 km^2)	No. of species	Species density (10^3 km^{-2})
Mediterranean climate regions			
Cape Floristic Region	90	8578	95.3
Cape Peninsula	0.47	2256	13.7
California Floristic Province	324	4452	13.7
SW Australia	320	3600	11.25
Greece	130	6000	30.8
Temperate Regions			
British Isles	308	1443	4.7
Eastern North America	3238	4425	13.7
Europe (Flora Europea)	10 000	10 500	1.05
Tropical Rainforests			
Panama	80	c. 8300	103.75
Malaysian Peninsula	130	c. 8000	61.5
Ivory Coast	320	c. 4700	14.7

Source: Bond & Goldblatt (1984); Cowling & Holmes (1992).

Most of the richness in the Cape flora is at the species level rather than at generic or family level (Linder *et al.* 1992). In other words, there are a number of genera with very large numbers of species but familial diversity is not unusually high. This pattern of diversity is repeated in insular floras such as that of Hawaii. The diversity of growth-form FTs is very low. There are many hundreds of shrub species, often from unrelated families, all from 0.5 to 1.5 m tall, with small, evergreen sclerophyll leaves. Similar convergence occurs in monocotyledonous groups, where grasses and sedges closely resemble leafless Restionaceae (Cowling & Holmes 1992; Linder *et al.* 1992). This growth-form mix of shrubs and grass-like plants (Restionaceae, Cyperaceae, Poaceae) remains essentially unchanged over rainfall gradients of <300 mm to >3000 mm p.a. (Cowling & Holmes 1992; Linder *et al.* 1992). There are almost no trees, in contrast to Australia's eucalypts and the conifers of Californian chaparral (Moll *et al.* 1980). Partly as a consequence, productivity changes little over large precipitation gradients (Fig. 9.1). This comparative monotony of growth form suggests high levels of 'redundancy' and a simple task for GCTE modellers in this system. The majority of species would probably be placed in the stress corner of Grime's C–S–R triangle (Grime 1988). However, fynbos plants sharing similar growth forms are highly diverse along other functional axes,

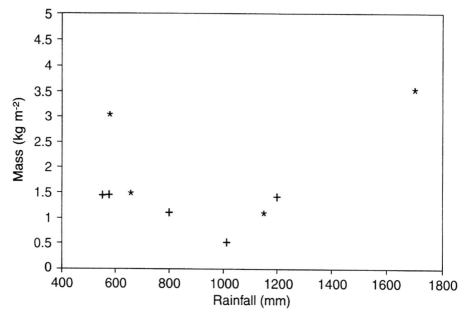

Figure 9.1 *Above-ground biomass vs. rainfall in a number of fynbos stands (from Stock & Allsopp 1992). Productivity is only weakly coupled to climate in this nutrient-poor system. Two post-burn ages are shown: 10–12 years (+) and 18–22 years (*).*

especially pollination and dispersal (Johnson 1992; Le Maitre & Midgley 1992) and nutritional mode (Lamont 1982). The floral diversity of fynbos is the focus of public interest in the biome and contributes to its horticultural fame. Clearly other functional attributes are needed to predict the fate of this diversity in the face of global change.

How will the distribution of fynbos change in response to climate change?

Fynbos is one of seven southern African biomes. Predicting the future distribution of these biomes seems far simpler than predicting the fate of the more than 22 000 vascular plant species that make up the South African flora. The most serious threat to the diversity of the Cape flora is its complete replacement by adjoining biomes. Despite its continental location, fynbos has a somewhat island-like position, occurring on nutrient-poor soils in climates with half or more of their rainfall falling in winter. Fynbos is bordered by arid Karoo shrublands, grasslands, spiny and succulent thickets, and forests. How will biome boundaries change with climate change?

Correlative approaches

Biome distributions correlate with climate and these correlations are usually the starting point for predictions of global change. The Holdridge Classification system is a widely used example. Southern African biomes, including fynbos, also correlate with climate (Campbell 1985; Rutherford & Westfall 1986; Cowling & Holmes 1992). These correlations have been used to predict future biome distribution. The most elegant attempt to date (Ellery *et al.* 1991) plotted biome distributions against three climatic indices:

1. the number of growth days calculated as the number of days where rainfall exceeds evaporative demand;
2. 'growth temperature', defined as the mean temperature on days when water was available for growth;
3. 'no-growth temperature', the corollary of the above for the dry season.

Using these three climate indices, Ellery *et al.* were able to correctly classify 21 out of 24 fynbos sites for which they had sufficient climatic information, and a very high proportion of sites for South African biomes in general. Fynbos climates were distinguished from grassland and non-succulent arid shrubland climates by growth temperatures lower than 'no-growth' temperatures (i.e. winter rainfall). Within winter rainfall areas, they were separated from succulent shrublands by longer growth seasons (higher rainfall), especially where summers were hot. If these correlations are causal, temperature increases and precipitation decreases following global climate change should lead to the spread of succulent shrublands into arid fynbos areas (Ellery *et al.* 1991). If rainfall shifted to summer, fynbos should be displaced by grassland, especially in the east. Precise prediction is complicated by the complex mountain topography and the poor network of climate stations.

However, the distribution of fynbos also coincides with nutrient-poor soils. For many years the correlation between fynbos and low-nutrient soils has been assumed to be causal (Specht & Moll 1983). If this is the case, climate change would have little impact on distribution of the biome.

Mechanistic approaches

It would be attractive to resolve the uncertainty about correlative vs. causal vegetation–climate relationships by developing models of climate–growth form interaction from first principles. Predictions of such mechanistic models could be compared with observed plant distribution to help distinguish between direct and indirect effects of climate. There are some promising beginnings in this area (Woodward 1987, 1993; Friend 1993). However, we are still a long way from predicting, say, the climate limits of grasslands vs. woodlands or deciduous vs. evergreen leaves with any confidence.

Experiments, whether natural or designed, are a powerful alternative for exploring causality. In fynbos, they cast doubt on both soils and climate as direct determinants of biome distribution. The low height of the vegetation is not a direct consequence of the nutrient-poor soils, nor of aridity (cf. Friend 1993). Fynbos is readily invaded by alien pine trees, *Acacia* species (all Australian) and *Hakea*. It is also invaded by broad-leaved native trees and shrubs with tropical affinities and, supposedly, higher nutrient requirements (Masson & Moll 1987; Manders *et al.* 1992). Tall closed forest occurs on soils physically identical to those under adjacent fynbos stands in some areas (van Daalen 1981). Successful experimental transplants of species from adjoining biomes, including forest, grassland and succulent and spinescent thickets, have shown that fynbos soils and climates are 'permeable' barriers (Manders *et al.* 1992; D. Euston-Brown & W. J. Bond, unpublished).

If neither soils nor climate directly determine fynbos boundaries with tree-dominated vegetation, what does? The answer seems to be fire. Fynbos is invaded by slow-growing forest trees in mesic areas, or by succulents and spiny shrubs in drier regions if fire is excluded for long periods (Moll *et al.* 1980; Manders *et al.* 1992; D. Euston-Brown & W. J. Bond, unpublished). The boundary with grassland is more uncertain. Grassiness increases with increasing summer rainfall, aspect and soil type to the east apparently independently of fire regime (Cowling 1984; Campbell 1985). However, experimental burning, in seasons unfavourable to shrubs, has been used to convert fynbos to grassland where these biomes abut (Trollope 1973). Ellery *et al.* (1991) acknowledge that their correlation between climate and biomes is indirect and must depend on the interrelationship between climate, plant growth and fire. Fire is a key process in maintaining fynbos boundaries, and many others in the world. In order to predict the effects of climate change on biome distribution, fire has to be taken into account (Fosberg *et al.* 1992).

Global change and fire regimes

Fire regimes are strongly influenced by fuels and weather. Fuel accumulation depends on system productivity and decomposition rates, both of which are strongly influenced by climate. Increasing carbon dioxide concentrations may also increase rates of fuel accumulation in fynbos and other mediterranean shrublands (Stock & Midgley 1995). Weather cycles (seasonal or multiannual) determine the moisture content of fuels and therefore fire frequency and intensity (Clark 1988; Swetnam & Bettancourt 1990). Changes in weather have been shown to cause changes in fire regimes, which in turn have caused changes in vegetation (Clark 1988, 1989; Swetnam & Bettancourt 1990).

   Most climate-change scenarios for the Cape region imply higher temperatures and possibly higher fuel accumulation rates leading to more frequent and more intense fires (van Wilgen *et al.* 1992). The new fire regimes would tend to favour fynbos over forest or more xeric thicket. Fynbos boundaries with these biomes should therefore remain unchanged or advance slightly. Biome boundaries with succulent arid shrublands at the xeric limits, or grasslands towards the summer rainfall margins in the east, appear to be more closely determined by climate (Ellery *et al.* 1991; Cowling & Holmes 1992). Changes in rainfall seasonality (to a summer rainfall regime) would probably have the most severe effect on the size of the biome, causing it to shrink to the west with grassland invading from the east (van Wilgen *et al.* 1992). GCTE scenarios for the region have estimated rainfall amount but not seasonality. In mediterranean climate regions the seasonality of rainfall is a key predictor of future biome change which deserves closer attention from climate modellers.

How will the diversity of fynbos change in response to climate change?

Unless rainfall seasonality changes, fynbos biome boundaries seem unlikely to be greatly affected by global climate change. What of loss of diversity within the biome? Again correlative approaches, comparable to Holdridge-type extrapolations for biome distributions, are possible. Linder (1991) correlated species richness of a representative sample of the Cape flora, measured at a landscape scale, with a variety of environmental variables. The best correlation was with total precipitation. If this correlation were causal (Linder is careful to argue that it may not be) then the predicted decline of total precipitation should cause a loss of species from high rainfall areas. Similar correlations between climatic variables and species richness have been reported at continental scales (Currie & Paquin 1987; Currie 1991) and also imply changes in biodiversity as a function of global climate change. One can also use a mechanistic approach, closer to the mechanistic gap-phase models used to predict succession in forest vegetation (Botkin *et al.* 1972; Shugart 1984), to assess sensitivity of biodiversity to global change. Grime (1988) and others have noted the poor relationship between the functional types of established plants and their regeneration phase. While we know relatively little about the physiology of growth in fynbos plants in relation to climate change, much more is known about the biology of regeneration in relation to changing fire regimes. If climate change causes changes in fire regime, then fire-response FTs can be used to predict effects on diversity.

Functional classifications of fire response

At least three aspects of fire are important for predicting vegetation response: frequency, intensity and fire season (Gill 1981). All three could be altered by global change because of changing fuel accumulation rates and altered weather patterns. To predict the impact of changing fire regimes on biodiversity, functional classifications that predict species response to all three components are needed.

Functional classifications for predicting fire frequency effects have existed for over a decade (Noble & Slatyer 1980). No similar system exists for predicting plant response to fire season and intensity. Following Shugart (1984), plants in fire-prone vegetation can be conveniently divided into four basic life histories based on whether they are killed by fire and whether they require fire for regeneration (Table 9.2). Typical population trajectories are shown in Figure 9.2.

Type I species

These species survive fire vegetatively by sprouting from protected buds but regeneration is stimulated by fire. They are good persisters and are not easily displaced. They are most vulnerable in their first fire cycle when seedlings have to grow to a sufficient size to survive the first fire. There is a trade-off between resource allocation to survival (thick bark, root reserves, etc.) and allocation to growth and reproduction so that Type I species will generally be much slower-growing and slower-maturing than Type II species (e.g. Hansen *et al.* 1991). Consequently they will be less mobile than Type II species in changing conditions.

Type II species

These species are either killed by fire or occur only as a dormant seed bank when burnt. Like Type I species, regeneration is stimulated by fire. Type II species are poor persisters and will be eliminated most quickly in changing environments. However, they can establish more rapidly than Type I species and may therefore prove more mobile. Both Type II and Type I species tend to have poor dispersal powers because dispersal is temporal (in seed banks) rather than spatial. A notable exception is species with canopy-stored seed-banks, such as members of the Proteaceae, whose seeds are released after fire and blown over the barren landscapes cleared by fire (Bond 1988). Because Type I and II species regenerate in the openings

Table 9.2 *Life history variation in fire-prone environments based on vegetative and reproductive responses to burning*

Reproductive response	Vegetative response	
	Survive fires	Killed by fires

Stimulated by fire	Type I
	Episodic recruitment
	Overlapping
	  generations
		Type II
		Episodic recruitment
		Non-overlapping
		  generations

Suppressed (unaffected)
  by fire
	Type III
	Continuous recruitment
	Overlapping
	  generations
		Type IV
		Continuous recruitment
		Variable

created by fire, seedling survival may be extremely sensitive to changing temperature or rainfall (e.g. Lamont *et al.* 1991).

Type III species

These species survive fire by sprouting from buds and do not require fire to regenerate. They are represented by broad-leaved trees and shrubs with floristic affinities to forest or subtropical thicket (e.g. *Heeria, Rhus, Cassine, Diospyros*). They are typically well dispersed by birds and should be highly mobile. They are excellent persisters with very high fire survival as sprouters. Their weak point is establishment. They are even more vulnerable than Type I species because seedlings arrive at a site later in the fire cycle but still need to accumulate sufficient resources to survive the next fire (see for example, Manders *et al.*, 1992). Thus Type III species require long fire frequencies to increase. Type III species can persist without fire and are the first phase of forest or thicket invasion. They risk being shaded out by Type IV species over long fire-free periods.

Type IV species

These species are killed by fire and regenerate without fire stimulus. This category is a very small and somewhat disparate group including annuals, short-lived perennials (e.g. *Anthospermum aethiopicum*) and long-lived trees (e.g. *Podocarpus latifolius*). The trees require long fire frequencies to establish and can only avoid death in the next fire by forming forest patches that do not burn. Type IV species do not form a significant component of fynbos diversity.

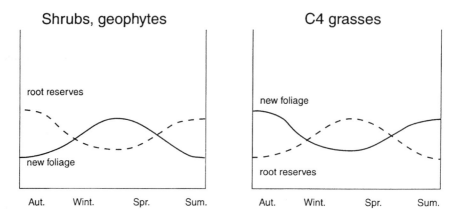

After fire in each season

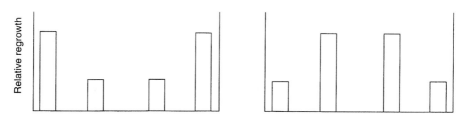

Figure 9.2   *A possible mechanism underlying observed response of sprouters (Types I and III) to fire season. Regrowth is poor after burning when reserves are depleted from new shoot extension. Fires at different phenological stages (top diagram) produce the response indicated in the lower diagram. Because of different phenologies, shrubs and geophytes differ from C4 grasses in their fire season response.*

Fire frequency effects

Short fire intervals can prevent regeneration of slow-growing and slow-maturing species (e.g. Noble & Slatyer 1980; Moore & Noble 1990). Because of allocation costs to survival, seedlings of Type I and Type III species are slow-growing and easily killed by short fire intervals, but both types are very resilient in the established phase. They will show high persistence but low mobility under changing climates.

Type II species (fire-stimulated non-sprouters) grow and mature rapidly from seeds but can be eliminated if fires occur before they have matured. Allocation trade-offs suggest that tall species will mature more slowly and will therefore be more vulnerable to short intervals between fires (cf. Loehle 1988). Local extinction of tall proteoid shrubs is not uncommon after short fire frequencies. Thus tall, Type II overstorey species may be eliminated by short fire frequencies as a result of climate change.

Fire season effects

Seasonal effects of fire on populations and communities can be substantial, especially for Type II (fire-stimulated non-sprouter) species, but their mechanistic base is still poorly understood. Fire season (and fire intensity) has been greatly changed in recent times by human intervention. Shifts in fire season may eliminate some groups of species but promote others. The best understood functional type includes tall shrubs with seed banks stored in the canopy, rather than in the soil. Populations of these serotinous species can be reduced – or increased – by an order of magnitude depending on fire season (Bond *et al.* 1984). The same functional type with similar fire responses occurs in Australian shrublands. The mechanisms underlying the fire season response are now fairly well understood (e.g. Bond 1984; Cowling & Lamont 1987; Midgley 1989; Bradstock & Bedward 1992). After fire, there is a delay between seed release and seed germination cued by low winter temperatures. Long delays, caused by post-burn temperatures too high for germination, lead to heavier losses to predators or pathogens. Despite intensive research, prediction remains difficult (e.g. Lamont *et al.* 1991). The difficulty of simple prediction is, I believe, a useful caution for those expecting precision from general functional classifications. In this particular system, the seasonal response depends on an *indirect* effect of fire (cf. Strauss 1991); the response of the species depends on the presence of other species, particularly seed predators, or the extent to which seeds are exposed to direct heat in summer (Bradstock & Bedward 1992). Wherever functional classification depends on a chain of *indirect* effects, prediction will be more difficult. Of course this implies sufficient understanding of the system to know when a function (or response) *is* indirect.

Sprouting species, both Type I and Type III, persist well through single fires but can be decimated by repeated burning at vulnerable seasons. This sensitivity of established plants to fire season contrasts with their resilience to fire frequency. Fire season effects on sprouters have been shown in fynbos (Trollope 1973; Le Maitre & Brown 1992) and also in African savannas (Trapnell 1959). The sensitivity to fire season seems to depend on the phenology of carbohydrate storage in relation to fire. Plants burnt just after allocation to new growth, when reserves have been depleted, suffer most (Fig. 9.3). Le Maitre & Brown (1992), for example, found a marked reduction in flowering in the fynbos geophyte *Watsonia borbonica* after defoliation when corms were depleted. This, in turn, led to poor seed set because of more intense seed predation. C_4 grasses and C_3 shrubs have different phenologies (summer-growing vs. winter-growing, respectively) and mixtures can be altered towards grasslands or shrublands by manipul-

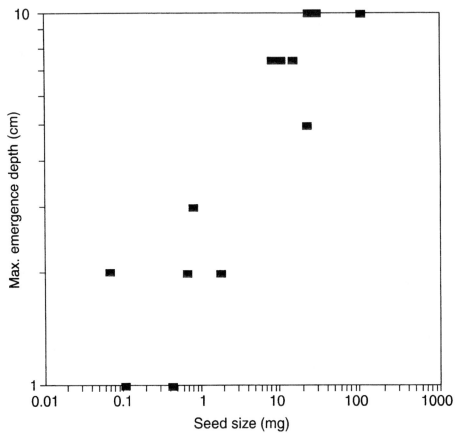

Figure 9.3   *Seed size vs. maximum emergence depths. Maximum seedling emergence depth is allometrically related to seed size. Data (from W. J. Bond & M. Honig, unpublished) are from 13 fynbos species, from several families, planted in sandy soil. Depth of emergence = $2.6W^{0.34}$, where W is seed mass in milligrams ($r^2=0.81$).*

ating fire season (Trollope 1973). The pattern needs further verification but a functional classification of fire seasonal response may simply collapse into shrubs and C3 graminoids vs. C4 grasses: shrubs favoured by summer/autumn burns, grasses by winter/early spring burns.

Fire intensity effects

Functional responses differ for Type II (non-sprouting) and Types I and III (sprouting) species. Responses of Type I and III species depend both on the amount of foliage burnt (topkill) and survival of meristematic tissue under the bark or below the soil for root crown sprouters. Some excellent models for predicting tree mortality from fire intensity have been developed in North American conifer forests (e.g. Peterson & Ryan 1986). Unfortunately no comparable system has been developed for plants that survive from buds lying at or just below the soil surface where topkill is complete. Variation in fire intensity is known to influence survival of sprouting shrubs (e.g. Moreno & Oechel 1991). However, a predictive understanding of fire-induced mortality of sprouting species has yet to be developed.

The effects of fire intensity on seed propagation are better understood. Seed germination in many fire-prone shrublands is heat stimulated (Jeffery *et al.* 1988; Keeley 1991; Bell *et al.* 1993) and germination can fail if fires are of low intensity (Shea *et al.* 1979; Auld 1986; Bond *et al.* 1990). Legumes are often heat-stimulated. In Western Australia, it is thought that low-intensity fires may have led to the elimination of acacias in jarrah forest understoreys (Shea *et al.* 1979). Very hot fires may also damage seed banks stored in surface soil layers. We are currently testing the hypothesis that these effects depend on seed size. There is a strong relationship between seed size and maximum emergence depth (Fig. 9.4). Intense fires appear to selectively eliminate small-seeded species, which have insufficient reserves to emerge from depths below the scorched surface layers (Fig. 9.4). The prediction would thus be that low-intensity burns would favour small-seeded species at the expense of large-seeded species, which would fail to get the necessary heat pulse. High-intensity burns would eliminate small-seeded species but favour deeply buried, typically myrmecochorous, large-seeded species. Elegant studies on Australian legumes support these basic arguments for large-seeded species (Auld 1986; Auld & O'Connell 1991).

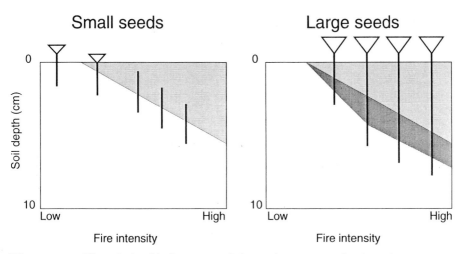

Figure 9.4   *The relationship between seed size and response to fire intensity. Lethal temperatures penetrate (shaded area) to progressively greater depths with increasing fire intensity. Small-seeded species can therefore be eliminated by high-intensity burns. Large-seeded species can emerge from greater soil depths. However, they may fail to germinate if soil temperatures fall below the range necessary to stimulate germination (dark shaded area).*

Since the legumes are the main nitrogen fixers, low-intensity fires may also have negative repercussions for the nitrogen budget in ecosystems (Shea *et al.* 1979). The putative effect of intensity on small-seeded species needs collaboration in the field (W. J. Bond & M. Honig, unpublished).

Predicting impacts on diversity

Diversity changes within the biome can be estimated from the number of species in different fire response FTs coupled with predicted impacts of climate change on the disturbance regime. The distribution of species across functional types is known only for a few genera and families and for a few types. Le Maitre & Midgley (1992) provide a useful summary. Using this, I have indicated approximate number of species in their sample of 4000 that would be influenced by different aspects of fire regime (Table 9.3).

Firstly, it is noteworthy that about half of the species in this sample are Type II and therefore poor persisters and vulnerable to climate change in the post-burn seedbed. Half are sprouters (Types I and III) which should resist change for decades (perhaps centuries), even if climate change makes regeneration impossible. Thus fynbos is likely to persist as a vegetation type long after the climate is no longer suitable for regeneration. Biodiversity of these remnants might, at worst, be halved.

Secondly, most fynbos species lack mobility. With the exception of serotinous species, seed dispersal of Type I and II species is short-distance and localized (Slingsby & Bond 1985). In addition, Type I species are poor regenerators because of slow seedling growth rates. Fynbos will expand much more slowly than forest or thicket, dominated by Type III and IV species.

Table 9.3 gives some indication of the response to changes in fire regime. Surprisingly few species (*ca.* 5%) may be affected by changes in fire frequency, at least in the short term. This is because of the scarcity of tall, and therefore slow-maturing, species. However, nearly a quarter of the sample is potentially sensitive to changes in fire intensity. This is because many species are both small-seeded and non-sprouters and therefore vulnerable to seed bank destruction by high fire intensities. Ironically intensity is one of the most neglected areas of fynbos fire research.

Finally, shifts in fire season, or rainfall seasonality, may have massive impacts on fynbos biodiversity. Changes in fire season are one of the few means of dislodging that half of the biodiversity which sprouts after fire. Changes in fire season would be driven by changes in rainfall seasonality and temperature. The direct effects of such climate change could be even more disruptive: for example, through changing the timing of germination

Table. 9.3 *The potential impact of climate change on diversity of a sample of the Cape flora*

Climate is assumed to change fire regimes which then threaten different fire response FTs. The most severe effects would be caused by changes in rainfall seasonality. These include changes to fire season but direct climate effects on the phenology of growth and reproduction, and pollinator and disperser activity (van Wilgen *et al* 1992; Bond 1995). Estimates of diversity within functional types are based on Table 6.1 in Le Maitre and Midgley (1992) and on information in P. Bond and Goldblatt (1984). Fire response types are described in Table 9.1. Species numbers are given for Proteaceae, Ericaceae, Asteraceae, Rutaceae, Fabaceae, the graminoid monocot families of Restionaceae and Cyperaceae (gram), and the geophytic Iridaceae and Liliaceae (*sensu lato*) (geo).

Variable	Effect of climate	Effect on functional type	Species number potentially affected									
			Prot	Eric	Aster	Rut	Fab	Brun	gram	geo	Total	%
No change	No change	No change	320	680	990	260	640	75	520	940	4418	100
Rate of change	fast change	Type II (non-sprouters): poor persisters, fast responders	256	612	495	130	480	35	90	0	2102	48
Rate of change	slow change	Type I (sprouters): good persisters, slow responders	64	68	495	130	160	40	423	940	2315	52
Fire frequency	interval reduced	Slow maturing non-sprouters lost (TYPE II)	192	—	—	—	24	—	<5	0	219	5
Fire intensity	higher intensity	Small-seeded non-sprouters lost (TYPE II)	0	612	153	0	0	5	40	0	807	18
Fire intensity	lower intensity	Large-seeded non-sprouters lost (TYPE II)	130	0	342	130	480	5	55	0	1142	26
Fire season	shift to summer rain	Winter burns promote grasses, impacting type I and II shrubs	Whole flora threatened by change in rainfall and fire seasonality									
Fire season	shift to winter rain	Summer burns promote shrubs at the expense of C4 grasses	Cape flora should expand with little loss of diversity									

in relation to the rainy season (van Wilgen *et al.* 1992), insect pollinator activity, and activity of dispersers such as ants (Bond 1995). The reproductive effects of global climate change are very poorly studied but would seem to have the most potential for impacting biodiversity in the Cape – and other mediterranean-type – floras.

Conclusions

I am sure I am not alone in feeling a sense of horror in attempting to predict vegetation change of this magnitude. It is tempting to merely extrapolate from present-day climate-vegetation patterns to future climate scenarios. In Cape fynbos, this approach is suspect (1) because distributions may be edaphically determined, and (2) because fire plays an obvious role in excluding trees, and possibly grassland. However, southern African biome distributions do coincide with physiologically interpretable climate boundaries (Ellery *et al.* 1991), supporting the notion of an equilibrium with current climate (and not, as in fynbos, with soil type). The relationship between climate, fire and vegetation and feedbacks between them deserves more attention from GCTE modellers to predict future rates and direction of biome changes.

Changes in biome distribution, unless extreme in magnitude, may have little impact on diversity within biomes. There is no obvious link between climate change and diversity. The most immediate effects may act through deflection of local succession. Unlike forests or herbaceous floras, fynbos species appear to have very similar growth forms with similar, slow growth rates. Thus climate change is less likely to influence comparative species performance and competitive ability and more likely to disrupt response to disturbances. Fires are by far the most common initiator of local succession, as they are in Australian and Californian shrublands, and fire response is therefore an appropriate focus for exploring impacts of global change.

The four basic fire life histories are a useful basis for fire-response FTs. By far the majority of fynbos species fall into the fire-stimulated sprouter and non-sprouter types. The four types differ in their persistence and mobility in the face of changing environments but it seems clear that, at least in the short term, persistent species will maintain fynbos physiognomy even if regeneration fails. Most fynbos species are not very mobile, and biome boundaries will change slowly.

An analysis of the differential response of species groupings, functional types, to the three components of fire regime suggests that:

1. only small numbers of species are at risk from increases in fire frequency;

2. many more species are potentially at risk from changes in fire intensity;
3. the greatest loss of biodiversity would probably follow changes in seasonality of rainfall with attendant changes in fire season.

These predictions are generalizations based on slender evidence and only applicable in the short term. They are based on the premise that the first and most severe effects of global change will be to the disturbance regime in fynbos systems. In the longer term, global change may be so disruptive to plant pollinator and disperser interactions, pathogens, decomposition and nutrient uptake processes, and growth and flowering phenology that prediction rapidly becomes impossible.

Is there any generality to the approaches used here? It seems probable that general mechanisms of fire response can be identified in at least the Gondwanan shrublands, Californian chaparral and possibly mediterranean maquis. Eucalypt and conifer woodlands, and even savanna, may also share similar basic fire-response mechanisms. Perhaps the imperative of predicting global impacts of climate change will act as a spur to identifying general patterns of plant fire responses from the vast, disparate and parochial literature on the topic.

Finally, it should be noted that, as in most centres of biodiversity, the more immediate threats to fynbos are conversion of land for agricultural and urban use (Rebelo 1992; Hobbs, Chapter 4, this volume), the continued spread of alien invasive weeds (MacDonald *et al.* 1986; Richardson *et al.* 1992), and the lack of a sustained will of people to conserve the remnants.

References

Auld, T. D. (1986) Population dynamics of the shrub *Acacia suaveolens* (Sm.) Willd.: fire and the transition to seedlings. *Australian Journal of Ecology*, **11**, 373–85.

Auld, T. D. & O'Connell, M. A. (1991) Predicting patterns of post-fire germination in 35 eastern Australian Fabaceae. *Australian Journal of Ecology*, **16**, 53–70.

Bell, D. T., Plummer, J. A. & Taylor, S. K. (1993) Seed germination ecology in south-western Western Australia. *Botanical Review*, **59**, 24–73.

Bond, P. & Goldblatt, P. (1984) Plants of the Cape Flora–a descriptive catalogue. *Journal of South African Botany, Suppl.*, **13**, 1–455.

Bond, W. J. (1984) Fire survival of Cape Proteaceae – influence of fire season and seed predators. *Vegetatio*, **56**, 65–74.

Bond, W. J. (1988) Proteas as 'tumbleseeds': wind-dispersal through the air and over soil. *South African Journal of Botany*, **54**, 455–60.

Bond, W. J. (1995) Effects of global change on plant-animal synchrony: implications for pollination and seed dispersal. In Moreno, J. M. & Oechel, W. C. (eds), *Global change and Mediterranean-type Ecosystems*, pp. 181–202. Ecological Studies no. 117. New York: Springer-Verlag.

Bond, W. J., Le Roux, D. & Erntzen, R. (1990) Fire intensity and regeneration of myrmecochorous Proteaceae. *South African Journal of Botany*, **56**, 326–31.

Bond, W. J., Vlok, J. & Viviers, M. (1984) Variation in seedling recruitment of Cape Proteaceae after fire. *Journal of Ecology*, **72**, 209–21.

Botkin, D. B., Janak, J. F. & Wallis, J. R. (1972) Some ecological consequences of a computer model of forest growth. *Journal of Ecology*, **60**, 849–72.

Bradstock, R. A. & Bedward, M. (1992) Simulation of the effect of season of fire on post-fire seedling emergence of two *Banksia* species based on long-term rainfall records. *Australian Journal of Botany*, **40**, 75–88.

Campbell, B. M. (1985) A classification of the mountain vegetation of the fynbos biome. *Memoirs of the Botanical Survey of South Africa*, **50**, 1–115.

Clark, J. S. (1988) Effect of climate change on fire regimes in northwestern Minnesota. *Nature*, **334**, 233–5.

Clark, J. S. (1989) Effects of long-term water balances on fire regime, north-western Minnesota. *Journal of Ecology*, **77**, 989–1004.

Cowling, R. M. (1984) A syntaxonomic and synecological study in the Humansdorp region of the fynbos biome. *Bothalia*, **15**, 175–227.

Cowling, R. M. & Holmes, P. M. (1992) Flora and vegetation. In Cowling, R. M. (ed.), *The Ecology of Fynbos*, pp. 23–61. Oxford University Press.

Cowling, R. M. & Lamont, B. B. (1987) Post-fire recruitment of four co-occurring *Banksia* species. *Journal of Applied Ecology*, **24**, 645–58.

Currie, D. J. (1991) Energy and large-scale patterns of animal and plant species richness. *American Naturalist*, **137**, 27–49.

Currie, D. J. & Paquin, V. (1987) Large-scale biogeographical patterns of species richness in trees. *Nature*, **329**, 326–7.

Ellery, W. N., Scholes, R. J. & Mentis, M. T. (1991) An initial approach to predicting the sensitivity of the South African grassland biome to climate change. *South African Journal of Science*, **87**, 499–503.

Fosberg, M. A., Goldammer, J. G., Rind, D.

& Price, C. (1992) Global change: effects on forest ecosystems and wildfire severity. In Goldammer, J. (ed.), *Fire in Tropical Biotas*, pp. 463–86. Berlin: Springer-Verlag.

Friend, A. D. (1993) The prediction and physiological significance of tree height. In Solomon, A. M. & Shugart, H. H. (eds), *Vegetation Dynamics and Global Change*, pp. 101–15. New York: Chapman and Hall.

Gill, A. M. (1981) Adaptive responses of Australian vascular plant species to fires. In Gill, A. M., Groves, R. H. & Noble, I. R. (eds), *Fire and The Australian Biota*, pp. 243–72. Canberra: Australian Academy of Sciences.

Good, R. (1974) *The Geography of Flowering Plants*. London: Longman.

Grime, J. P. (1979) *Plant Strategies and Vegetation Processes*. Chichester: Wiley.

Grime, J. P. (1988) The C-S-R model of primary plant strategies – origins, implications and tests. In Gottlieb, L. D. & Jain, S. (eds), *Evolutionary Plant Biology*, pp. 371–93. London: Chapman and Hall.

Hansen, A., Pate, J. S. & Hansen, A. P. (1991) Growth and reproductive performance of a seeder and a resprouter species of *Bossiea* as a function of plant age after fire. *Annals of Botany*, **67**, 497–509.

Harper, J. L. & White, J. (1974) The demography of plants. *Annual Review of Ecology and Systematics*, **5**, 419–63.

Heinselman, M. A. (1973) Fire in the virgin forests of the Boundary Waters Canoe Area, Minnesota. *Quaternary Research*, **3**, 329–82.

Janzen, D. H. (1988) Insect diversity of a Costa Rican dry forest: why keep it, and how? *Biological Journal of the Linnean Society*, **30**, 343–56.

Jeffery, D. J., Holmes, P. M. & Rebelo, A. G. (1988) Effects of dry heat on seed germination in selected indigenous and alien legume species in South Africa. *South African Journal of Botany*, **54**, 28–34.

Johnson, S. D. (1992) Plant-animal relationships. In Cowling, R. M. (ed.), *The Ecology of Fynbos*, pp. 175–205. Oxford University Press.

Keeley, J. E. (1991) Seed germination and life history syndromes in the California chaparral. *Botanical Review*, **57**, 81–116.

Lamont B. B. (1982) Mechanisms for enhancing nutrient uptake in plants with particular reference to mediterranean South Africa and western Australia. *Botanical Review*, **48**, 597–689.

Lamont, B., Connell, S. W. & Bergl, S. M. (1991) Seed bank and population dynamics of *Banksia cuneata*: the role of time, fire and moisture. *Botanical Gazette*, **152**, 114–22.

Lawton, J. H. & Brown, V. K. (1993) Redundancy in ecosystems. In Schulze, D. E. & Mooney, H. A. (eds), *Biodiversity and Ecosystem Function*, pp. 255–70. (Ecological Studies, 99.) Berlin: Springer-Verlag.

Le Maitre, D. C. & Brown, P. J. (1992) Life cycles and fire-stimulated flowering in geophytes. In van Wilgen, B. W. *et al.* (eds), *Fire in South African Mountain Fynbos*, pp. 145–60. (Ecological Studies. 93.) Berlin: Springer-Verlag.

Le Maitre, D. C. & Midgley, J. J. (1992) Plant reproductive ecology. In Cowling, R. M. (ed.), *The Ecology of Fynbos: Nutrients, Fire and Diversity*, pp. 135–74. Oxford University Press.

Leishman, M. R. & Westoby, M. (1992) Classifying plants into groups on the basis of associations of individual traits – evidence from Australian semi-arid woodlands. *Journal of Ecology*, **80**, 417–24.

Linder, H. P. 1991. Environmental correlates of patterns of species richness in the southwestern Cape Province of South Africa. *Journal of Biogeography*, **18**, 509–18.

Linder, H. P., Meadows, M. E. & Cowling, R. M. (1992) History of the Cape flora. In Cowling, R. M. (ed.), *The Ecology of Fynbos: Nutrients, Fire and Diversity*, pp. 113–34. Oxford University Press.

Loehle, C. (1988) Tree life history strategies: the role of defenses. *Canadian Journal of Forestry*, **18**, 209–22.

MacDonald, I. A. W., Kruger, F. J. & Ferrar, A. A. (eds) (1986) *The Ecology and Control of Biological Invasions in Southern Africa*. Cape Town: Oxford University Press.

Manders, P. T., Richardson, D. M. & Masson, P. H. (1992) Is fynbos a stage in succession to forest? Analysis of the perceived ecological distinction between two communities. In van Wilgen, B. W. *et al.* (eds), *Fire in South African Mountain Fynbos*, pp. 81–107. (Ecological Studies 93.) Berlin: Springer-Verlag.

Masson, P. H. & Moll, E. J. (1987) The factors affecting forest colonisation of fynbos in the absence of fire at Orange Kloof, Cape Province, South Africa. *South African Forestry Journal*, **143**, 5–10.

Midgley, J. J. (1989) Season of burn of serotinous fynbos Proteaceae: a critical review and further data. *South African Journal of Botany*, **55**, 165–70.

Moore, A. D. & Noble, I. R. (1990). An individualistic model of vegetation stand dynamics. *Journal of Environmental Management*, **31**, 61–81.

Moll, E. J., McKenzie, B. & McClachlan, D. (1980) A possible explanation for the lack of trees in the fynbos, Cape Province, South Africa. *Biological Conservation*, **17**, 221–8.

Moreno, J. M. & Oechel, W. C. (1991) Fire intensity and herbivory effects on postfire resprouting of *Adenostoma fasciculatum* in southern California chaparral. *Oecologia*, **85**, 429–33.

Noble, I. R. & Slatyer, R. O. (1980) The use of vital attributes to predict successional changes in plant communities subject to recurrent disturbances. *Vegetatio*, **43**, 5–21.

Peterson, D. L. & Ryan, K. C. (1986) Modelling postfire conifer mortality for long range planning. *Environmental Management*, **10**, 797–808.

Rebelo, A. G. (1992) Preservation of biotic diversity. In Cowling, R. M. (ed.), *The Ecology of Fynbos: Nutrients, Fire and Diversity*, pp. 309–44. Oxford University Press.

Richardson, D. M., MacDonald, I. A. W., Holmes, P. M. & Cowling, R. M. (1992) Plant and animal invasions. In Cowling, R. M. (ed.), *The Ecology of Fynbos: Nutrients, Fire and Diversity*, pp. 271–308. Oxford University Press.

Rutherford, M. C. & Westfall, R. H. (1986)

Biomes of southern Africa – an objective classification. *Memoirs of the Botanical Survey of South Africa*, **54**, 1–98.

Shea, S. R., McCormick, J. & Portlock, C. C. (1979) The effects of fires on regeneration of leguminous species in the northern jarrah (*Eucalyptus marginata* Sm) forest of Western Australia. *Australian Journal of Ecology*, **4**, 195–205.

Shugart, H. H. (1984) *A Theory of Forest Dynamics*. New York: Springer-Verlag.

Slingsby, P. & Bond, W. J. 1985. The influence of ants on the dispersal distance and seedling recruitment of *Leucospermum conocarpodendron* (**Proteaceae**). *South African Journal of Botany*, **51**, 30–4.

Smith, T. M., Shugart, H. H., Woodward, F. I. & Burton, P. J. (1993) Plant functional types. In Solomon, A. M. & Shugart, H. H. (eds), *Vegetation Dynamics and Global Change*, pp. 272–92. London: Chapman & Hall.

Specht, R. L. & Moll, E. J. (1983) Heathlands and sclerophyll shrublands – an overview. In Kruger, F. *et al.* (eds), *Mediterranean Type Ecosystems – the Role of Nutrients*. pp. 41–65. Berlin: Springer-Verlag.

Stock, W. D. & Allsopp, N. (1992) Functional perspective of ecosystems. In Cowling, R. M. (ed.), *The Ecology of Fynbos: Nutrients, Fire and Diversity*, pp. 241–59. Oxford University Press.

Stock, W. D. & Midgley, G. (1995) Ecosystem response to elevated CO_2 : nutrient availability and nutrient cycling. In Moreno, J. M. & Oechel, W. C. (eds), *Global Change and Mediterranean-type Ecosystems*, pp. 326–42. Ecological Studies no. 117. New York: Springer-Verlag.

Strauss, S. Y. (1991) Indirect effects in community ecology: their definition, study and importance. *Trends in Ecology and Evolution*, **6**, 206–10.

Swetnam, T. W. & Bettancourt, J. L. (1990) Fire-southern oscillation relations in the southwestern United States. *Science*, **249**, 961–1076.

Trapnell, C. G. (1959) Ecological results of woodland burning experiments in northern Rhodesia. *Journal of Ecology*, **47**, 129–68.

Trollope, W. S. W. (1973) Fire as a method of controlling macchia (fynbos) vegetation on the Amatole mountains of the eastern Cape. *Proceedings of the Grassland Society of Southern Africa*, **8**, 35–41.

van Daalen, J. C. (1981) The dynamics of the indigenous forest-fynbos ecotone in the southern Cape. *South African Forestry Journal*, **119**, 14–23.

van Wilgen, B. W., Bond, W. J. & Richardson, D. M. (1992) Ecosystem management. In Cowling, R. M. (ed.), *The Ecology of Fynbos: Nutrients, Fire and Diversity*, pp. 345–71. Oxford University Press.

Walker, B. H. (1991) Biodiversity and ecological redundancy. *Conservation Biology*, **6**, 18–23.

Woodward, F. I. (1987) *Climate and Plant Distribution*. Cambridge University Press.

Woodward, F. I. (1993) Leaf responses to the environment and extrapolation to larger scales. In Solomon, A. M. & Shugart, H. H. (eds.) *Vegetation Dynamics and Global Change*, pp. 71–100. New York: Chapman and Hall.

Wright, H. A. & Bailey, A. W. (1982) *Fire Ecology: United States and Southern Canada*. New York: Wiley.

⑩ Defining functional types for models of desertification

J. F. Reynolds, R. A. Virginia and W. H. Schlesinger

Introduction

As one progresses from semi-arid to arid conditions, vegetation often changes from a graminaceous growth-form to a variety of short arborescent growth-forms, known informally as 'shrubs'. Found on several continents, these transition zones are thought to be particularly sensitive to future climate change owing to the close proximity of contrasting life forms (e.g. shrubs vs. grasses) and to the differential physiological response (e.g. C_3 vs. C_4 photosynthesis) of the plant species often involved (IGBP 1990). Past and future changes in the border between semi-arid grasslands and desert shrublands and the potential for an increasing area of arid lands – desertification – offer a useful index of environmental change, including both human impacts and regional climatic changes. Historical losses of semi-arid grasslands and expansion of desert shrublands are well known in southern New Mexico and west Texas (Buffington & Herbel 1965; Archer et al. 1988).

Future global climate change is likely to cause dramatic shifts in the distribution and extent of deserts, but our ability to predict the magnitude of such shifts is limited. Most equilibrium models of future global climate predict a contraction of the area of desert lands (Smith et al. 1992), but a transient increase in the extent of mid-continental drought and desert vegetation may dominate the dynamics of dryland vegetation during the next century (Rind et al. 1990).

In this chapter we describe our use of ecosystem functional types (EFTs) in the development of models of desertification in the Jornada Basin of southern New Mexico. Our use of EFTs is hierarchical and based on our hypothesis that changes in ecosystem function at the transition between arid and semi-arid regions are best understood in the context of the spatial and temporal distribution of soil resources. That is, during the process of desertification of productive grasslands, what was previously a relatively uniform distribution of water, nitrogen (N), and other soil resources is replaced by an increase in both their spatial and temporal heterogeneity. These conditions promote the invasion of shrubs, furthering this heterogeneity since soil resources tend to accumulate under shrubs while wind and water remove resources from intershrub spaces, transporting them across the landscape. A conceptual model showing the various linkages and feedbacks between ecosystem properties at the patch and landscape scales to regional and global biogeochemistry is given in Fig. 10.1. Although our work is based on local studies in New Mexico, we believe it is applicable to other areas in the world and may be useful in the development of global–scale models of desertification.

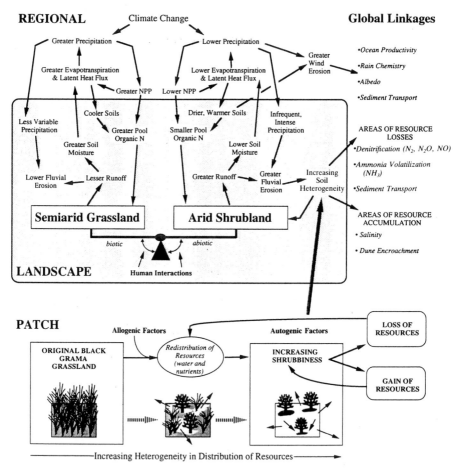

Figure 10.1   *Linkages and feedbacks at various spatial and temporal scales during the process of desertification. A relatively uniform distribution of soil resources in productive grasslands is replaced by an increase in both their spatial and temporal heterogeneity followed by an invasion of shrubs. As abiotic factors predominate over biotic ones, the balance tips in favour of the development of arid ecosystems. Modified from Schlesinger* et al. *(1990).*

Desertification of the Jornada Basin

During the past 100 years, ecosystems of the Jornada Basin of southern New Mexico have changed dramatically. Large areas of what was previously black grama (*Bouteloua eriopoda*) grasslands have been replaced by communities dominated by shrubs, especially creosote bush (*Larrea tridentata*) and mesquite (*Prosopis glandulosa*) (Buffington & Herbel 1965; York & Dick-Peddie 1969; Hennessy *et al.* 1983) (Fig. 10.2). The invasion of desert shrubs into former grassland areas in southern New Mexico is often attributed to one or more of the following *allogenic* factors: climatic change (Neilson 1986), human fire suppression (Humphrey 1958), overgrazing (Archer 1989; Conley *et al.* 1992; Grover & Musick 1990), a proliferation of rodents due to the human elimination of natural predators (Brown & Heske 1990), and rising atmospheric CO_2 concentrations (Mayeux *et al.* 1991; Idso 1992; Johnson *et al.* 1993). It is likely that several of these allogenic factors have worked in concert to favour desert shrubs over grasses, and it is unlikely that an elimination of any one of these would allow an immediate return of the historical grasslands (Laycock 1991).

We suggest that the invasion and persistence of desert shrubs in semiarid grasslands is augmented by several *autogenic* factors that operate at the patch scale (Fig. 10.1). Shrub dominance leads to greater spatial heterogeneity of soil properties because the infiltration of rainfall to the soil profile is confined to the area beneath shrub canopies, while barren intershrub spaces generate overland flow, soil erosion by wind and water,

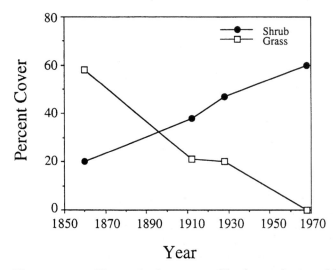

Figure 10.2 *Changes in the percent of land area dominated by shrubs (mainly mesquite) and grassland vegetation on the Jornada Experimental Range from 1858 through 1962. Adapted from Buffington & Herbel (1965).*

and nutrient losses from the ecosystem (Rostagno 1989; Takar *et al.* 1990; Ward *et al* 1990; Stockton & Gillette 1990). The cycling of plant nutrients is progressively confined to the zone beneath shrubs, leading to the development of 'islands of fertility' that characterize many deserts dominated by shrub vegetation (Charley & West 1975; Virginia & Jarrell 1983; MacMahon & Wagner 1985; Lajtha & Schlesinger 1986). These islands of fertility become preferred sites for the regeneration of shrubs (Goldberg & Turner 1986; McAuliffe 1988). Our model suggests that these vegetation-induced changes in the distribution of soil resources produce a positive feedback, augmenting the persistence and regeneration of shrubs in 'desertified' habitats.

Autogenic factors also include the activities of animals that help augment local soil heterogeneity, potentially aiding the invasion of grasslands by shrubs. Mun & Whitford (1990) report significantly higher nutrient concentrations in the mounds of banner-tailed kangaroo rats than in grassland soils a short distance away. Brown & Heske (1990) suggest that kangaroo rats may reduce grass cover in semi-arid regions, largely as a result of feeding, digging and other activities that act to concentrate soil resources in local areas.

Recognition of the causes and consequences of changes in vegetation affects our ability to manage arid and semi-arid ecosystems. We suggest that any process that increases the spatial heterogeneity of semi-arid soils will encourage the invasion of native grasslands by shrubs. For instance, increased soil heterogeneity caused by grazing cattle is analogous to, but more extreme than, the effect of native animals (Afzal & Adams 1992). Conversely, the re-establishment of native shrubs on desert lands that have been 'homogenized' by human impacts is slow (Carpenter *et al.* 1986). In many areas of southern Arizona, for example, shrubs are unable to reinvade areas of desert that have been abandoned from earlier attempts at agriculture (*New York Times*, April 21, 1992). Cultivation of these sites has destroyed the original islands of fertility that would be preferred sites for the regeneration of shrubs (Goldberg & Turner 1986).

A simple comparison between the spatial heterogeneity of soils in the black grama grassland and shrubland communities at the Jornada Long Term Ecological Research (LTER) site in southern New Mexico was derived from samples taken at intervals of 1 m along a transect 100 m long in each community (Fig. 10.3). For each soil resource, the coefficient of variation was significantly greater in creosote bush and mesquite shrublands than in grassland. Further sampling showed that the heterogeneity in shrublands is associated with the 'islands of fertility' that accumulate around shrubs (Fig. 10.4).

Spatial variation in soil nutrients is associated with spatial variation in soil microbial populations and soil microfauna that promote nutrient cycling in shrub deserts. Gallardo & Schlesinger (1992) found that the spatial

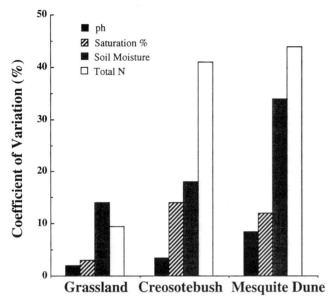

Figure 10.3 *Coefficient of variation associated with mean values for soil properties. Based on 200 samples from each community. From Schlesinger* et al. *(1990), with permission.*

coefficient of variation for soil microbial biomass was 50–80% in shrubland habitats compared with 30–40% in grassland habitats at the Jornada LTER site. Soil microbial biomass was significantly greater under shrubs than in the shrub interspaces, a pattern also seen in semi-arid shrubland habitats in Argentina (Mazzarino *et al.* 1991). Freckman & Mankau (1986) found that soil nematodes were significantly more abundant under shrubs than in the barren areas between shrubs in a hot desert environment of southern Nevada.

Spatial variation in soil resources is reflected in local spatial variation in net primary production in grassland and shrubland functional types in southern New Mexico. Despite similar levels of net primary production (NPP) in these communities, the coefficient of variation for NPP at a scale of 1 m^2 is significantly higher in shrubland than in grassland habitats (L. Huenneke, personal communication). Thus, desertification is not associated with a loss of plant productivity but with a change in its local distribution on the landscape.

One might expect that such changes could affect higher trophic levels, but differences in the local distribution of NPP do not appear to be associated with differences in the density and diversity of birds and lizards in grassland and shrubland habitats in southern New Mexico. In shrublands these species are often found perching or hiding in the shrub islands, but the high spatial variation of soil nutrients and NPP in shrublands is

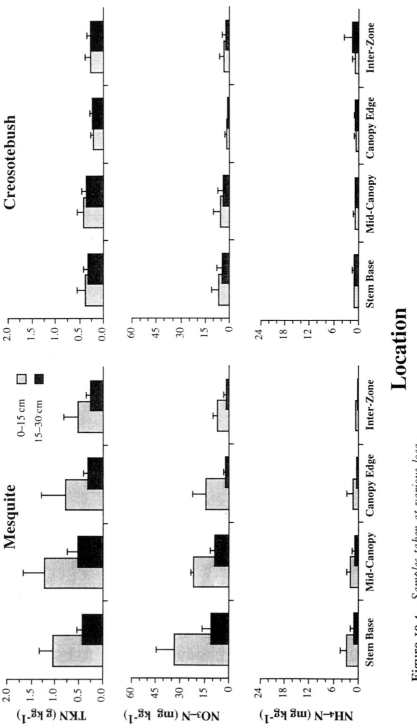

Figure 10.4 *Samples taken at various locations from the base of a shrub to the bare interzone between shrubs illustrate the island of fertility associated with shrubs.*

apparently not significant to the overall population density of large mobile organisms. Even the spatially heterogeneous shrubland habitat must appear 'fine-grained' to these species.

Ecosystem functional types (EFTs)

Gitay & Noble (Chapter 1, this volume) argue that in order for plant functional types to have extrapolative and predictive power, they should represent a simplification that describes similar responses, and the responses must be due to the same mechanisms. Using the distribution of soil resources as an index of ecosystem structure and function, we define patch, patch mosaic, and regional EFTs for modelling desertification at the landscape and regional scales (Table 10.1). Our definitions of these EFTs represent an attempt to be consistent with these criteria.

Patch EFTs

A patch represents a fine-scale unit of land that is internally consistent for the purpose at hand, such as for the basis of a model of ecosystem function (Woodmansee 1987; IGBP 1990). We delineate patch size as a function of the size of a single type of plant (grass clump or shrub) growing on a homogeneous soil type (e.g. sandy loam soil). For our purposes, patch EFTs range in size from *ca.* 1 m² to 10 m². We identify two general types of vegetation patches: *grass* and *shrub island*. In southern New Mexico, a grass patch is dominated by black grama (*Bouteloua*) whereas shrub island patches are dominated by shrubs such as mesquite and creosote bush. A shrub island patch is composed of a single shrub, representing a 'hot spot' or an 'island' of biological activity within a matrix of relatively barren soil. The behaviour of patch EFTs is determined by the functional and structural properties of the plants (grasses and shrubs), the interactions of the plants with their immediate abiotic and biotic environment (see Table 10.1), and the strength of autogenic factors operating at this scale (Fig. 10.1).

Patch mosaic EFTs

A unit of land consisting of contiguous patch EFTs forms a *patch mosaic* EFT. Patch mosaics range in size from *ca.* 1 ha to 1 km², although delineation of the size is arbitrary. We define three types of patch mosaics for modelling desertification: *grass* (composed solely of grass patch EFTs), *mixed*, and *shrub island* (composed solely of shrub island EFTs) (Fig. 10.5).

    The behaviour of a patch mosaic EFT is a function of its composition, configuration and dominant flowpaths (Table 10.1). Composition is the total

Table 10.1 *Ecosystem functional types used in models of desertification*

	Typical ecological scales of interest		Internal distribution of soil resources	Ecosystem functional types (EFTs)	Dominant functional and structural characteristics of EFT	Typical ecological model outputs
	Spatial	Temporal				
Patch	*ca.* 1–10 m²	Hours, days, weeks	Uniform	Grass	High canopy cover Shallow-rooted Tight coupling of biological processes (e.g. photosynthesis) to soil moisture	Primary production (photosynthesis, respiration) Growth, allocation and herbivory Phenology, reproduction and mortality Decomposition and nutrient dynamics Water use and balance
				Shrub island	Low canopy cover Shallow and deep-rooted Coupling of biological processes to soil moisture is species-dependent	

Landscape	ca. 1 ha–1 km²	Months, years	Uniform	Grass patch mosaic	High infiltration rate of rainfall Infrequent horizontal transport of water and nutrients Biotic processes confined to upper soil layers Topographic position (determines input and outputs of nutrients due to horizontal transport) Standing biomass	Water budgets and dynamics (e.g. overland flow, infiltration, run-on, run-off, etc.) Spatial patterns of evapotranspiration Fluvial patterns and processes (including movement of soil, nutrients, detritus) Spatial patterns of trace gas fluxes (e.g. ammonia volatilization) Net productivity
			Variable: depends on composition of patch EFTs	Mixed patch mosaic	Frequent horizontal transport of water and nutrients Patch composition Patch configuration Directional flowpaths for water movement Standing biomass	

Table 10.1 (*cont.*)

	Typical ecological scales of interest		Internal distribution of soil resources	Ecosystem functional types (EFTs)	Dominant functional and structural characteristics of EFT	Typical ecological model outputs
	Spatial	Temporal				
			Heterogeneous	Shrub island patch mosaic	Cycling of nutrients confined to zone beneath shrubs Effective infiltration of rainfall confined to area under shrub canopies Frequent horizontal transport of water and nutrients Patch composition Patch configuration Directional flowpaths for water movement Low canopy cover Standing biomass	
Regional	*ca.* 50–100 km^2	Years, decades, centuries	Variable, depends on degree of desertification	Vegetation functional types	Per cent plant cover Areas of resource accumulation Areas of resource losses	Cover (LAI) Deciduousness/ Phenology Vegetation height and patchiness Trace gas fluxes

number and the proportion of patch EFTs. Configuration includes a variety
of spatial characteristics of the patch EFTs in the mosaic, such as spatial
distribution (regular, clumped, etc.), shapes formed by neighbouring
patches (irregular, circular, random, etc.), and contrast (similarity of a patch
to its contiguous neighbours). These characteristics are important in
determining the behaviour of a particular mosaic, for example the effect of
vegetation on hydrological flow (Pickup 1985; Turner & Gardner 1991),
seedling establishment (Montaña *et al.* 1990), and exchanges of water,
organic matter, propagules, nutrients, sediments, etc. (Sklar & Costanza
1991).

In the conceptual model of the Jornada Basin by Wondzell *et al.* (1987),
the term *landscape* is used to indicate a land unit that may comprise several
patch mosaics (Fig. 10.5). In the case of landscapes, distinct geomorphic
surfaces (e.g. alluvial fans, piedmonts) or topographic features (e.g. water-
sheds) often form natural boundaries.

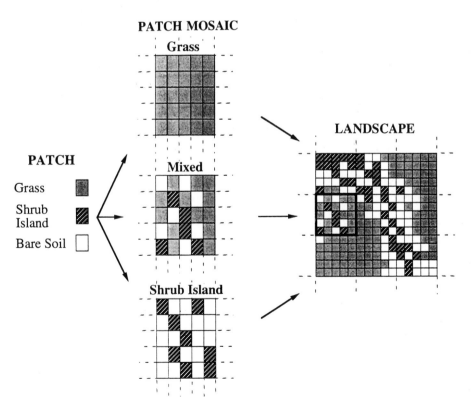

Figure 10.5 *A landscape is composed of patch mosaic EFTs; the behaviour
of the landscape is a function of its composition (number and proportion of patch
EFTs) and configuration (spatial distribution, shapes, contrast, etc.).*

Regional EFTs

We define a region to be from about one half to a full grid cell ($1° × 1°$) in size, containing many complex landscapes. At this coarse scale, our definition of regional EFTs is analogous to the general *Vegetation Functional Types* of Walker (Chapter 5, this volume), that is, semi-arid grassland and arid shrublands.

The patch, patch mosaic and regional EFTs possess heterogeneity at different spatial and temporal scales. However, to be useful as functional types in modelling they must represent distinct degrees of heterogeneity directly relevant to understanding and predicting ecosystem behaviour at different scales (Kolasa & Rollo 1991). We suggest that the dominant structural and functional characteristics of these EFTs (Table 10.1) allow this prediction. The ultimate test of their usefulness is one of our ongoing research goals.

Models of desertification

In this section we describe how the patch, patch mosaic and regional EFTs may be used to develop a predictive model for desertification.

Patch dynamics

Patch dynamics and responses are dominated by plant–soil–atmosphere interactions that occur relatively rapidly. Differences in functional and structural characteristics of dominant arid-land species lead to the development of distinct patch types. For example, mesquite (*Prosopis*) and creosote bush (*Larrea*), which now dominate extensive areas of the Sonoran and Chihuahuan Deserts, differ in their structure, phenology, and physiology. Mesquite is a deciduous, deeply rooted legume capable of symbiotic nitrogen fixation; it produces leaf litter with a low $C : N$ ratio and can contribute fixed N to the soil (Simpson 1977; Shearer *et al.* 1983). *Larrea*, in contrast, is an evergreen xerophyte with a relatively shallow root system and litter that has a much higher $C : N$ ratio (Lopez *et al.* 1979). These two plants differ in key functional traits (mode of N acquisition, nutrient- and water-use efficiency, root system architecture, etc.), which result in different patterns of response and persistence (Table 10.2).

Such differences are useful in understanding the functional differences in shrub island EFTs. To illustrate this, we ran a series of simulations with the *Patch AridLands Simulator* (PALS) (Reynolds *et al.* 1992) to examine the relationships between N-mineralization (N-min) rates, runoff–infiltration characteristics and primary production in creosote bush and

Table 10.2 *Key functional traits of two dominant shrubs of the Southwestern deserts of the US*

Characteristic	Creosote bush	Mesquite
Resource availability		
Water	Low	High
Nitrogen	Low	High
Resource use efficiency		
Water	High	Low
Nitrogen	High	Low
Structure/growth		
Leaf phenology	Evergreen	Deciduous
Relative growth rate	Low	High
Rooting depth	Shallow	Deep

mesquite shrub island patches. We hypothesize that during the process of island formation (i.e. structure), changes in water (e.g. the amount and distribution of infiltration) and nutrients (e.g. rates of N-mineralization) have an important impact on the functional response of a patch EFT to its environment. In the models, differing degrees of island formation were mimicked by choosing values for parameters in PALS that control the rates of N-min (l=low, m=medium, and h=high) and infiltration (L=low, M= medium, and H=high). For example, recently established shrubs on bare or compacted soils were assigned low N-min potentials and low infiltration rates; well-developed islands were given high levels of each. All other parameters and initial states were held constant for a one-year simulation in a 3×3 factorial of these combinations using 1986 Jornada LTER daily climatic data as driving variables. The relative responses of these two patch EFTs are shown in Fig. 10.6, where for each infiltration level there are three levels of N-min, representing the different N-min potentials.

There was an increase in net primary production (NPP) and soil N mineralized with increases in infiltration in the creosote bush patch. NPP was more strongly related to total N mineralized than to amount of infiltration. This is probably because changes in the hydrological parameters in PALS affected infiltration for only a small fraction of the total number of rain events that occurred during 1986; thus differences in the water budgets at different infiltration rates were not as great as would be the case with a greater variation in total rainfall amounts. Several different combinations of N-min and infiltration caused similar amounts of N-mineralization but showed differences in production. For example, the m,L island had 266 g of total shrub production compared with 219 g in the l,H island for the same amount of nitrogen mineralization (Fig. 10.6A). This probably occurred because the m,L island had both greater water-use efficiency (WUE) and nitrogen-use efficiency (NUE) than the l,H island (Fig. 10.6B).

These resource-use efficiencies are not model parameters but instead are a function of the timing and amount of growth, N availability and rainfall. For each level of N-min potential, increasing infiltration resulted in

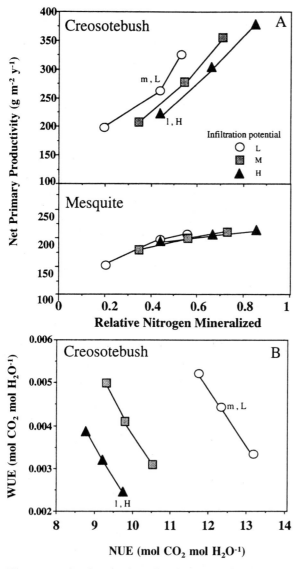

Figure 10.6 *Output from Patch AridLands Simulator showing the interactions of N-mineralization and infiltration potentials on NPP in creosote bush and mesquite Patch EFTs.*

decreases in both NUE and WUE. In contrast, within each of the infiltration potentials (low, medium, high), increasing N-min potential caused decreases in NUE, but increases in WUE.

Increases in soil water and N had much less effect in the mesquite shrub island EFT (Fig. 10.6A). With the exception of the lowest levels of N-min, this deep-rooted, N-fixing shrub was only slightly responsive to these driving variables. Under the conditions of these simulations, the mesquite patch was effectively 'saturated' with soil water and N.

We are using the PALS model to develop response surfaces for grass and shrub island EFTs under a wide range of environmental conditions. For example, differences in soil types significantly affect the patterns shown in Fig. 10.6. These results can, in turn, be subjected to various analyses to search for patterns and thresholds that can be summarized for use in simple model formulations of patch mosaics (e.g. Reynolds *et al.* 1993).

Landscape dynamics

Landscape properties are critical to modelling desertification. When shrub-lands replace grasslands, landscape processes change. Shrub encroachment may lead to the formation of erosion zones (Pickup 1985), which affect fluvial transport across the landscape. Greater overland flow during rain-storms lowers the recharge of soil moisture over large areas of the land-scape, but it augments the infiltration of moisture to the soil beneath ephemeral streambeds, seasonal lakes, and other local areas where water accumulates (Stafford Smith & Morton 1990). As suggested by Noy-Meir (1985), when rainfall in arid lands is redistributed by local run-off, those parts of the landscape that are a source of run-off will have limited vegeta-tion, while local areas that receive 'run-on' will support significantly greater amounts of plant biomass, especially shrubs. The same amount of precipi-tation may result in a higher overall regional productivity than if it were distributed evenly across the landscape. Schlesinger & Jones (1984) observed significant lower shrub biomass in areas of the Mojave Desert that are dependent only upon incident rainfall, as a result of the diversion of overland flow from upslope. In a number of arid zones the existence of dense vegetation stripes alternating with bare areas is attributed to hetero-geneity in the redistribution of rainwater (Montaña *et al.* 1990). Thus, the greater redistribution of moisture in arid shrublands compared with semi-arid grasslands contributes regional heterogeneity to the distribution of soil resources, promoting the establishment and persistence of shrubs.

We hypothesise that there is a system threshold of stability (i.e. the balance of the seesaw or 'teeter totter' in Fig. 10.1) between semi-arid grassland and shrubland ecosystems that is predictable from the dynamics of patch mosaic EFTs. This is shown as a conceptual model in Fig. 10.7, where patch mosaics of various compositions and configurations that occur during desertification are shown, ranging from grass (I) to mixed (II–V) to shrub island (VI). For each mosaic, the probability that it will persist is shown both for current climatic conditions and conditions of increasing

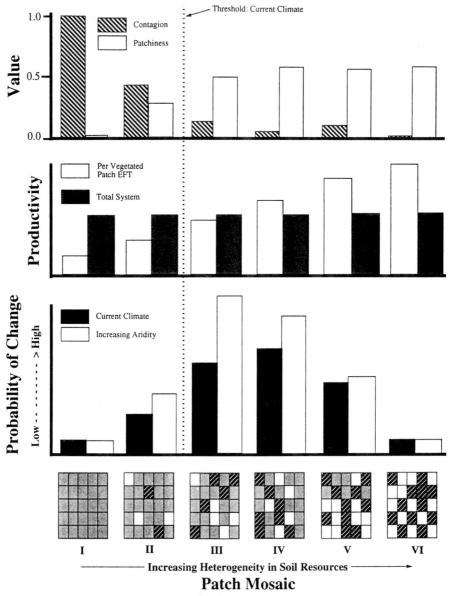

Figure 10.7 *Hypothesized relationships between patch mosaic EFTs and probability of change and productivity. Landscape indices computed from larger maps (2500 cells) of mosaic. Shading as in Fig. 10.5.*

aridity. These probabilities are a function of the composition and configuration of specific patch mosaics shown; these same probabilities apply to regional dynamics, i.e. the balance of the 'teeter totter', if the region is dominated by landscapes of that particular patch mosaic EFT. In most cases regional stability will be a function of mosaics in various stages of desertification.

Our conceptual model predicts that if a region is dominated by grass patch mosaics, it will persist as a grassland, both under current climatic conditions and with increasing aridity (Fig. 10.7). Patch mosaic II (also a grassland) represents a configuration of grass and shrub island EFTs that, under current climate, will also persist (although with a lower probability than mosaic I since there is some shrub encroachment). However, with increasing aridity, the probability that this region will undergo some change is slightly higher, because shrub growth and establishment could be favoured over grasses under these conditions. These systems are relatively stable, because the feedback mechanisms operating at the patch scale in our desertification model (Fig. 10.1) are strong and hence the distribution of resources (mainly water and nitrogen) will remain relatively homogeneous. In contrast, there is a high probability that mixed mosaic III will not persist in its present state. The various autogenic factors that contribute greater redistribution of resources are operable, and these will lead to a greater redistribution of resources and the further invasion of the mosaic by shrubs. Under increasing aridity, there is a still greater probability that it will not persist. Thus, we suggest that a threshold of stability exists between mixed mosaics II and III.

The shrub island patch mosaic, dominated by shrub island EFTs and bare soil, represents a highly desertified landscape. This system is stable (low probability of change) under both current climatic conditions and increasing aridity. Mosaics III–V represent landscapes at increasing stages of desertification. Under today's climatic conditions, these landscapes are not stable but are changing (at different rates) towards a configuration similar to the shrub island mosaic. We predict that the rates of desertification of mosaics III–V might be increased under conditions of increasing aridity. Since desertification is not so much associated with a loss of plant productivity as with a change in its local distribution on the landscape, our model predicts that total system production will remain unchanged in spite of differences in patch EFT production (Fig. 10.7).

Is it possible to quantify the system threshold that tips the balance in favour of increasing desertification? Can knowledge about small-scale patch structure and function help us to better understand the larger-scale process of landscape desertification? Because our model predicts that desertification is largely driven by increasing heterogeneity of resources, we believe that an understanding of the mechanisms of patch formation and persistence (using patch EFTs) along with the use of landscape properties (using patch mosaic EFTs) may allow us to extrapolate to a regional scale. For example, the relationship between the composition and configuration of the patch mosaics and two landscape indices, contagion and patchiness, is shown in Fig. 10.7. Contagion measures the extent to which the patch types are clumped (Li & Reynolds 1993) and patchiness measures the contrast of neighbouring patches (contrast between similar patches is low) (Romme 1982). With increasing shrub encroachment, contagion decreases whereas patchiness reaches a plateau, in which the contrast between grasses and bare soil is roughly equal to that of shrubs and bare soil (cf. Hook *et al.* 1991). We suggest that there may be a relation between the composition and configuration of patch mosaics EFTs and their stability (Fig. 10.7).

Regional dynamics

Given their differing physiognomy, grasslands and shrublands are easily differentiated in remote sensing (Warren & Hutchinson 1984; Prince & Justice 1991). Recognition of the striking differences in ecosystem function between semi-arid grasslands and desert shrublands – as distinctive functional types at this scale – may allow us to predict the significance of such shifts to global ecosystem function (Schlesinger *et al.* 1990). For example, at the Jornada Experimental Range, the total content of organic carbon in grassland habitats (2303 g m^{-2}; 92% in soil) exceeds that found in mesquite habitats (2040 g m^{-2}; 88% in soil), implying that the shift from grassland to shrubland has been associated with a loss of CO_2 to the atmosphere.

If we recognize semi-arid grasslands and desert shrublands as 'vegetation functional types' we find that they differ in a variety of ecosystem properties (Table 10.1; Fig. 10.1); a change in the regional extent of desert shrubland has the potential to cause significant changes in global properties. For example, south of the US–Mexican border, a greater loss of grass cover due to overgrazing is associated with greater surface albedo and higher maximum summer air temperatures than in comparable areas in Arizona (Balling 1988, 1991). Losses of plant cover in desertified habitats are associated with greater rates of wind erosion, transporting soil materials and their nutrients globally (Talbot *et al.* 1986; Swap *et al.* 1992), with effects on atmospheric planetary albedo (Ackerman & Chung 1992), rain chemistry (Gillette *et al.* 1992; Durand *et al.* 1992) and oceanic productivity (Martin & Gordon 1988). The global influence of a greater expanse of desert land during the last glacial is shown by a greater deposition of dust in the Vostok ice core from Antarctica (DeAngelis *et al.* 1987; Petit *et al.* 1990).

Conclusion

Changes in desert landscapes and the global consequences of desertification will be more easily predicted if we understand the underlying mechanisms that drive local vegetation change and the consequences of such change. Recognition of the causes and consequences of changes in functional types at the patch, landscape, and regional scales is critical to our ability to manage arid and semi-arid ecosystems. Our recognition of a hierarchy of ecosystem functional types (EFTs) is based on our hypothesis that changes in ecosystem function at the transition between arid and semi-arid regions are best understood in the context of the spatial and temporal distribution of soil resources. These EFTs embrace functional heterogeneity at different spatial and temporal scales, which is critical for understanding and predicting ecosystem behaviour.

Acknowledgements

We are grateful to our colleagues in the Jornada LTER, Laura Huenneke, Wes Jarrell, Gary Cunningham, and Walt Whitford, who contributed to these ideas, and to Paul Kemp, Daryl Moorhead, and Harbin Li who assisted in the development of PALS. This research was supported by DOE Grant DE-FG05–92ER61493 and NSF grant DEB 89051 and it is a contribution to the Jornada LTER under NSF grants BSR 88–11160 and DEB 92–40261.

References

Ackerman, S. A. & Chung, H. (1992) Radiative effects of airborne dust on regional energy budgets at the top of the atmosphere. *Journal of Applied Meteorology*, **31**, 223–33.

Afzal, M. & Adams, W. A. (1992) Heterogeneity of soil mineral nitrogen in pasture grazed by cattle. *Soil Science Society of America Journal*, **56**, 1160–5.

Archer, S. (1989) Have southern Texas savannas been converted to woodlands in recent history? *American Naturalist*, **134**, 545–61.

Archer, S., Scifres, C., Bassham, C. R. & Maggio, R. (1988) Autogenic succession in a subtropical savanna: conversion of grassland to thorn woodland. *Ecological Monographs*, **58**, 111–27.

Balling, R. C. (1988) The climatic impact of a Sonoran vegetation discontinuity. *Climatic Change*, **13**, 99-109.

Balling, R. C. (1991) Impact of desertification on regional and global warming. *Bulletin of the American Meteorological Society*, **72**, 232–4.

Brown, J. H. & Heske, E. J. (1990) Control of a desert-grassland transition by a keystone rodent guild. *Science*, **250**, 1705–7.

Buffington, L. C. & Herbel, C. H. (1965) Vegetational changes on a semidesert grassland range from 1858 to 1963. *Ecological Monographs*, **35**, 139–64.

Carpenter, D. E., Barbour, M. G. & Bahre, C. J. (1986) Old-field succession in Mojave desert scrub. *Madrono*, **33**, 111–22.

Charley, J. L. & West, N. E. (1975) Plant-induced soil chemical patterns in some shrub-dominated semidesert ecosystems of Utah. *Journal of Ecology*, **63**, 945–64.

Conley, W., Conley, M. R. & Karl, T. R. (1992) A computation study of episodic events and historical context in long-term ecological processes. Climate and grazing in the northern Chihuahuan desert. *Coenoses*, **7**, 55–60.

DeAngelis, M., Barkov, N. I. & Petrov, N. V. (1987) Aerosol concentrations over the last climatic cycle (160 kyr) from an Antarctic ice core. *Nature*, **325**, 318–21.

Durand, P., Neal, C. & LeLong, F. (1992) Anthropogenic and natural contributions to the rainfall chemistry of a mountainous area in the Cevennes National Park (Mont-Lozere, southern France). *Journal of Hydrology*, **130**, 71–85.

Freckman, D. W. & Mankau, R. (1986) Abundance, distribution, biomass and energetics of soil nematodes in a northern Mojave Desert ecosystem. *Pedobiologia*, **29**, 129–42.

Gallardo, A. & Schlesinger, W. H. (1992) Carbon and nitrogen limitations of soil microbial biomass in desert ecosystems. *Biogeochemistry*, **18**, 1–17.

Gillette, D. A., Stensland, G. J., Williams, A. L., Barnard, W., Gatz, D., Sinclair, P. C. & Johnson, T. C. (1992) Emissions of alkaline elements, calcium, magnesium, potassium and sodium, from open sources in the contiguous United States. *Global Biogeochemical Cycles*, **6**, 437–57.

Goldberg, D. E. & Turner, R. M. (1986) Vegetation change and plant demography in permanent plots in the Sonoran desert. *Ecology*, **67**, 695–712.

Grover, H. D. & Musick, H. B. (1990) Shrubland encroachment in southern New Mexico, U. S. A., An analysis of desertification processes in the American Southwest. *Climatic Change*, **17**, 305–30.

Hennessy, J. T., Gibbens, R. P., Tromble, J. M. & Cardenas, M. (1983) Vegetation changes from 1935 to 1980 in mesquite dunelands and former grasslands of southern New Mexico. *Journal of Range Management*, **36**, 370–4.

Hook, P. B., Burke, I. C. & Lauenroth, W. K. (1991) Heterogeneity of soil and plant N and C associated with individual plants and openings in North American shortgrass steppe. *Plant and Soil*, **138**, 247–56.

Humphrey, R. R. (1958) The desert grassland, A history of vegetational change and an analysis of causes. *Botanical Review*, **24**, 193–252.

IGBP (International Geosphere–Biosphere Programme) (1990) *A Study of Global Change: The Initial Core Projects, Report No. 12*. Stockholm: International Council of Scientific Unions.

Idso, S. B. (1992) Shrubland expansion in the American Southwest. *Climatic Change*, **22**, 85–6.

Johnson, H., Polley, H. W. & Mayeux, H. S. (1993) Increasing CO_2 and plant-plant interactions: Effects on natural vegetation. In Rozema, J., Lambers, H., van de Geign, S. C. & Cambridge, M. L. (eds) *CO_2 and the Biosphere*, pp. 157–70. Dordrecht: Kluwer Academic Publishers.

Kolasa, J. & Rollo, C. D. (1991) The heterogeneity of heterogeneity: A glossary. In Kolasa, J. & Pickett, S. T. A. (eds), *Ecological Heterogeneity*, pp. 1–23. New York: Springer-Verlag.

Lajtha, K. & Schlesinger, W. H. (1986) Plant response to variations in soil nitrogen availability in a desert shrubland community. *Biogeochemistry*, **2**, 29–37.

Laycock, W. A. (1991) Stable states and thresholds of range condition in North American rangelands. A viewpoint. *Journal of Range Management*, **44**, 427–33.

Li, H. & Reynolds, J. F. (1993) A new contagion index to quantify spatial patterns of landscapes. *Landscape Ecology*, **8**, (3), 155–62.

Lopez, E. C., Mabry, T. M. & Tavizon, S. F. (1979) *Larrea*. Saltillo, Mexico: Centro de Investigacion en Quimica Aplicada.

MacMahon, J. A. & Wagner, F. H. (1985) The Mojave, Sonoran and Chihuahuan deserts of North America. In Evenari, M., Noy-Meir, I. & Goodall, D. W. (eds). *Hot Deserts and Arid Shrublands, 12A*, pp. 105–202. Amsterdam: Elsevier Scientific Publishers.

Martin, J. H. & Gordon, R. M. (1988) Northeast Pacific iron distribution in relation to phytoplankton productivity. *Deep Sea Research*, **35**, 177–96.

Mayeux, H. S., Johnson, H. B. & Polley, H. W. (1991) Global change and vegetation dynamics. In *The Proceedings of the National Noxious Range Weed Conference*, pp. 62–74. Boulder, Colorado: Westview Press.

Mazzarino, M. J., Oliva, L., Abril, A. & Acosta, M. (1991) Factors affecting nitrogen dynamics in a semiarid woodland (Dry Chaco, Argentina), *Plant and Soil*, **138**, 85–98.

McAuliffe, J. R. (1988) Markovian dynamics of simple and complex desert communities. *American Naturalist*, **131**, 459–90.

Montaña, C., Lopez-Portillo, J. & Mauchamp, A. (1990) The response of two woody species to the conditions created by a shifting ecotone in an arid ecosystem. *Journal of Ecology*, **78**, 789–98.

Mun, H.-T. & Whitford, W. G. (1990) Factors affecting annual plant assemblages on banner-tailed kangaroo rat mounds. *Journal of Arid Environments*, **18**, 165–73.

Neilson, R. P. (1986) High-resolution climatic analysis and Southwest biogeography. *Science*, **232**, 27–34.

Noy-Meir, I. (1985) Desert ecosystem structure and function. In Evenari, M., Noy-Meir, I. & Goodall, D. W. (eds), *Hot Deserts and Arid Shrublands, 12A*, pp. 93–103. Amsterdam: Elsevier Scientific Publishers.

Petit, J. R., Mounier, L., Jouzel, J., Korotkevich, Y. S., Kotlyakov, V. I. & Lorius, C. (1990) Paleo-climatological and chronological implications of the Vostok core dust record. *Nature*, **343**, 56–8.

Pickup, G. (1985) The erosion cell – A geomorphic approach to landscape classification in range assessment. *Australian Rangeland Journal*, **7**(2), 114–21.

Prince, S. D. & Justice, C. O. (eds). (1991) Coarse resolution remote sensing of the Sahelian environment. *International Journal of Remote Sensing, Special Issue*, **12** (6), 1133–421.

Reynolds, J. F., Hilbert, D. W. & Kemp, P. R. (1993) Scaling ecophysiology from the plant to the ecosystem: A conceptual framework. In Ehleringer, J. E. & Field, C. B. (eds), *Scaling Physiological Processs*, pp. 127–40. San Diego: Academic Press.

Reynolds, J. F., Hilbert, D. W., Chen, J.-L., Harley, P. C., Kemp, P. R. & Leadley, P. W. (1992) *Modeling the Response of Plants and Ecosystems to Elevated CO_2 and Climate Change*. DOE/ER-60490T-H1. Springfield, Virginia: National Technical Information Service, US Dept of Commerce.

Rind, D., Goldberg, R., Hansen, J., Rosenzweig, C. & Ruedy, R. (1990) Potential evapotranspiration and the likelihood of future drought. *Journal of Geophysical Research*, **95**, 9983–10004.

Romme, W. H. (1982) Fire and landscape diversity in subalpine forests of Yellowstone National Park. *Ecological Monographs*, **52**, 199–221.

Rostagno, C. M. (1989) Infiltration and sediment production as affected by soil surface conditions in a shrubland of Patagonia, Argentina. *Journal of Range Management*, **42**, 382–5.

Schlesinger, W. H. & Jones, C. S. (1984) The comparative importance of overland runoff and mean annual rainfall to shrub communities of the Mojave Desert. *Botanical Gazette*, **145**, 116–24.

Schlesinger, W. H., Reynolds, J. F., Cunningham, G. L., Huenneke, L. F., Jarrell, W. M., Virginia, R. A. & Whitford, W. G. (1990) Biological feedbacks in global desertification. *Science*, **247**, 1043–8.

Shearer, G., Kohl, D. H., Virginia, R. A., Bryan, B. A., Skeeters, J. L., Nilsen, E. T., Sharifi, M. R. & Rundel, P. W. (1983) Estimates of N2-fixation from variation in the natural abundance of 15N in Sonoran Desert ecosystem. *Oecologia, Berlin*, **56**, 365–73.

Simpson, B. B. (1977) *Mesquite: its biology in two desert ecosystems.* US/IBP Synthesis Series. Stroudsburg, Pennyslvania: Dowden, Hutchinson & Ross.

Sklar, F. H. & Costanza, R. (1991) The development of dynamic spatial models for landscape ecology: A Review and prognosis. In Turner, M. G. & Gardner, R. H. (eds), *Quantitative Methods in Landscape Ecology*, pp. 239–88. New York: Springer-Verlag.

Smith, T. M., Leemans, R. & Shugart, H. H. (1992) Sensitivity of terrestrial carbon storage to CO_2-induced climate change: Comparison of four scenarios based on general circulation models. *Climatic Change*, 21, 367–84.

Stafford Smith, D. M. & Morton, S. R. (1990) A framework for the ecology of arid Australia. *Journal of Arid Environments*, 18, 255–78.

Stockton, P. H. & Gillette, D. A. (1990) Field measurement of the sheltering effect of vegetation on erodible land surfaces. *Land Degradation and Rehabilitation*, 2, 77–85.

Swap, R., Garstang, M., Greco, S., Talbot, R. & Kallberg, P. (1992) Saharan dust in the Amazon basin. *Tellus*, 44B, 133–49.

Takar, A. A., Dobrowolski, J. P. & Thurow, T. L. (1990) Influence of grazing, vegetation life-form, and soil type on infiltration rates and inter-rill erosion on a Somali rangeland. *Journal of Range Management*, 43, 486–90.

Talbot, R. W., Harris, R. C., Browell, E. V., Gregory, G. L., Sebacher, D. I. & Beck, S. M. (1986) Distribution and geochemistry of aerosols in the tropical North Atlantic troposphere: Relationship to Saharan dust. *Journal of Geophysical Research*, 91, 5173–82.

Turner, M. G. & Gardner, R. H. (eds) (1991) *Quantitative Methods in Landscape Ecology*, p. 536. New York: Springer-Verlag.

Virginia, R. A. & Jarrell, W. M. (1983) Soil properties in a mesquite-dominated Sonoran desert ecosystem. *Soil Science Society of America Journal*, 47, 138–44.

Ward, T. J., Krammes, J. S. & Bolton, S. (1990) A comparison of runoff and sediment yields from bare and vegetated plots using rainfall simulation. In Riggins, R., Jones, E. B., Singh, R. & Rechard, P. (eds), *Watershed Planning and Analysis in Action*, pp. 245–55. New York: American Society of Civil Engineers.

Warren, P. L. & Hutchinson, C. F. (1984) Indicators of rangeland change and their potential for remote sensing. *Journal of Arid Environments*, 7, 107–26.

Wondzell, S. M., Cunningham, G. L. & Bachelet, D. (1987) A hierarchical classification of landforms: Some implications for understanding local and regional vegetation dynamics. In Moir, W. & Aldon, E. A. (eds), *Proceedings of the Symposium on Strategies for Classification and Management of Natural Vegetation for Food Production in Arid Zones*, pp. 15–23. Denver, Colorado: US Forest Service.

Woodmansee, R. G. (1987) Ecosystem processes and global change. In Ross, T., Woodmansee, R. G. & Risser, P. G. (eds), *Spatial and Temporal Variability of Biospheric and Geospheric Processes*, pp. 11–27. (SCOPE Report.) New York: Wiley & Sons.

York, J. C. & Dick-Peddie, W. A. (1969) Vegetation changes in southern New Mexico during the past hundred years. In McGinnies, W. G. & Goldman, B. J. (eds), *Arid Lands in Perspective*, pp. 157–66. Tucson: University of Arizona Press.

11 Plant functional types in temperate semi-arid regions

O. E. Sala, W. K. Lauenroth and R. A. Golluscio

Introduction

Grouping plants into categories as a way of simplifying some of the complexities of nature has been one of the aims of biology and ecology since the very beginning. Alexander von Humboldt (1806), early in the nineteenth century, distinguished 16 groups of plants. Subsequently, many authors have attempted different classifications of plants (see Barkman (1988) for a thorough review of this topic). These classifications have varied in their scope, in the characters taken into account, and in the methods used. Some classifications have been broad, attempting to include all plant types, such as Raunkiaer's (1907) or Grime's (1988); others have been narrow and, for example, directed only at aquatic plants. Some classifications use only morphological characters and others include functional characters, such as phenology or bud height. Barkman (1988) made the distinction between growth forms, which were groups based only upon morphological characteristics, and life forms, which were based upon morphological and/or physiological adaptations to a certain ecological factor. The techniques authors have used to arrive at groups have also changed through time. Early groupings were often the result of the author's experience, whereas at present objective multivariate methods are preferred (Leishman & Westoby 1992; Golluscio & Sala 1993).

In order to model the response of ecosystems to global change, and to assess their contribution to present changes in the composition of the atmosphere and in the climate, it will be necessary to reduce the number of elements in the models. It will not be possible to model every species nor every ecosystem. The units that aggregate several species with a common behaviour in the ecosystem are the functional types (FTs). Species within each FT should have a similar function in the ecosystem, and they should play a common ecological role. Functional types will be a critical hierarchical level at which to develop global change models.

This chapter describes research that involved defining and testing hypotheses about the relationships among FTs in the Patagonian steppe and extrapolating results to a regional scale. The objective of this chapter is to present (1) a set of hypotheses about the partitioning of water among FTs in the Patagonian steppe; (2) a test through manipulative experiments of these hypotheses; (3) a set of regional-scale hypotheses about the distribution of FTs and the occurrence of vegetation units, based upon the results of previous items; and (4) a test of these regional-scale hypotheses.

Functional types of arid regions

The functional types were defined *a priori* based upon morphological and phenological characteristics. We tested specific hypotheses relating to the kind of resources used by each FT, which in turn have major implications about their competitive interactions and their role in arid ecosystems.

The functional types of arid regions are grasses, shrubs, forbs and succulents. The conceptual model that relates these four FTs focuses on water relations because water is the critical resource explaining both the structure and the dynamics of these ecosystems. Water is partitioned among the four FTs along two axes: the depth from which each FT is able to absorb water, and the residence time of water in the soil. Residence time is the period of time during which water remains within the range of water potential available for plants (Fig. 11.1).

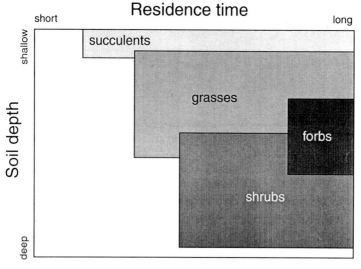

Figure 11.1   *Conceptual model of the partitioning of water resources among the four functional types of the Patagonian steppe: grasses, shrubs, succulents and forbs. The axes of the model are the depth from which each functional type is able to absorb water, and the residence time of water, which is the period of time during which water remains within the range of water potential available for plants.*

The model is based upon the following hypotheses:

1. Grasses absorb most of the water from the upper layers of the soil. They utilize frequent and short-duration pulses of water availability.
2. Shrubs absorb most of the water from the lower layers of the soil. They utilize infrequent and long-duration pulses of water availability.
3. Succulents absorb water from the shallowest layers of the soil. They utilize pulses of the shortest duration.
4. Forbs overlap with grasses and shrubs regarding the region of the soil where they absorb water. They utilize pulses of the longest duration. These pulses are infrequent, but they usually yield a quantity of water sufficient to complete a growth cycle of the forbs.

Our model is related to those proposed by Walker *et al.* (1981) and Walker & Noy-Meir (1982) in that it is based upon ideas first put forward by Walter (1971) as an explanation for the existence of savannas in tropical regions. Walter suggested that 'only in the tropics where both summer rain and a deep, loamy sand coincide, are they [grasses and woody plants] found existing in a state of ecological equilibrium'.

Walter (1971) suggested in his two-layer hypothesis for savannas that woody plants and grasses compete for water in the surface layers of the soil, but woody plants have exclusive access to a source of water relatively deep underground. Knoop & Walker (1985) tested this hypothesis for southern African savannas through removal experiments.

Based upon our hypothetical model we can make deductions, which may be tested in the field. The deductions are: (1) that the removal of shrubs will not increase the water status or productivity of grasses and will increase the availability of deep soil water; and (2) that the removal of grasses will not increase the water status or the productivity of shrubs and will increase the availability of water in the upper layers of the soil. The experiments consisted of removing grasses or shrubs and monitoring primary production, plant water status and soil water potential at different locations in the soil profile. Production was measured by using a harvest technique, plant water status by the pressure chamber technique (Scholander *et al.* 1965), and soil water potential with themocouple hygrometers (Spanner 1951). A complete description of the experimental design and the methods is given by Sala *et al.* (1989).

The experiments were carried out in the Occidental District of the Patagonian steppe (Soriano 1956). The five-year average above-ground net primary production (ANPP) of this ecosystem was 60 g m^{-2} a^{-1}; total canopy cover was 49% (Golluscio *et al.* 1982; Fernández-A. *et al.* 1991). Grasses, shrubs and forbs account, respectively, for 64%, 33% and 3% of canopy cover and 53%, 43% and 4% of ANPP. Temperatures range between 1 °C in July and 15 °C in January. The annual precipitation average, over a period of 37 years, was 136 mm. It was concentrated during the winter months, resulting in a recharge of the profile at this time.

The experimental results showed that the removal of shrubs did not result in an increase in the productivity of grasses (Fig. 11.2). Removal of grasses resulted in a small increase in the productivity of shrubs. The increase in shrub productivity was much smaller than the decrease in total productivity (grass productivity) that resulted from grass removal. The removal of grasses freed resources, which were used by shrubs. The efficiency with which one FT used the resources, water and nutrients liberated by removal of the other FT was calculated as

$$\text{Efficiency} = \Delta FT_1 / \text{ANPP } FT_2 \tag{11.1}$$

where ΔFT_1 is the change in the productivity of FT_1 as the result of removing FT_2, and ANPP FT_2 is the production of the experimentally removed FT in the control plots. The efficiency ranged between 0 for the removal of shrubs and 25% for the removal of grasses. These results partly support the overall hypothesis that these FTs use different resources.

Removal of shrubs did not result in any change in the water status of grasses (Fig. 11.3A). The lack of response in leaf water potential agreed with

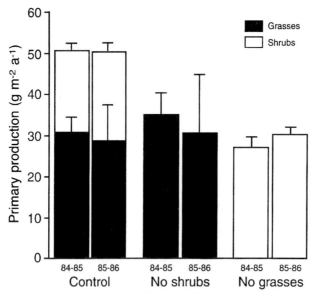

Figure 11.2 *Above-ground net primary production (+ standard error) of grasses and shrubs in the Patagonian steppe for treatments in which grasses or shrubs were selectively removed, and the control where both functional types were present (after Sala et al. 1989).*

the lack of response observed in the production of grasses as a result of the removal of shrubs. By contrast, removal of grasses resulted in an increase in leaf water potential of shrubs. This increase occurred on only a few sample dates: two during the first year, one during the second year, and never during the last year (Fig. 11.3B). These infrequent and small increases in the shrub water status may account for the small increase in production observed as a result of grass removal.

Soil water potential showed a clear seasonal pattern (Fig. 11.4). The entire soil profile was wet by the end of the winter or beginning of the spring and dried out throughout the spring and summer, when all layers attained very low soil water potential values. The differences among treatments were only evident during the drying period. In the upper layers, removal of grasses resulted in an increase in soil water potential, but the

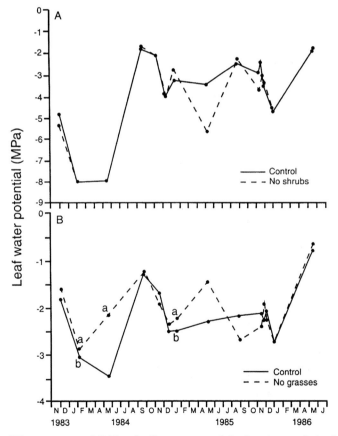

Figure 11.3 *Midday leaf water potential of grasses and shrubs in the Patagonian steppe for the control, and treatments in which grasses or shrubs were selectively removed. Different letters indicate significant differences (p<0.05) between treatment and the control. Absence of letter for a date indicates that differences were not significant (after Sala et al. 1989).*

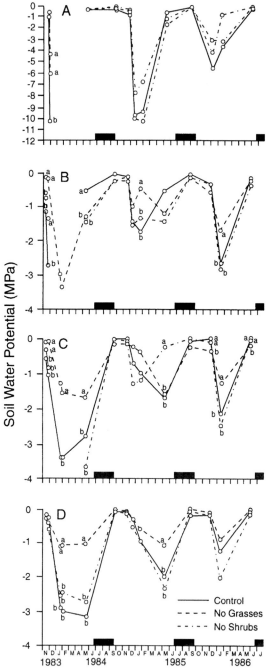

Figure 11.4   *Soil water potential at (A) 5, (B) 15, (C) 30 and (D) 60 cm depth in the Patagonian steppe for the control, and treatments in which grasses or shrubs were selectively removed. Different letters indicate significant differences (p<0.05) among treatments for a given date. Absence of letter for a date indicates that differences were not significant (after Sala et al. 1989).*

removal of shrubs did not result in any changes in soil water. In lower layers, removal of grasses also resulted in an increase in soil water potential but removal of shrubs did not result in the hypothesized increase in soil water availability, suggesting that they absorb from even deeper layers.

These experimental data support the overall hypothesis that shrubs and grasses in the Patagonian steppe use different water resources. Shrubs absorb water exclusively from the lower layers. Grasses indeed take up most of the water from the upper layers of the soil. However, grass removal resulted in a small increase in soil water in deeper layers, in shrub leaf water status and in shrub production. This may indicate that grasses influence the input of water to lower layers and/or that they are also able to absorb from lower soil layers.

These results have important implications for the nutrient economy of both FTs. A characteristic of arid and semi-arid regions is that soil organic matter and nutrient availability are concentrated in the layers very near the soil surface (Clark 1977; Cole *et al.* 1977; Schimel & Parton 1986). Therefore, if shrubs and grasses absorb their nutrients from the same location from which they absorb water, shrubs will be at a competitive disadvantage since they will be absorbing from a nutrient-poor layer. By contrast, grasses will have the advantage of absorbing from the richest layer of the soil. How do shrubs cope with these circumstances? Do they have lower requirements than grasses? Do they meet a larger fraction of their requirements via internal retranslocation?

Biogeographical model of the distribution of arid functional types

The objective of this section is to develop a biogeographical model of the distribution of temperate grasslands and shrublands. The model is based upon our current understanding about the functioning of grasses and shrubs, and about the resources used by each of these FTs. This is an attempt to extrapolate our experimental results about FTs to the scale of vegetation units. The approach is to match abiotic requirements of FTs with resource availability to estimate the probability of occurrence of grasslands, shrublands or a mixture. Finally we will discuss the influence of biotic factors, which range from competition to human intervention.

Generalizing the results obtained for the Patagonian steppe and for the southern African savannas, we will propose the potential conditions for temperate semi-arid sites to support grasses, woody plants or a mixture. These conditions can be assessed with information about the seasonality of precipitation and the texture of the soil.

If grasses have an advantage at locations where soil water is stored near the surface, two easily assessed characteristics can be used to evaluate the potential of that location to support grasses. First, soil texture will have a large influence on the location at which water is stored. In general, fine-textured soils will store more water near the surface than coarse-textured soils (Fig. 11.5). Given an identical precipitation regime, a site with a silty soil will have a larger proportion of its total water storage in the surface layers (0–30 cm) than will a sandy or gravelly soil. This occurs because the storage capacity (cm^3 cm^{-1}) is higher for silty soils.

Although soil texture is important in determining the location of water stored in the soil, the seasonality of precipitation must also be taken into account. We propose to account for the effects of seasonality of precipitation by evaluating the overlap between the wet season and the warm season. Because grass roots densely occupy the top soil layers, we assume

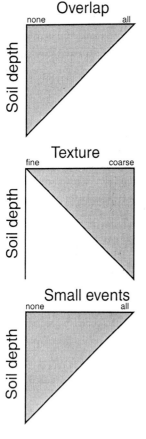

Figure 11.5   *Conceptual model of the effects of the overlap between the warm and wet seasons, soil texture, and the proportion of precipitation received as small events on the depths in the soil at which water is available to plants.*

that precipitation that occurs during the time of year when air temperatures are favourable for plant growth will favour grasses. Precipitation falling during the cold portion of the year will have a higher probability of being stored in deep layers, where grasses are less effective, and therefore should favour woody plants.

Two independent soil water modelling exercises support the idea that seasonality controls the distribution of available water in the soil profile. In the North American shortgrass steppe, where most of the precipitation occurs during the warm season, water availability is skewed towards the upper layers (Sala *et al.* 1992). The 4–15 cm soil layer was the layer that had the highest frequency of being wet (soil water potential > −1 MPa) over a 33 year period (Fig. 11.6). By contrast, in the South American Patagonian steppe, where 70% of the precipitation occurs during the cold months, the layer with the highest probability of being wet is the deepest layer (Paruelo & Sala 1995).

The combination of the soil texture and precipitation seasonality variable results in an explanation of grass–woody plant relationships (Fig. 11.7). Areas with maximum precipitation during the warmest portion of the year should support grasslands on all sites except those with very coarse-textured soils. Areas with maximum precipitation during the coldest portion of the

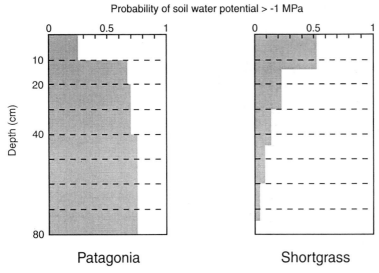

Figure 11.6 *Probability of soil water potential being greater than −1 MPa for different soil layers for the South American Patagonian steppe, which has mainly winter precipitation, and for the North American shortgrass steppe, which has predominantly summer precipitation. Results were obtained using two daily simulation models run for periods of 19 (Patagonia) and 33 years (short-grass) (after Sala* et al. *1992; Paruelo & Sala 1995).*

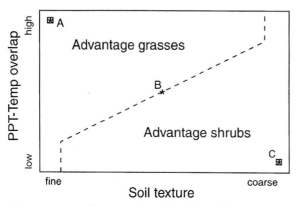

Figure 11.7 *Conceptual model of the effects of soil texture, and the overlap between the warm and wet seasons, upon the relative importance of grasses and shrubs. The model based exclusively on abiotic variables suggests that vegetation is composed exclusively of either grasses or woody plants only close to points A and C. Other points are mixtures of grasses and shrubs. The influence of biotic factors will be maximum near point B.*

year should support shrublands except on those sites with very fine-textured soils. Are mixtures of grasses and woody plants possible with this explanation? Yes: we would expect a vegetation composed exclusively of either grasses or woody plants only in very close proximity to either points A and C. Other points in the space represent mixtures of grasses and shrubs. As an interesting side issue we suggest that the ease with which one can manipulate a vegetation type decreases as one moves from the dividing line towards either A or C.

The remainder of this chapter is concerned with relations between grasslands and shrublands, why they occur where they do, and how humankind's uses of grasslands and shrublands change their structure and FT composition.

Consider the following ordinary non-linear differential equations:

$$dG/dt = r_g\, G(1 - G/G_{max}) \tag{11.2}$$

$$dS/dt = r_s\, S(1 - S/S_{max}) \tag{11.3}$$

for the case in which the dynamics of the vegetation is determined entirely by climatic and soil variables. The maximum biomass of each FT (S_{max} or G_{max}) is an expression of the suitability of the site. Taking grasses as an example, we can write

$$G_{max} = f(PPT,\ T,\ ST,\ Bg_{max}) \tag{11.4}$$

where PPT is monthly precipitation, T is monthly temperature, ST is soil texture and Bg_{max} is the maximum standing crop of grasses to be expected per unit of annual precipitation. If we assume a linear decline in the suitability of a site to support grasses as one moves from point A to point C in Fig. 11.7, the following equation applies:

$$G_{max} = (0.1 + 0.45X_1, + 0.45X_2)Bg_{max} \qquad (11.5)$$

We can ignore the scale factor Bg_{max} by considering G_{max} to be a proportion of the maximum biomass. The variable X_1 in Eqn 11.5 is an expression of overlap between the wet and the warm seasons based upon monthly data of precipitation and temperature. We use the product moment correlation coefficient expression:

$$X_1 = \frac{CORR(PPT, T) + 1}{2} \qquad (11.6)$$

$CORR(PPT, T)$ is the correlation coefficient between monthly average precipitation and monthly average air temperature. X_1 has a range from zero, when PPT and T are perfectly negatively correlated, to 1 when PPT and T are perfectly positively correlated.

The variable X_2 in Eqn 11.5 is an effect of soil texture. The function has the same shape as the function relating the effect of soil texture to water-holding capacity

$$X_2 = (2 - e^{KST}) \qquad (11.7)$$

where K is a scaling factor and ST is a soil texture variable. X_2 has a range from 1 for very fine-textured soils to zero for very coarse-textured soils.

The parameters r_g and r_s in Eqns 11.1 and 11.2 are the intrinsic rates of increase of either grass or shrub biomass. They are taken to be constants in this analysis, although one could argue for their dependence upon information similar to that used to calculate the maximum standing crop parameters. In that case, variable r values may be interpreted as changes in species composition, within FTs, as sites become more or less favourable for the particular FT.

The behaviour of this simple model over time will result in $G = G_{max}$ and $S = S_{max}$, assuming the simulation is carried out for a sufficient period of time and that the seasonal distributions of precipitation and temperature do not change. The interesting aspects of the analysis of the model revolve around the calculations of G_{max} and S_{max} and the associated equilibrium solutions G^* and S^*. How well does the model predict climatically controlled steady-state biomass? This is a question of how good are G_{max} and S_{max}.

Sims et al. (1978) presented soil, climate and FT composition data to which the calculations of G_{max} and S_{max} can be compared (Table 11.1). It is important to note that the model predicts the relative contribution of

Table 11.1. *Comparison of predictions from our model with data from ten North American sites*

Site	Soil texture	Per cent grasses	Per cent shrubs	X_1	X_2	G_{max}	S_{max}
Richland, WA (Ale)	Silt loam	65	29	0.09	0.91	0.55	0.45
National Bison Range, MT (Bridger)	Cobbly silt loam	75	0	0.74	0.91	0.84	0.16
Bangtail Ridge, MT (Bridger)	Silt loam	56	0	0.84	0.91	0.89	0.11
Cottonwood, SD (Cottonwood)	Silty clay loam	97	<1	0.88	0.99	0.94	0.06
Dickinson, ND (Dickinson)	Loamy fine sand	75	<1	0.92	0.57	0.77	0.23
Hays, KS (Hays)	Loam	85	1	0.95	0.68	0.83	0.17
Jornada Exp. Range, NM (Jornada)	Loamy fine sand	44	14	0.83	0.57	0.73	0.27
Pawhuska, OK (Osage)	Silty clay loam	96	0	0.89	0.99	0.95	0.05
Amarillo, TX (Pantex)	Silty clay loam	80	0	0.94	0.99	0.97	0.03
Central Plains Exp. Range, CO (Pawnee)	Fine sandy loam	52	11	0.92	0.57	0.77	0.23

Source: Sims *et al.* (1978).

grasses and shrubs to total cover ($G_{max} + S_{max} = 1$), whereas the observed data represent percentage cover of each FT and do not add up to 1 because bare soil cover is not presented. Model predictions for shrubs were highest for the sites with the largest shrub components; predictions for grasses were highest for sites with the largest proportions of grasses (Fig. 11.8 and 11.9). The model predicts the shrub contribution better than the grass contribution. The grass predictions are approximately related to the observed cover values except for the Ale site in Washington where predictions were much lower than the observations. This cannot be considered a stringent test of the model, since it contains information only about climatic and soils variables and ignores biotic interactions, management and historical information. The reason for carrying out this comparison is to show general correspondence.

Our analysis recognizes the potential for competition between grasses and shrubs as one of several biotic interactions that influence the balance between the two FTs rather than the dominant interaction. The degree to

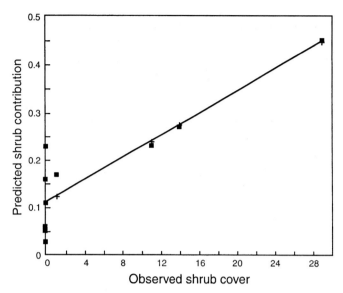

Figure 11.8 *Comparison between predicted and observed shrub cover for ten North American sites. Predictions are the result of the model presented here and represent the fraction of plant cover accounted for by shrubs. The observed values represent ten independent measurements of shrub cover from Sims et al. (1978). Observed cover values represent the fraction of soil area covered by shrubs; since total plant cover in most arid and semi-arid regions is not 100% the sum of shrub plus grass cover is less than 100%.*

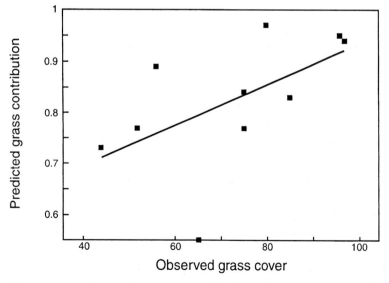

Figure 11.9 *Comparison between predicted and observed grass cover for ten North American sites. Predictions are the result of the model described here; observed values are those reported by Sims et al. (1978).*

which competition between FTs will determine the balance between grasses and shrubs in semi-arid regions is related to the local environment. In terms of Figure 11.7, the importance of competition between shrubs and grasses in determining whether a site will be occupied predominantly by grasses or by shrubs will be least for sites located near points A and C. The importance of competition will be greatest for sites near B.

The outcome of competition will be difficult to predict near point B because here all of the biotic influences will have their maximum impact. To understand why this is so we need to examine the characteristics implied in Figure 11.7. A simplified view of this model reveals an explanation of the behaviour of grasslands and shrublands near points A, B and C.

The significance of the points A and C is clear from this perspective. Point A represents a site with very fine-textured soils and a perfect correlation between the warmest and wettest months. According to our model, this is the epitome of a grassland site. By contrast, point C is located on a coarse-textured soil at a site that receives essentially all of its precipitation during the coldest months. This is a site that we would expect to be dominated by shrubs. How will the vegetation at sites A and C be influenced by biotic factors, including manipulation by humankind?

To answer this question, we need to use the concept of substitution efficiency developed in Eqn 11.1. This is the number of units of one FT that can be supported at a particular site as a result of removing a unit of the other FT. This is both FT- and site-dependent. According to the model, point C represents a very unfavourable site for grasses and consequently we would expect a small substitution efficiency. The probability of replacement of one FT by the other or the replacement rate is related to the distance of the site from point B in Figure 11.7. The further the site is from point B, the lower will be the efficiency of substitution and, therefore, the probability of replacement.

Let us now return to the question of the importance of biotic factors in determining the structure of the vegetation. Because the substitution efficiency of grasses for shrubs at point C or shrubs for grasses at point A is low, we would expect these two sites to show a maximum degree of independence from biotic factors. For instance, heavy herbivory on a site such as A would be expected to produce a vegetation composed of unpalatable grasses rather than one dominated by shrubs. In this case the effect of applying a force to the existing vegetation may result in a shift in species composition within the grass FT (or perhaps a shift in the relative frequency of tolerant genotypes). An analogous response may be expected for a shrubland represented by point C. Herbivory or perhaps herbicide treatment to control shrubs will result in either a change in species composition among shrubs or a short-lived vegetation dominated by grasses or other forbs. In addition to their degree of independence from biotic effects, sites represented by points A and C should be expected to show a minimum of year-to-year fluctuation in FT composition.

Previously we suggested that sites near point B should have behaviour quite different from those represented by points A and C. These sites occupy the boundary between the grassland and shrubland regions. Rather than a sharp edge separating the two, we suggest that there is a region in parameter space with a degree of indeterminacy for vegetation structure. It is here that biotic and other non-climatic influences should have their maximum impact. Sites in this region will be characterized by alternative vegetation structures depending upon their recent history. This is the region for which bifurcation models with bimodal behaviour are applicable.

Such sites have contributed to the idea that past heavy use by domestic herbivores has resulted in the conversion of many grasslands into shrublands. While this explanation may be exactly correct for sites in the vicinity of point B, it is clear that our model does not support such a deduction in the general case. The generality from the model is that the further the site is from the dividing line (point B), the smaller will be the difference between the alternative states of the system. In other words, the greater the distance from point B, the more similar the alternative steady states will be. Our previous discussion makes clear the notion that at point A or C the alternative states will have converged to a point.

Conclusion

Grasses are better suited to utilizing water stored in the upper layers of the soil than shrubs, which are more effective at utilizing water stored in the deep layers of the soil. Location of soil water determines the proportion of FTs, shrubs and grasses. Characteristics of the environment such as soil texture and seasonality of precipitation are the major controls of the distribution of water in the soil profile. Sites that have fine-textured soils and in which the wet season occurs in synchrony with the warm season represent the most favourable conditions for grasses. By contrast, shrubs dominate on coarse-textured soils and in regions in which precipitation does not occur during the warm season.

The probability of replacement of one FT (grasses) by the other (shrubs) increases as conditions become more different from the optimum. The model suggests that the boundary between grassland and shrubland regions is not sharp but implies a degree of indetermination for vegetation structure. It is in these regions, where the probability of replacement is high, that biotic factors have their maximum impact.

Acknowledgements

We thank T. B. Kirchner for helpful discussion regarding the development of the model.

References

Barkman, J. J. (1988) New systems of plant growth forms and phenological plant types. In Werger, M. J. A. *et al.* (eds), *Plant Forms and Vegetation Structure*, pp. 9–44. The Hague: SPB Academic Publishing.

Clark, F. E. (1977) Internal cycling of ^{15}nitrogen in shortgrass prairie. *Ecology*, **58**, 1322–33.

Cole, C. V., Innis, G. S. & Steward, J. W. B. (1977) Simulation of phosphorus cycling in semiarid grasslands. *Ecology*, **58**, 1–15.

Fernández-A., R. J., Sala, O. E. & Golluscio, R. A. (1991) Woody and herbaceous aboveground production of a Patagonian steppe. *Journal of Range Management*, **44**, 434–7.

Golluscio, R. A., León, R. J. C. & Perelman, S. B. (1982) Caracterización fitosociológica de la estepa del Oeste de Chubut: su relación con el gradiente ambiental. *Boletín de la Sociedad Argentina de Botánica*, **21**, 299–324.

Golluscio, R. A. & Sala, O. E. (1993) Plant functional types and ecological strategies in Patagonian forbs. *Journal of Vegetation Science*, **4**, 839–46.

Grime, J. P. (1988) The C–S–R model of primary plant strategies – origins, implications and tests. In Gottlieb, L. D. & Jain, S. K. (eds), *Plant Evolutionary Biology*, pp. 371–93. London: Chapman and Hall.

Humboldt, A. von (1806) *Ideen zu einer Physiognomik der Gewächse*. Stuttgart: Cotta.

Knoop, W. T. & Walker, B. H. (1985) Interactions of woody and herbaceous vegetation in a southern African savanna. *Journal of Ecology*, **83**, 235–53.

Leishman, M. R. & Westoby, M. (1992) Classifying plants into groups on the basis of associations of individual traits – evidence from Australian semi-arid woodlands. *Journal of Ecology*, **80**, 417–24.

Paruelo, J. M. & Sala, O. E. (1995) Water losses in the Patagonian steppe: A modelling approach. *Ecology*, **76**, 510–20.

Raunkiaer, C. (1907) *Planterigets livsformer og deres Betydning for Geografien*. Copenhagen: Munksgaard.

Sala, O. E., Golluscio, R. A., Lauenroth, W. K. & Soriano, A. (1989) Resource partitioning between shrubs and grasses in the Patagonian steppe. *Oecologia*, **81**, 501–5.

Sala, O. E., Lauenroth, W. K. & Parton, W. J. (1992) Long term soil water dynamics in the shortgrass steppe. *Ecology*, **73**, 1175–81.

Schimel, D. S. & Parton, W. J. (1986) Microclimatic controls on nitrogen mineralization and nitrification in shortgrass steppe soils. *Plant and Soil*, **93**, 347–57.

Scholander, P. F., Hammel, H. T., Bradstreet, E. D. & Hemmingsen, E. A. (1965) Sap pressure in vascular plants. *Science*, **148**, 339–46.

Sims, P. L., Singh, J. S. & Lauenroth, W. K. (1978) The structure and function of the western North American grasslands. 1.

Abiotic and vegetational characteristics. *Journal of Ecology*, **66**, 251–85.

Soriano, A. (1956) Los distritos florísticos de la Provincia Patagónica. *Revista de Investigaciones Agrícolas*, **10**, 323–47.

Spanner, D. C. (1951) The Peltier effect and its use in the measurement of suction pressure. *Journal of Experimental Botany*, **2**, 145–68.

Walker, B. H., Ludwing, D., Holling, C. S. & Peterman, R. M. (1981) Stability of semi-arid savanna grazing systems. *Journal of Ecology*, **69**, 473–98.

Walker, B. H. & Noy-Meir, I. (1982) Aspects of stability and resilience of savanna ecosystems. In Huntley, B. J. & Walker, B. H. (eds), *Ecology of Tropical Savannas*, pp. 577–90. Berlin: Springer-Verlag.

Walter, H. (1971) *Natural Savannas. Ecology of Tropical and Subtropical Vegetation*. Edinburgh: Oliver and Boyd.

⑫ Interactions between demographic and ecosystem processes in a semi-arid and an arid grassland: a challenge for plant functional types

W. K. Lauenroth, D. P. Coffin, I. C. Burke and R. A. Virginia

Introduction

Categorizing plants into groups is an obvious way to deal with the complexity of multispecies communities, and a wide variety of criteria have been employed in the actual categorization (Raunkiaer 1934; Box 1981; Prentice *et al.* 1992). In North American grasslands the idea of functional groups of plants arose from the US International Biological Program as a combination of physiological, morphological and life-span characteristics (Sims *et al.* 1978) (Table 12.1). The utility of such a categorization is most obvious when a number of grassland sites, each dominated by different species, are compared (Sims *et al.* 1978).

The idea of functional groups or functional types suggests that the variability in key characteristics is greater between types than within. Is it possible to construct a classification of plant types that meets that criterion for all characteristics? In our opinion the answer is no. We propose that the utility of a particular functional-type classification is limited to the range of questions that lead to the classification. In the context of global change, the degree to which aggregation of plants into functional types is appropriate depends upon the spatial scale of the question (Fig. 12.1). For instance, to address questions in the central grassland region of the United States at the county, state and even grassland-type scale it is inappropriate to utilize functional types. At these relatively small spatial scales, answers to questions about the ecological effects of global change will need to explicitly address effects on the species that are important to ecosystem structure and function. For example, in eastern Colorado, USA, any answer to a question about the effects of global change on the shortgrass steppe that does not mention *Bouteloua gracilis* will be of limited value. In eastern Colorado and throughout much of the Great Plains *Bouteloua gracilis* is a dominant or co-dominant species in addition to being the most grazing- and drought-resistant grass species in the ecosystem.

Table 12.1 *Functional groupings of plants commonly used in North American grasslands*

Morphology	Life span	Phenology[1]
Grasses[2]	Annual	Cool
		Warm
	Perennial	Cool
		Warm
Forbs[3]	Annual	Cool
		Warm
	Perennial	Cool
		Warm
Shrubs/Low trees	Perennial	Cool
		Warm
Succulents[4]	Perennial	Cool

[1] Plants were categorized as warm or cool season corresponding with the season of maximum growth. Warm season plants are essentially C_4 pathway type and cool season plants C_3 pathway types.

[2] Includes members of the Cyperaceae.

[3] Forbs are herbs not in the families Poaceae or Cyperaceae.

[4] CAM pathway type.

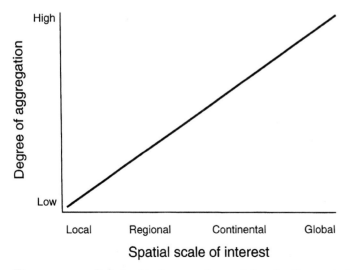

Figure 12.1 *Relationship between the spatial scale of a particular global change problem and the degree of aggregation that will be appropriate to solve the problem.*

Understanding the fate of *B. gracilis* is critical to understanding the response of the ecosystem. Analyses that combined *B. gracilis* with other species in a C4 perennial grass functional group would not provide the information required by ecosystem managers. As the spatial scale at which questions are being asked increases, the appropriateness of aggregating plants into functional types also increases. To a large degree such aggregation is a practical necessity.

The overall objective of this chapter is to focus on two dominant species in North American grasslands with the aim of exploring some aspects of variability between functional groups in population dynamics and effects on biogeochemistry. Each of the species is categorized as a C4 perennial grass. *Bouteloua gracilis* (H.B.K.) Lag. ex Griffiths is the dominant species of the shortgrass steppe (Fig. 12.2). *Bouteloua eriopoda* (Torr.) Torr. is the dominant species of desert grassland communities in the south-western US (Fig. 12.2). Both are bunchgrasses and are key forage species for livestock management. In additional to their similarities, they have important life-history differences that have important implications for their population dynamics. *Bouteloua gracilis* has a very long life span and recruits infrequently from seeds. By contrast, *B. eriopoda* has a short life span and recruits frequently by vegetative spread. Our specific objectives are to:

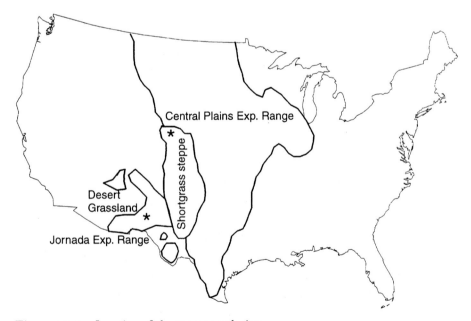

Figure 12.2 *Location of the two research sites.*

1. explore how differences in population dynamics affect the ways in which *B. gracilis* and *B. eriopoda* dominate sites in their respective grassland types;
2. investigate the spatial distribution of carbon and nitrogen in soils associated with individual plants of each species; and
3. evaluate the actual and potential interactions between population dynamics and ecosystem properties.

Methods

Site descriptions

Two sites were selected to represent the grassland types dominated by *B. gracilis* and *B. eriopoda*. The Central Plains Experimental Range (CPER) is located in north-central Colorado, North America, in the precipitation shadow of the Rocky Mountains and approximately 40 km south of Cheyenne, Wyoming (40°49'N, 104°46'W) (Fig. 12.2). Average annual precipitation was 321 mm over the 52 years from 1939 to 1990 with a standard deviation of 98 mm (Lauenroth & Sala 1992). Average annual temperature was 8.6 °C with a standard deviation of 0.6 °C over the same time period. The climate of the CPER is typical of mid-continental semi-arid sites in the temperate zone of North America except for the large influence of the Rocky Mountains 60 km to the west. Maxima in precipitation and temperature occur in June, July and August and minima occur in December, January and February.

Vegetation at the CPER is representative of the northern portion of the shortgrass steppe (Lauenroth & Milchunas 1992). Most locations, regardless of past grazing history, are dominated by *B. gracilis*, which accounts for approximately 90% of both basal cover and forage production (Milchunas *et al.* 1989). Associated species include *Buchloe dactyloides* (Nutt.) Engelm, *Opuntia polyacantha* Haw, *Sphaeralcea coccinea* (Pursh) Rydb. and *Carex eleocharis* Bailey.

The Jornada Experimental Range (JER) is located in south-central New Mexico approximately 37 km north of Las Cruces (32°57'N, 106°51'W) (Fig. 12.2.). Average annual precipitation from 1914 to 1968 was 206 mm with a standard deviation of 77 mm (Dittberner 1971). Average annual temperature was 14.9 °C over the same 54 year period. The climate of the JER is representative of mid-continental subtropical arid sites in North America. Maxima in temperature occur in June, July and August and maxima in precipitation occur in July, August and September. Minima in temperature occur in December, January and February; minima in precipitation occur in March and April.

The vegetation of the JER is a mixture of grasslands and shrublands (Schlesinger *et al.* 1990). The grasslands are dominated by *Bouteloua eriopoda* with varying amounts of *Sporobolus flexuosus* (Thurb.) Rydb., *Aristida* sp. and *Yucca elata* Engelm. The shrublands are dominated by either *Larrea tridentata* (DC) Coville or *Prosopis juliflora* (Swartz.) DC. var. *glandulosa* (Torr.) Cockerell.

A feature shared by both sites, and one that is common to all arid and semi-arid grasslands dominated by bunchgrasses, is a small-scale spatial structure that we refer to as the plant–interspace mosaic (Fig. 12.3). The mosaic consists of two phases: plants and the bare ground separating them. The largest fraction of the bare spaces are underlain by the roots of the dominant bunchgrasses (Milchunas & Lauenroth 1989; Hook *et al.* 1994). Therefore few of them represent safe sites for seedlings (Aguilera & Lauenroth 1993). In the many arid and semi-arid ecosystems, this plant–interspace mosaic has important implications for the spatial distribution of nitrogen and water (Charley & West 1977; Burke 1989; Schlesinger *et al.* 1990; Hook *et al.* 1991; Burke *et al.* 1995).

Population dynamics

We assessed the population dynamics of *B. gracilis* and *B. eriopoda* by using a simulation model. The model, STEPPE, simulates the recruitment, growth and death of individual plants on a 0.1 m^2 plot (Coffin & Lauenroth 1990). Estimates of parameters for recruitment and mortality were obtained from field data. Since our interests were in the effects of recruitment and mortality on the dynamics of the two species, we used the same growth rate parameter for both species.

Recruitment

Recruitment by seedlings is an infrequent event for both species. The probability of seedling establishment was estimated to be 0.0140 yr^{-1} for *B. gracilis* (Lauenroth *et al.* 1994) and 0.295 yr^{-1} for *B. eriopoda* (Wright 1972). Although both species are clonal plants, they differ substantially in their

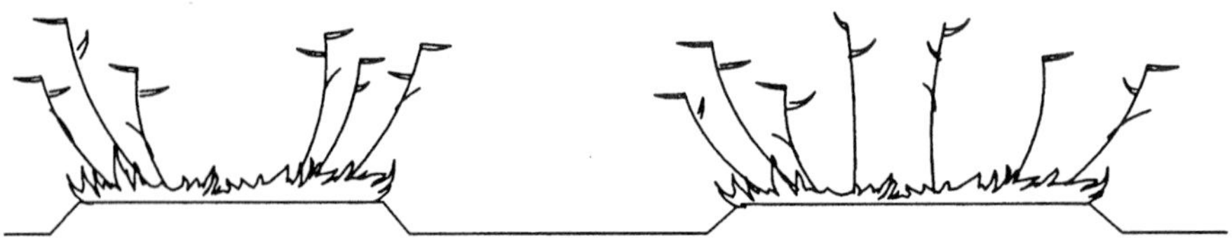

Figure 12.3. *Graphic representation of the plant–interspace mosaic characteristic of arid and semiarid grasslands.*

abilities to spread vegetatively. *Bouteloua gracilis* has a very low capability to expand either plant size or population* size by vegetative spread; the only mechanism *B. gracilis* has is tillering, which occurs very slowly (Samuel 1985). By contrast, *B. eriopoda* is very effective at spreading vegetatively (Wright 1972). Flowering culms have the capability to produce roots at nodes, and under favourable growing conditions a new individual results. The probability of recruitment of new individuals by vegetative spread for *B. eriopoda* was estimated to be 0.5 per year (Wright 1972).

Mortality

The two species also differ substantially in terms of mortality. *B. gracilis* is a very long-lived species, whereas *B. eriopoda* is a relatively short-lived species. Coffin & Lauenroth (1988) evaluated causes of mortality of *B. gracilis* and found that small-scale disturbances were the most frequent cause of death. From their results we estimated that 0.5% of a cohort survives 400 years. Dittberner (1971) investigated the life span of *B. eriopoda* using information from chart quadrats. Based upon his data we estimated that 0.5% of a cohort survives 30 years.

Experimental simulations

The simulation model was run for a 50 m^2 area at each of the two sites (CPER and JER). Since the plot size for the model is 0.1 m^2, 500 individual plots were simulated for each site. On each plot, the model kept track of the size and age of each individual. The soils for the simulated plots were both sandy loams, which are the most frequent soil textures at both sites. Stochastic weather data were generated for each site from historical data. To eliminate the confounding effects of interspecific competition, all species other than the two perennial grasses were eliminated from the model. Intraspecific competition was operational during the simulations.

The model was run to a steady-state biomass for each species and then for an additional 50 years to collect population data for subsequent analyses. Data for each species consisted of the size and age of each individual over the 50 years for the 50 m^2 area. All individuals that achieved less than 10% of maximum size were ignored in the analysis; these were seedlings, most of which only survived one year. The other plants were separated into three categories: juveniles were those individuals greater than 10% of maximum size but less than 50%; adults were those greater than 50% but less than 75% of maximum size; mature individuals were those that achieved more than 75% of maximum size.

---

* Our use of the term *population* refers to a number of physiologically distinct rather than genetically distinct individuals.

Soil carbon and nitrogen

Total soil carbon and nitrogen data from under and between individuals were available from each site. Data for *B. gracilis* were from Burke *et al.* (1995). Cores 5 cm in diameter were collected from 12 locations in the Pawnee National Grasslands in north-eastern Colorado. One of the locations sampled was the CPER. At each site, 24 cores (12 under *B. gracilis* and 12 between) were collected to a depth of 10 cm. The cores were composited in pairs so that each site had 12 independent subsamples (6 under and 6 between). Data for *B. eriopoda* were from R. A. Virginia (unpublished). Three locations were sampled at the JER; at each location, 10 cores each 5 cm in diameter were collected (5 under *B. eriopoda* and 5 between). Samples were collected to a depth of 20 cm and divided in the field into 0–10 cm and 10–20 cm increments. We report data only for the 0–10 cm depth for comparative purposes. Data were statistically analysed using analysis of variance (SAS Institute 1988). The data were analysed separately for the shortgrass steppe and the desert grassland. Each analysis tested for the effects of plant microsite (under and between individual plants) on soil carbon and nitrogen.

Results

Population dynamics

The two species had important differences in both the dynamics and the structures of their populations. Populations of *B. gracilis* had less year-to-year variability ($CV = 1\%$) than those of *B. eriopoda* ($CV = 5\%$) (Fig. 12.4). Over the 50 years of the simulation, most of the individuals of *B. gracilis* that were alive were mature individuals (Fig. 12.4a). Less than 15% of the live individuals were adults and still fewer were juveniles. The pattern for *B. eriopoda* was the exact opposite. Most individuals of *B. eriopoda* were juveniles (Fig. 12.4b). Fewer, but still a substantial number, were adults, and a very small fraction were mature individuals. Average density (number per square metre) of individuals of *B. eriopoda* was almost twice that of *B. gracilis*, despite the fact that simulated basal cover was 3–4 times greater for *B. gracilis* than for *B. eriopoda* (Fig. 12.5). The differences in density reflect the differences in size structure of the two populations, whereas the differences in basal cover are a function of the differences in water availability between the two ecosystem types.

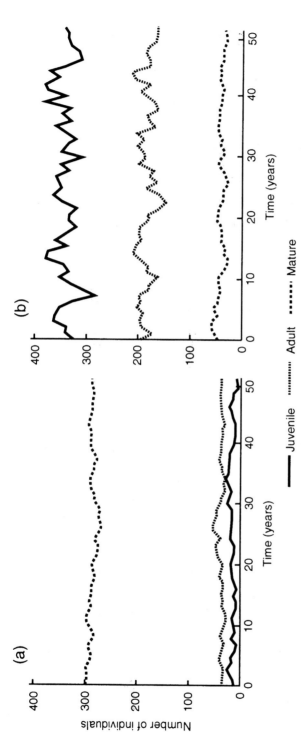

Figure 12.4 *Dynamics of numbers of live plants of* Bouteloua gracilis *(a) and* B. eriopoda *(b) over the 50 simulated years.*

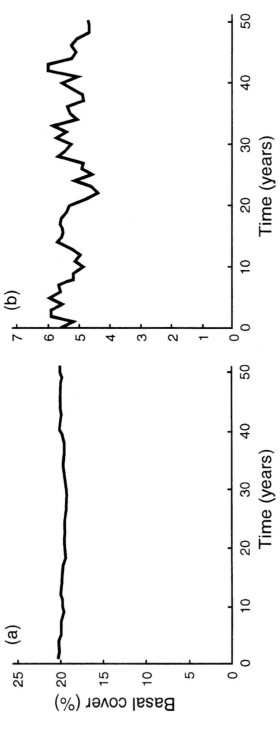

Figure 12.5 *Dynamics of basal cover (%) of Bouteloua gracilis (a) and B. eriopoda (b) over the 50 simulated years.*

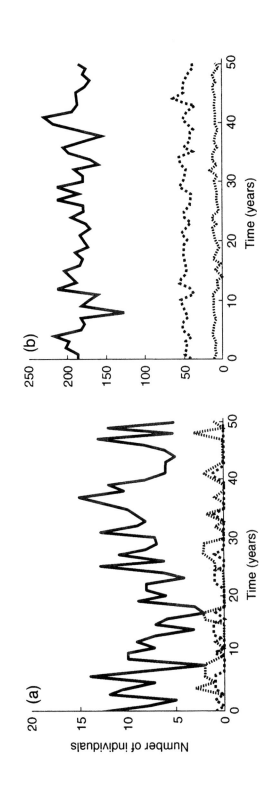

Figure 12.6 *Dynamics of numbers of dead plants of Bouteloua gracilis (a) and B. eriopoda (b) over the 50 simulated years.*

—— Juvenile .... Adult ▪▪▪▪▪▪ Mature

The patterns of plant mortality over the 50 years of simulation reinforce the differences between the two species in live plants. For both populations, most of the plants that died were juveniles (Fig. 12.6). The average number of juvenile deaths per year was 200 individuals for *B. eriopoda* (4 m^{-2}) and 9 for *B. gracilis* (0.18 m^{-2}). This result reflects the large differences between the two species in recruitment. More adults died each year than matured for *B. eriopoda*. This reflects the small numbers of individuals reaching the mature category. Very few adults or mature individuals of *B. gracilis* died during the 50 years of observation.

A summary of average population structure for both live and dead plants makes the differences between the two species clear (Fig. 12.7). The populations of live plants were qualitatively different between the two species. *B. eriopoda* populations consisted almost entirely of juvenile and adult plants. By contrast, populations of *B. gracilis* consisted of mostly large and/or old individuals, with very few juveniles or adults. The size–age composition of dead plants was the same for the two species, with important quantitative differences.

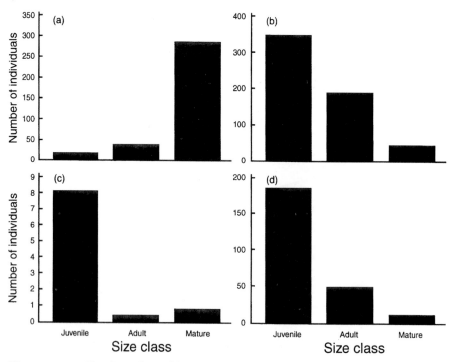

Figure 12.7 *Total numbers of live and dead plants of* Bouteloua gracilis *(a, live; c, dead) and* B. eriopoda *(b, live, d, dead) for the 50 simulated years.*

Soil carbon and nitrogen

In addition to its clear above-ground features, the plant–interspace mosaic structure of arid and semi-arid grasslands is associated with some important below-ground characteristics. The root systems of these two species can be conceptualized as a cone that widens with depth in the soil (Fig. 12.8). This below-ground shape is known for *B. gracilis* from sampling at the CPER (Coffin & Lauenroth 1990; Lee & Lauenroth 1994). Because of morphological similarities among bunchgrasses, we assume the same holds for *B. eriopoda*. Within the area delineated by the cone, root biomass decreases exponentially with depth (Sims *et al.* 1978). The result of these two characteristics is that most of the root biomass of the two bunchgrasses is directly under the plant and within 20 cm of the soil surface.

Shortgrass steppe soils at the CPER had approximately three times more organic carbon and nitrogen than desert grassland soils at the JER (Fig. 12.9). Soils directly below individual *B. gracilis* plants had significantly more carbon and nitrogen than soils from adjacent interspaces. The differences between 'under' and 'between' microsites at the CPER was approximately 200 g C m^{-2} and 20 g N m^{-2}. The C : N ratio of these soils was approximately 10:1.

Soils under and between individuals of *B. eriopoda* at the JER had statistically the same amount of carbon and nitrogen. Numerically, the soils underneath individual plants had more carbon and nitrogen than those under

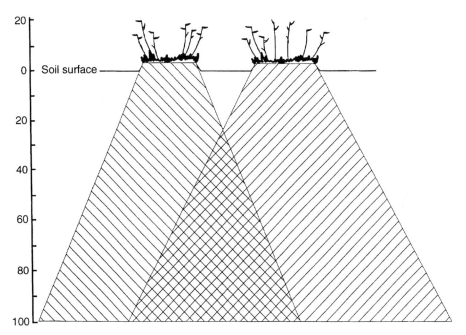

Figure 12.8 *Graphic representation of the plant–interspace mosaic, emphasizing the relationship between above- and below-ground plant biomass.*

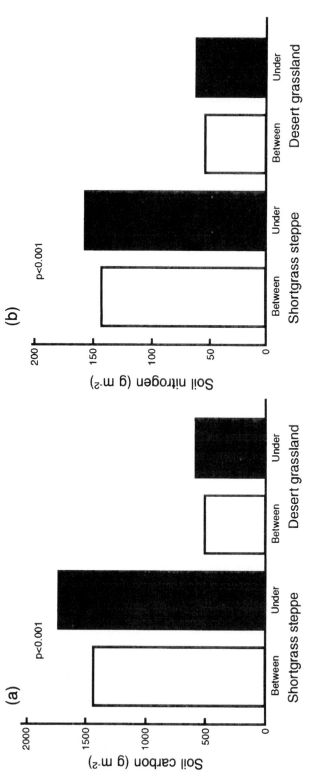

Figure 12.9 *Amounts of soil carbon and soil nitrogen under and between plants at the shortgrass steppe (a) and desert grassland (b) sites.*

interspaces, suggesting that the same spatial processes were operating at both sites.

Discussion

Similarities in physiology (C4 photosynthesis), life span (perennial) and morphology (bunchgrass) argue for placing *B. gracilis* and *B. eriopoda* into the same functional type: C4 perennial bunchgrass. Furthermore, the fact that they are each dominant species and key forage plants in their respective grasslands provides additional temptation to generalize about their roles in North American grasslands. What we have demonstrated in this analysis is that, despite their similarities, demographic differences lead to large and important differences between the species with respect to their population dynamics and in their effects on ecosystem properties.

Bouteloua gracilis is the dominant species throughout the shortgrass steppe (Lauenroth & Milchunas 1992). At the CPER, *B. gracilis* is long-lived and recruits infrequently (Coffin & Lauenroth 1988; Lauenroth *et al.* 1994). Consequently, stands of *B. gracilis* are largely composed of very old individuals (Fig. 12.7a). Is 400 years a good estimate of maximum life span? With the small amount of data available on the topic, we have to admit that we don't know. We are quite confident that *B. gracilis* lives a very long time, but whether that means 150, 200 or 400 years, we cannot be certain. Based upon what we know about recruitment (Lauenroth *et al.* 1994) and growth rates, we can estimate the minimum life span that *B. gracilis* must achieve to maintain current numbers and sizes of plants observed at the CPER under the current disturbance regime.

Evaluation of the relationship between life span and steady-state biomass ($g\ m^{-2}$) was accomplished by running the simulation model for a range of different mortality rates. The model was run to steady-state (450 years) and then for 50 additional years to collect the data for each mortality rate. Results for the final 50 years were averaged to come up with a single estimate of either steady-state biomass or steady-state numbers of individuals in each of the three size–age classes. The results are expressed in terms of maximum life span. The relationship between maximum life span for *B. gracilis* and steady-state biomass suggested maximum sensitivity for life spans of less than 200 years (Fig. 12.10). Density of the three size–age classes of individuals supported the biomass results (Fig. 12.11). The largest decreases in the density of mature individuals and adults occurred for life spans of less than 200 years. The maximum increase in juveniles coincided with the decrease in mature individuals and adults, suggesting

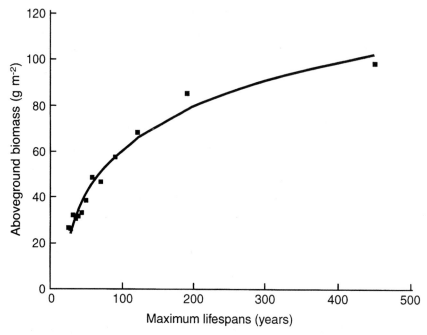

Figure 12.10   *Relationship between the maximum age of individuals of* Bouteloua gracilis *and the simulated steady-state above-ground biomass.*

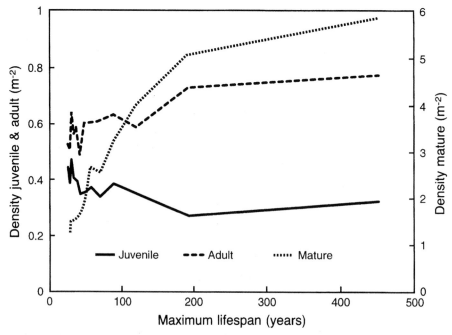

Figure 12.11   *Relationship between the maximum age of individuals of* Bouteloua gracilis *and the simulated steady-state density (individuals per square metre).*

the importance of competition by established individuals in the survival of seedlings (Aguilera & Lauenroth 1993). Our conclusion from this analysis is that our results for *B. gracilis* are not very sensitive to our estimate of maximum life span. Reducing our estimate by 50% would have a very small effect on the results.

B. eriopoda is the dominant species of the grassland phase of the vegetation in the desert grassland region (Sims *et al.*, 1978; Pieper *et al.* 1983). Stands of *B. eriopoda* at the JER are composed primarily of very young individuals (juveniles). Mature individuals are encountered very infrequently (Fig. 12.7b). This size–age structure is the consequence of high rates of mortality and frequent recruitment by vegetative spread.

The differences in demography between these two species of *Bouteloua* appear to have had important effects on the spatial patterns of storage of soil carbon and nitrogen in the two grasslands. Populations of long-lived individuals of *B. gracilis* are associated with strong spatial patterning of both carbon and nitrogen at the scale of individual plants. By contrast, the populations of short-lived individuals of *B. eriopoda* are associated with a generally uniform distribution of soil carbon and nitrogen. Are there processes by which individual plants can influence the accumulation of carbon and nitrogen in soils? The answer with respect to carbon is very straightforward (Fig. 12.12). Plants fix atmospheric carbon, which gets translocated to above- and below-ground organs. Since the biomass ratio of an individual bunchgrass is in the range of 10–20 g of below-ground material per gram of above-ground material, most of the fixed carbon is being deposited directly in the soil. Furthermore, since most of the roots of individual bunchgrasses are found directly beneath the plants and within 20 cm of the soil surface (Coffin & Lauenroth 1991; Lee & Lauenroth 1994), it is not surprising that plants have the ability to modify the amount of carbon stored in the soil. The explanation for how plants accumulate significant quantities of soil nitrogen beneath them is in many ways a consequence of the deposition of carbon. Since carbon and nitrogen are found in plant material in a relatively constant proportion (C : N ratio), accumulation of carbon necessarily results in the accumulation of nitrogen, even though some nitrogen may be extracted from senescing organs. We postulate that the source of the additional nitrogen is the interspace area (Fig. 12.12). The balance of uptake, distribution to plant organs, and deposition in dead material favours a movement of nitrogen from the areas between individual plants, resulting in plants occupying relatively nitrogen-rich patches in a background of nitrogen-poor soil. In the shortgrass steppe, this means that plants occupying 25–35% of the soil surface are depleting the remaining 65–75% of the area.

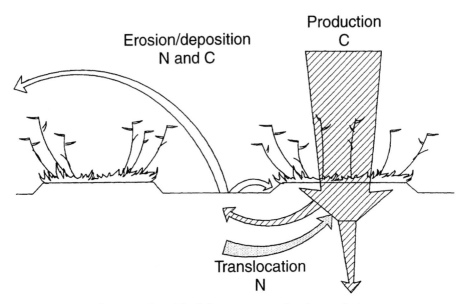

Figure 12.12   *Conceptual model of the movement of carbon and nitrogen under and between individual bunchgrasses.*

The explanation for why *B. gracilis* has a significant effect on soils while *B. eriopoda* does not may be related to their longevity. Although both are perennials, *B. eriopoda* is relatively short-lived compared with *B. gracilis. B. eriopoda* does not occupy a location long enough to have a large effect on carbon or nitrogen storage. This explanation is supported by the fact that the data for *B. eriopoda* showed a non-significant trend towards more carbon and nitrogen beneath than between plants. This suggests that if they lived long enough the effects would accumulate. How long does a bunchgrass need to occupy a location in order to influence its carbon and nitrogen status? The answer may be different for the two grasslands. The semi-arid shortgrass steppe has three times more soil carbon and nitrogen than the arid desert grassland. It is possible that an individual of *B. gracilis* can influence the soil beneath it in a shorter amount of time because of greater water availability which would support higher activity rates for a longer period of time each year. Burke *et al.* (1995) reported, from studies of abandoned agricultural fields, that the small-scale pattern in carbon and nitrogen associated with individual plants of *B. gracilis* could be established in as little as 50 years. Since essentially no individuals of *B. eriopoda* live that long, it should not be surprising that the results for the desert grassland are different from those for the shortgrass steppe.

Conclusions

The differences in demography between two species of plant of the same 'functional type' confer critical differences in ecosystem structure and dynamics. These attributes should result in significant differences in how these two ecosystems respond to disturbances or to climate change. The dynamic populations of *B. eriopoda* in the desert grassland with high recruitment and high mortality rates may recover very rapidly from short-term pulse disturbances that result in the death of individuals. They will also be very susceptible to any environmental change that interferes with recruitment. If climate change alters the vegetative recruitment of individuals, desert grasslands dominated by *B. eriopoda* should rapidly change to dominance by another species. A controversy exists in the desert grassland region about the cause of a well-documented change in the vegetation over the past century (Buffington & Herbel 1965; Gibbens & Beck 1988). Two forces, climate change and domestic livestock grazing, have been postulated as the causes of the change from a grassland vegetation dominated by *B. eriopoda* to a shrubland vegetation dominated by either *Larrea tridentata* (creosote bush) or *Prosopis glandulosa* (mesquite) (Neilson 1986; Schlesinger *et al.* 1990). The results of our simulations do not provide any information to help distinguish between these alternative explanations. Either climate change or long-term grazing could result in decreasing recruitment of *B. eriopoda* and rapid change to a different vegetation. Wright & Van Dyne (1976, 1981) reported results from both field data and simulations that concurred that the major effect of livestock grazing on *B. eriopoda* was a decrease in recruitment.

The long-lived populations of *B. gracilis* in the shortgrass steppe with very low recruitment and mortality rates should recover more slowly than the desert grassland from short-term pulse disturbances but change more slowly as a result of long-term forces such as livestock grazing or climate change. Evidence from short-term pulse disturbance experiments supports the idea of slow recovery by *B. gracilis* (D. P. Coffin & W. K. Lauenroth, unpublished). The shortgrass steppe has not undergone a change in vegetation comparable to that in the desert grasslands. All evidence suggests that these ecosystems dominated by *B. gracilis* are very resistant to change by domestic livestock grazing (Milchunas *et al.* 1989; Lauenroth *et al.* 1994) and major droughts (Weaver & Albertson 1956).

The consequences of disturbances for carbon and nitrogen balance will be different for the two ecosystems. In the desert grassland, disturbances that result in the mortality of individual *B. eriopoda* plants will increase the vulnerability of the system to losses of carbon and nitrogen through erosion, but losses of material from directly under the plant will have the same significance as losses of material from between plants. In the shortgrass steppe, disturbances that cause mortality of *B. gracilis* will also increase the vulnerability of the system to losses, but the material directly under plants represents a more important loss of both carbon and nitrogen than material from between plants. Because of the slight (2–3 cm) elevation of the soil surface associated with individual plants (Fig. 12.3), this material is very susceptible to redistribution by wind when the individual dies. Large-scale disturbances could result in the removal of a large quantity of carbon and nitrogen from the affected area.

The concept of functional types is important and will be absolutely necessary if ecologists are to be able to cut through the enormous complexity of the biosphere to understand, in a general sense, how ecosystems will respond to global change. Our analysis provides evidence that classification of plants into functional types will depend upon the question that is being asked, the spatial extent of the analysis, and, in some cases, information that goes well beyond that which has been used in the past (e.g. functional groups; Sims *et al.* 1978). In the case of desert grasslands and the shortgrass steppe in North America, combining *B. gracilis* and *B. eriopoda* into a single functional type, i.e. C4 perennial bunchgrass, would obscure important differences in the way the two ecosystems will respond to climate change. At scales up to regional such a distinction will be very important. At continental and global scales the differences may not be significant given all of the other sources of variability in the analysis.

Acknowledgements

The Central Plains Experimental Range and the Jornada Experimental Range are administered by the USDA Agricultural Research Service and both are National Science Foundation funded Long Term Ecological Research sites. Support for this work was provided by grants from the National Science Foundation (BSR 90-11659) and the Colorado State Experiment Station (1-50661). We thank Paul Hook for figures 12.3 and 12.12.

References

Aguilera, M. O. & Lauenroth, W. K. (1993) Seedling establishment in adult neighbourhoods: intraspecific constraints in the regeneration of the bunchgrass *Bouteloua gracilis*. *Journal of Ecology*, 81, 253–61.

Box, E. O. (1981) *Macroclimate and Plant Forms: An Introduction to Predictive Modelling in Phytogeography*. The Hague: Junk.

Buffington, L. C. & Herbel, C. H. (1965) Vegetation changes on a semidesert grassland range. *Ecological Monographs,* **35,** 139–64.

Burke, I. C. (1989) Control of nitrogen mineralization in a sagebrush steppe landscape. *Ecology,* **70,** 1115–26.

Burke, I. C., Lauenroth, W. K. & Coffin, D. P. (1995) Soil organic matter recovery in semiarid grasslands: implications for the Conservation Reserve Program. *Ecological Applications,* **5,** 793–801.

Charley, J. L. & West, N. E. (1977) Plant-induced soil chemical changes in some shrub-dominated semi-desert ecosystems of Utah. *Journal of Ecology,* **63,** 945–64.

Coffin, D. P. & Lauenroth, W. K. (1988) The effects of disturbance size and frequency on a shortgrass plant community. *Ecology,* **69,** 1609–17.

Coffin, D. P. & Lauenroth, W. K. (1990) A gap dynamics simulation model of succession in a semiarid grassland. *Ecological Modelling,* **49,** 229–66.

Coffin, D. P. & Lauenroth, W. K. (1991) Effects of competition on spatial distribution of roots of blue grama (*Bouteloua gracilis* (H.B.K.) Lag. ex Griffiths). *Journal of Range Management,* **44,** 68–71.

Dittberner, P. L. (1971) *A demographic study of some semidesert grassland plants.* Masters Thesis, New Mexico State University, Las Cruces, NM. 77 pp.

Gibbens, R. P. & Beck, R. F. (1988) Changes in grass basal area and forb densities over a 64-year period on grassland types of the Jornada Experimental Range. *Journal of Range Management,* **41,** 186–92.

Hook, P. B., Burke, I. C. & Lauenroth, W. K. (1991) Heterogeneity of soil and plant N and C associated with individual plants and openings in the North American shortgrass steppe. *Plant and Soil,* **138,** 247–56.

Hook, P. B., Lauenroth, W. K. & Burke, I. C. (1994) Spatial patterns of plant cover and roots in a semiarid grassland: Abundance of canopy openings and regeneration gaps. *Journal of Ecology,* **82,** 485–94.

Lauenroth, W. K. & Milchunas, D. G. (1992) The shortgrass steppe. In Coupland, R. T. (ed.), *Ecosystems of the World, Vol 8a,* *Natural Grasslands,* pp. 183–226. Amsterdam: Elsevier Scientific Press.

Lauenroth, W. K., Milchunas, D. G., Dodd, J. L., Hart, R. H., Heitschmidt, R. K. & Rittenhouse, L. R. (1994) Effects of grazing on ecosystems of the Great Plains. In Vavram, M. & Laycock, W. A. (eds), *Ecological Implications of Livestock Herbivory in the West,* pp. 69–100. Denver, Colorado: Society for Range Management.

Lauenroth, W. K. & Sala, O. E. (1992) Long term forage production of North American shortgrass steppe. *Ecological Applications,* **2,** 397–403.

Lauenroth, W. K., Sala, O. E., Coffin, D. P. & Kirchner, T. B. (1994) Establishment of *Bouteloua gracilis* (H.B.K.) Lag. ex Griffiths in the shortgrass steppe: A simulation analysis. *Ecological Applications,* **4,** 741–9.

Lee, C. A. & Lauenroth, W. K. (1994) Spatial distributions of grass and shrub root systems in the shortgrass steppe. *American Midland Naturalist,* **132,** 117–23.

Milchunas, D. G. & Lauenroth, W. K. (1989) Three-dimensional distribution of vegetation in relation to grazing and topography in the shortgrass steppe. *Oikos,* **55,** 82–6.

Milchunas, D. G., Lauenroth, W. K., Chapman, P. L. & Kazempour, M. K. (1989) Effects of grazing, topography, and precipitation on the structure of a semiarid grassland. *Vegetatio,* **80,** 11–23.

Neilson, R. P. (1986) High-resolution climatic analysis and southwest biogeography. *Science,* **232,** 27–34.

Pieper, R. D., Anway, J. C., Ellstrom, M. A., Herbel, C. H., Packard, R. L., Pimm, S. L., Raitt, R. J., Staffeldt, E. E. & Watts, J. G. (1983) *Structure and Function of North American Desert Grassland Ecosystems.* New Mexico State University Agricultural Experiment Station Special Report 39.

Prentice, I. C., Cramer, W., Harrison, S. P., Leemans, R., Monserud, R. A. & Solomon, A. M. (1992) A global biome model based on plant physiology and dominance, soil properties and climate. *Journal of Biogeography,* **19,** 117–34.

Raunkiaer, C. (1934) *The Life Forms of Plants and Statistical Plant Geography.* (Translated

by Carter, Fausboll and Tansley.) Oxford
University Press.

Samuel, M. J. (1985). Growth parameter differences between populations of blue grama. *Journal of Range Management*, **38**, 339–42.

SAS Institute (1988) *SAS/STAT User's Guide*, Release 6.03 edn. Cary, North Carolina: SAS Institute.

Schlesinger, W. H., Reynolds, J. F., Cunningham, G. L., Huenneke, L. F., Jarrell, W. M., Virginia, R. A. & Whitford, W. G. (1990) Biological feedbacks in global desertification. *Science,* **247**, 1043–8.

Sims, P. L., Singh, J. S. & Lauenroth, W. K. (1978) The structure and function of ten western North American grasslands: I. Abiotic and vegetational characteristics. *Journal of Ecology*, **66**, 251–85.

Weaver, J. E. & Albertson, F. W. (1956) *Grasslands of the Great Plains*. Lincoln: Johnsen Publ. Co.

Wright, R. G. (1972) Computer processing of chart quadrat maps and their use in plant demographic studies. *Journal of Range Management*, **25**, 476–8.

Wright, R. G. & Van Dyne, G. M. (1976) Environmental factors influencing semidesert grassland perennial grass demography. *The Southwestern Naturalist*, **21**, 259–74.

Wright, R. G. & Van Dyne, G. M. (1981) Population age structure and its relationship to the maintenance of a semidesert grassland undergoing invasion by mesquite. *The Southwestern Naturalist*, **26**, 13–22.

13 Plant functional types in African savannas and grasslands

R. J. Scholes, G. Pickett, W. N. Ellery and A. C. Blackmore

Introduction

We define a plant functional type (PFT) as a set of co-varying parameter ranges for plant attributes related to resource acquisition, growth, reproduction, dispersal and response to environmental stress. For a plant to belong to a particular functional type, its attributes must fall within the defined range for all the required parameters. The PFT concept differs from more traditional ways of classifying plants in that it groups together those plants with similar ecological properties, rather than those plants which necessarily look similar, or have similar evolutionary origins (Menaut & Noble 1988). The functional type concept has been implicit in both folk and scientific taxonomies for a long time (e.g. Knight & Loucks 1968), but has recently gained renewed emphasis and respectability by the need to predict the response of plants to substantial and rapid environmental change at the global scale. The conventional botanical typology, based on inferred genetic and evolutionary relationships, is ill-suited to this purpose since it has too many taxa at the species level to be applied at a continental or global scale. The more aggregated levels of the taxonomy, such as the family, contain little explicit ecological information.

There are observational and theoretical reasons to believe that PFTs exist beyond the fact that they have been defined to exist. Any given environment usually contains several plants with widely different attributes, showing that there is seldom a unique solution to a given set of environmental challenges. This is consistent with the finding that complex, non-linear, highly linked systems (such as plant metabolism) have multiple stable states. However, the number of growth forms and life histories is not infinite: in fact, it is usually substantially less than the number of species. This is because not all combinations of attributes are feasible, or viable in the presence of competing or predating organisms. Plants in a given environment are faced with the same constraints and opportunities for trade-offs. They therefore tend to converge on a limited number of attribute package deals, sometimes referred to as 'strategies' or 'functional guilds' (Grime 1979; Noble & Slatyer 1980).

This chapter reviews the research that has been done on plant functional type classifications in the savannas and grasslands of southern Africa. On the basis of this work, we propose a functional typology for African savannas and grasslands. The typology may be applicable to grasslands and savannas on other continents, but our poor knowledge of non-African vegetation does not permit us to claim confidently that it does.

Approaches to functional type classifications

There are two fundamental approaches to identifying PFTs. Environment-based, or 'top-down', approaches identify the main environmental determinants of plant performance or presence, and then segment determinant axes into ranges. This segmentation need not be arbitrary: it can be as a result of considering the observed limits of species. The PFTs are then the factorial combinations of ranges on the various axes. Individual plants are slotted into a pre-existing category, according to their preferred range in nature. Some combinations may remain unoccupied since they are biologically unfeasible. The difficulty with this approach lies in identifying the environmental determinants in a useful and predictive way, now and for future conditions. The factorial nature of this approach can rapidly generate more PFTs than there are species, which rather defeats the purpose of the exercise.

The plant-based, or 'bottom-up', approach identifies plant attributes thought to be important in determining growth, reproduction and survival, and quantifies them for a number of species. This attribute-by-species matrix is then searched for covarying clusters, typically by applying sophisticated multivariate techniques. These clusters are defined as PFTs. The problem with this approach is that attributes need to be identified which are simultaneously sensitive, predictive and measureable. Unless *a priori* simplifying assumptions are made, a very large set of difficult-to-measure attributes needs to be accumulated. Addition of a new species after the initial classification may require that a new class be created to accommodate it.

We have experimented with both of these approaches in the savannas and grasslands of southern Africa, and summarize our findings in this chapter. The working typology that we propose is a combination of both approaches plus the experiences, insights and operational practices of researchers and land managers in the region. It could be termed a 'pragmatic' classification, since it has a specific purpose in mind: the management of savanna and grassland vegetation for human benefit in a changing environment.

Vegetation functional types

Vegetation functional types (VFTs) can be loosely defined as those areas of the vegetated land surface which have similar ecological attributes, such as composition in terms of plant functional types, structure (i.e. distribution of leaf area with height), phenology (i.e. distribution of leaf area over time), and potential biomass and productivity. The concept of vegetation functional types corresponds closely to the concept of a biome, and is closely related to the concept of plant functional types. An operational question is: should VFTs be defined top-down (as biomes have been in the past) or bottom-up, as the consequences of the coincidence of PFTs? Many studies have shown biome boundaries to correlate closely with climate indices, but is it reasonable to consider a biome to have climate tolerances in the same sense that PFTs have environmental tolerance ranges? This dilemma, in various forms, has always bedevilled plant community concepts. Are there emergent properties at the community level which are more than just the aggregate behaviours of the individual species? The pragmatic view is that as long as plants tend to co-occur (i.e. if their environmental tolerance limits tend to coincide) then it is useful and parsimonious to give a name and attribute ranges to that composite entity. There are several mechanisms that could cause the realized or fundamental tolerance limits of coexisting plants (or PFTs) to cluster, rather than to spread randomly along an environmental axis. Biophysical constraints, such as the freezing point of water, lead to natural discontinuities in some axes. Plants alter their local environment and may therefore maintain it within the environmental range of other plants. Finally, interactions between taxa may require or preclude coexistence.

African vegetation ecologists have generally taken a pragmatic view, and described communities and biomes even though they may believe that plants are fundamentally individualistic. We believe that there is evidence that PFTs are not independently distributed, and therefore VFT concepts such as the biome have an objective as well as a subjective reality. For instance, many of the trees of the savanna and forest biomes can grow successfully within the climatic boundaries of the grassland biome, but do not unless fire is excluded. Regular, hot fires are only possible because of the accumulation of grass fuel. This interaction means that the distribution of trees is not independent of the distribution of grasses. The observed grassland VFT boundary will not emerge spontaneously from the PFT distributions of trees and grasses unless this interaction is considered.

Environment-based approaches

Gross morphology has often been used as a surrogate for hard-to-measure functional attributes. One of the most successful schemes is that of Raunkiaer (1934), which is based on the position of the perennating organ (see reviews by Adamson 1939 and Cain 1950). The scheme originated in temperate and boreal vegetation, where the main environmental drivers were proposed to be survival during periods of low temperature, and competition for light at other times. It works quite successfully in more tropical vegetation (e.g. Rutherford & Westfall 1986), for different reasons. To survive fires, the perennating organ must either be below the soil surface, or above the flame zone. To survive mammalian herbivores, it must similarly be too low or too high to be eaten. Ellery *et al.* (1991) began with the assumption that the biomes of southern Africa, as defined by Rutherford & Westfall (1986) based on life-form spectra, represented meaningful functional types. They then developed a decision-tree model, which successfully predicted the current distribution of the biomes, based on climate data. They found that at least three climatic axes were needed, although they could be calculated in a number of different ways: a measure of water availability to plants, of temperature during the growing period, and of temperature during the non-growing period.

Grasslands

The fundamental distinction observed by southern African graziers is between 'sweet' and 'sour' *veld* (a colloquial term for rangeland). Cattle grazing 'sweet' grasslands gain mass all year round, but cattle on 'sour' grasslands lose mass in winter, when the protein in the dry foliage drops below the threshold needed to sustain ruminant digestion. Ellery (1992) developed a hypothesis that relates the occurrence of sweet and sour grasslands to the carbon : nutrient balance of the grasses. The hypothesis is that this balance is in turn controlled by a complex (but predictable) interplay of soil, water availability and parent material. Thus dry, warm sites favour nutrient assimilation over carbon assimilation, and are 'sweet'; the reverse is true for cool, wet sites, which are 'sour'. These climatic generalizations can be overridden by differences in soil fertility: fertile soils are 'sweeter' than infertile soils under the same climate (Fig. 13.1).

According to this analysis, 'sweetness' and 'sourness' are fundamental properties of the environment rather than of the species. In support of this view, fertilization, or removal of foliage by grazing or fire, results in 'sour' grasses becoming 'sweeter'. However, with some possible exceptions,

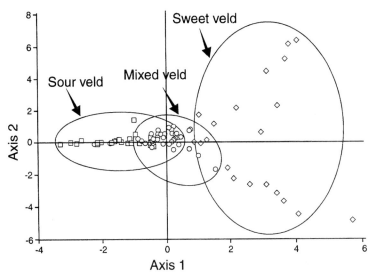

Figure 13.1 *The first two axes of a detrended correspondence analysis of species composition of South African grasslands, with the local grazing-preference categories of 'sweet' and 'sour' veld overlaid. The first axis is associated with aridity (wet on the left, dry on the right) and the second with temperature (cool at the bottom, warm on top).*

distinct suites of species are associated with 'sweet' and 'sour' environments. We argue that this pattern of species fidelity is the result of the interaction between competition between grasses and selective herbivory in different environments.

A weakness of top-down approaches is the need to decide, *a priori*, what the key environmental determinants are. This makes the classification very sensitive to changes in theory. For instance, a functional classification of grasslands which is still reflected in the vegetation map of South Africa is the distinction by Acocks (1975) between 'pure' and 'false' grasslands. 'Pure' grasslands are held to be a climatic climax, and 'false' grasslands are seral to forest, maintained in sub-climax condition by repeated human-induced fire disturbance. The distinction between these types has been increasingly questioned (Feely 1987; Ellery & Mentis 1992). Fire is now regarded as an intrinsic factor in grassland ecology, rather than an imposed disturbance. Grasslands and their fire regimes predated the arrival of pastoral inhabitants in South Africa. All grasslands become woodier as fires are excluded. The wetter 'pure' grasslands merely do so faster and more completely.

Savannas

The IUBS Responses of Savannas to Stress and Disturbance programme (Frost *et al.* 1986) proposed four key determinants of savanna form and function: water availability, nutrient availability, fire and herbivory. Water and nutrients are 'primary determinants' in the sense that they constrain the expression of fire and herbivory. In reality, each of these determinants is itself composed of a number of axes. For example, water availability has at least a magnitude and a seasonality. Fire has frequency, intensity and seasonality components. Herbivory has at least an intensity and type (type distinguishes between mammals and invertebrates, browsers and grazers, and different grazing feeding behaviours) and can also have frequency and seasonality, for instance in the case of migratory herbivores. 'Nutrients' includes nitrogen and phosphorus at a minimum, and potentially up to 20 other essential and toxic elements. Thus even this simplified scheme has at least nine axes. If each axis has a minimum of two levels (high and low), there are 2^9 (512) potential PFTs.

We simplified this array by working only with the primary determinants, water and nutrients, and limited them to a single dominant axis each. For water availability, this axis was defined as the mean number of days per year when soil water availability was sufficient to permit plant growth at its maximum potential rate. This was estimated by using a simple water balance model based on long-term mean monthly rainfall and evaporative demand. The model accounts for some seasonal effects, for instance the greater effectiveness of rain falling in the cooler times of the year. The remaining seasonal effects were ignored on the assumption that all the savannas and grasslands we were dealing with had a similar seasonal pattern of water availability (wet in the summer and dry in the winter). Nutrient availability was combined into a single value ('fertility') using a variety of approaches (Pickett 1994). A simple 'expert system' model based on parent material, landscape age and profile depth was found to have essentially the same capacity to predict grass growth as more complex schemes for integrating multiple-element soil analyses. Soils were categorized into five fertility classes, ranging from 'very low' through 'low', 'average' and 'high' to 'very high'.

Pickett (1994) searched for VFTs within the space defined by the availability of water and nutrients. The approach consisted of grouping proposed PFTs into VFTs, and then seeing if they occupied predictable sectors of the environmental space. Structural and functional attributes of the woody (125 spp) and graminoid (78 spp) components of 19 semi-arid savanna communities were collected in the field, from the literature, and from discussion with experienced field personnel. The choice of structural, vegetative and life history attributes was based on features of the vegetation currently thought to reflect the way savannas function (Huntley 1982; Frost *et al.* 1986). A combination of the top-down and bottom-up approaches was used. The attribute set for a community consisted of the attributes of its component species, weighted by their dominance in the community. A sites-by-attributes matrix was ordinated and classified, for single attributes and for combinations of attributes. The woody plant and graminoid components of the vegetation were considered separately, to prevent their obvious differences from completely dominating the ordination procedure.

The 19 savanna woody plant communities were grouped into two broad (and partly overlapping) VFT groups, and eight VFT subgroups (Fig. 13.2) on the basis of similarities and differences in the vertical distribution of browse volume, palatability, spinescence, and the predominant leaf type and leaf size of the vegetation. One super-group contains plants with compound leaves with leaflets smaller than 5 mm, highly spinescent and palatable. This group occurs only on fertile soils where the water availability in this context is still relatively arid in a global sense. The other super-group contains plants with larger leaves, either compound or simple, non-spinescent and unpalatable, and occurs on infertile soils. There is a trend towards decreasing palatability with increasing water supply. The overlap between super-groups occurs in the low-nutrient, low-water part of the environmental space. We suggest that this is because in highly water-constrained environments, plants have a limited capacity (or demand) for nutrients. Soils that would be infertile in other contexts therefore have an adequate nutrient supply in this context. Within both groups there is a tendency towards taller canopies with increasing water supply.

The woody plant VFTs (especially A,C,D and F) reflect trends that have been predicted for savanna structure and function in relation to available water and nutrients (Huntley 1982; Frost *et al.* 1986), for instance to increasing height and biomass with increasing water availability. VFTs B,E and H are exceptions to the general trends. The graminoid components of the savanna communities did not group into VFTs in the same way as the woody plants. The ecological meaning of the graminoid VFTs was poorly defined in comparison with the woody plant VFTs. The woody plant component, because of its long response time, probably reflects the average water and nutrient conditions of a site better than the grass component, which has a short response time and is susceptible to erratic spatiotemporal fluctuations of, for example, moisture supply and disturbance.

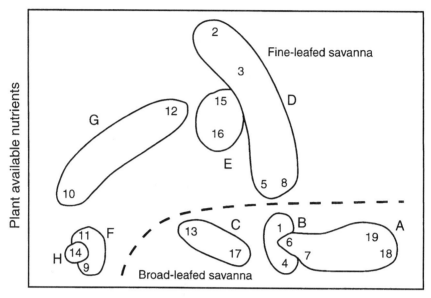

Figure 13.2 *Vegetation functional types in South African savannas. The dashed line separates the two 'supergroups' described in the text. The solid lines enclose subgroups.*

Supergroups 1: broad-leafed savannas. A: large, simple-leaved, unpalatable non-spinescent, tall. B: as A except abundance of moderately to highly palatable species and intermediate spinescence. C: as A except predominance of palatable species and slightly higher occurrence of compound leaves.

Supergroup 2: fine-leafed savannas. D: tall, compound-leaved, highly palatable, intermediate to high spinescence, most leaf above 2 m. E: similar to D except for intermediate-sized leaves and low to intermediate spinescence. F: small, compound-leaved, highly palatable and spinescent, low stature.

Overlap region. G: small, predominantly simple-leaved, low or high palatability, low spinescence, low stature. Most of the browse volume in these communities was concentrated below 1.5 m. H: similar to G except predominantly large leaves and moderately palatable species.

Plant-based approaches

Grasslands

Historically, a combination of floristic, habitat, structural and functional criteria have been used to distinguish vegetation types within the grassland biome in South Africa. Ellery (1992) recorded the species composition in 101 South African grassland sites in a wide range of environments. He scored each species with respect to 18 attributes (Table 13.1). The importance of each attribute class at each site was then calculated according to the fraction of the grass biomass at the site which exhibited the character.

Table 13.1 *The attributes considered important for a functional classification of southern African grasses*

Attributes	Classes
Phenology	annual, biennial, perennial, facultative perennial
Photosynthetic pathway	C_3, C_4 NADP–ME, C_4 NAD–ME, C_4 PCK, C_3 and C_4
Growth form	tufted, stoloniferous, rhizomatous, all forms, tufted stoloniferous, tufted rhizomatous
Reproduction	obligate vegetative, obligate seeder, vegetative and sexual
Seed size	small, medium, large
Seed number	many, average, few
Seed dispersal	awned, non-awned, barbed/hooked, plumed
Leaf pubescence	glabrous, slightly hairy, moderately hairy, densely hairy
Leaf position	strongly basal, moderately basal, few raised, many raised
Height	0–100, 100–500, 50–1000, >1000 mm
Flowering time	early, early-middle, middle, middle-late, late, early–middle–late
Leaf length	<100, <250, <500, <1000 mm
Leaf shape	expanded, slightly folded, strongly folded, slightly rolled, strongly rolled
Leaf width	<3, <5, <8, <12, <20 mm
Defence	none, growth form, wiry, terpenes, tannins, 'mtshiki', other
Woodiness	absent, slight, moderate, strong
Geophytism	absent, slight, moderate, strong
Palatability	unpalatable, 'sour', intermediate, 'sweet', mtshiki

The sites-by-species and sites-by-attributes matrices were subjected to Detrended Correspondence Analysis (DCA).

Most of the variance in the sites-by-species was accounted for by the first two axes. The first axis was associated with water availability and fire frequency (which are highly positively correlated), and the second axis with temperature during the growing season. The first two DCA axes of the sites-by-attributes matrix accounted for less variance. Defence, flowering time, reproduction and growth form were the characters best separated on the first two axes.

Wetter sites tend to have tufted, moderately to strongly geophytic, obligately sexual, large- and few-seeded, early- or prolonged-flowering grasses, while arid sites had rhizomatous or stoloniferous grasses, strongly woody-based, small- and many-seeded and late-flowering with an NAD–ME photosynthetic pathway. Cool sites had more C_3 species; hot sites favoured the

PCK variant of C4. Tannin-based defences were on the wetter sites, and the 'sweet' grasses on the drier sites.

Savannas

Blackmore (1992) measured a range of ecophysiological parameters on savanna plants at 15 sites in South Africa. The parameters that proved useful in discriminating differences between plant types were leaf morphology (sclerophylly, succulence and turgid : dry mass ratio were found to be highly correlated), leaf size and photosynthetic rate parameters (maximum photosynthetic rate and dark respiration were correlated). Intraseasonal and post-disturbance variation in parameters, particularly in the photosynthetic attributes, presents a practical limitation to this approach. For grasses, lamina breaking strength was predictive, but also showed intraseasonal variability. Photosynthetic quantum efficiency, light compensation point, and bark thickness were not useful. It was not easy to find water relations parameters that were both predictive and practicable to measure in the field on large numbers of samples.

Among the grass plants, most of the ecophysiological attributes were related to the water availability environmental axis, whereas there was not a clear pattern for either water or nutrients among the woody plants. On the other hand, ordination of the attribute–species matrix for woody plants revealed clusters that were at least intuitively reasonable; this was not true for grasses. In the field, there were not great differences in the maximum photosynthetic rates of C3 and C4 plants, and other physiological parameters also did not separate along C3–C4 lines.

It was concluded that the development of a PFT classification for savannas based on ecophysiological measurements was still a long way off. This is partly due to our incomplete understanding of the ecophysiology of most savanna plants and the difficulty of measuring the parameters, but also because of the predominance of weak, but pervasive intercorrelations between parameters.

A pragmatic proposal

We offer the following PFT typology for southern African savannas and grasslands, as a working hypothesis (Table 13.2). Its purpose is to capture the main functional attributes of these rangelands, for the purpose of making regional-scale generalizations. Many species that grow in southern Africa will find no appropriate place in our scheme: this is because under current conditions they do not make a significant contribution to biomass, primary or secondary production. We propose these types largely based on our experience and on conventional wisdom in the region, rather than as the direct product of a rigorous survey process.

Conclusions

The purpose of the studies described above was to see if there were emergent structural and/or functional properties of savanna and grassland plants and communities that reflected current environmental conditions. At a very general level of detail (about 17 PFTs for the entire region) it is possible to make reasonably robust predictions. This level of resolution is probably appropriate to support the next generation of global-change models. The environment-based approach to PFT and VFT classification ('top-down') has been more successful in southern Africa than the 'bottom-up' plant-based approach. The latter is hindered by the vast number of species involved, the rudimentary state of our ecophysiological knowledge of wild plants, and technical difficulties in measuring appropriate parameters. The environmental axes of plant water availability and available nutrients are strongly related to many of the broad structural and functional attributes of savannas and grasslands in the region, and the distribution of the PFTs we have proposed. In grasslands, temperature during the growing season is also important, and non-growing season temperature is needed to separate VFTs in southern Africa. The usefulness of these VFTs and PFTs as tools for the management of vegetation in a changing environment has yet to be proven.

References

Acocks, J. P. H. (1975) Veld types of South Africa. *Memoirs of the Botanical Survey of South Africa*, **40**, 126 pp.

Adamson, R. D. (1939) The classification of life-forms of plants. *Botanical Review*, **5**, 546–61.

Blackmore, A. C. (1992) *The variation of ecophysiological traits of savanna plants, in relation to indices of plant available moisture and nutrients*. MSc thesis, University of Witwatersrand, Johannesburg.

Cain, S. A. (1950) Life forms and phyto-climate. *Botanical Review*, **16**, 1–32.

Ellery, W. N. (1992) *Classification of vegetation of the South African grassland biome*. PhD thesis, University of Witwatersrand, Johannesburg.

Ellery, W. N. & Mentis, M. (1992) How old are South Africa's grasslands? In *Proceedings of the conference on forest-savanna boundaries*, Glasgow.

Ellery, W. N., Scholes, R. J. & Mentis, M. T. (1991) Differentiation of the grassland biome of South Africa based on climatic indices. *South African Journal of Science*, **87**, 499–503.

Feely, J. M. (1987) The early farmers of Transkei, southern Africa. *Cambridge Monographs in African Archaeology*, No. 24.

Table 13.2 *Seventeen broad PFTs found in southern African savannas*

The difference between 'trees' and 'shrubs' is based on height, not on species. Therefore one species can have both tree and shrub forms. Trees are sufficiently tall not to have the above-ground buds damaged by fire, and to escape most mammalian herbivory.

Size and longevity	Phenology	Water	Herbivory	Nutrients	'PFT nick name'
Woody plants Long-lived (>5 years)	Drought deciduous	(Mostly fine-leafed)	Thorny (hydrolysable tannins)	Fast growth and nutrient uptake. May be N-fixing.	Thorn shrub
					Thorn tree
Large (>4 m² area of influence) High fraction of secondary growth		(Mostly broad-leafed)	Indigestible (condensed tannins)	Inherently slow growth and uptake capacity	Broad-leafed shrub
					Broad-leafed tree
	Evergreen	Sclerophyllous	Indigestible (fibre and tannins)	Inherently slow growth and uptake capacity	Evergreen shrub
					Evergreen tree
		Succulent	Often palatable, sometimes tannins, resins or alkaloids		Succulent shrub
					Succulent tree

		Phreatophytic	Tall, or unpalatable		Phreatophyte
Graminoids Small (<1m^2), short-lived (<2 years per tiller) Monocots, buds basal. Mainly vegetative reproduction	Wet season deciduous	Water storage organ	Toxic	Inherently slow growth	Geoxylic suffrutex
		Xerophytes	High fibre	Root sheaths common on sandy soil	Wire grasses
		Mesophytes	Low digestiblity	Seasonally low N	Tuft grasses
			Lawn-forming	Continuously adequate N	Creeping grasses
Forbs As above, but dicots with terminal buds mainly sexually reproducing	Ephemeral/ Annual	Avoid drought as seeds	Toxic (alkaloids, etc.)		Ephemeral forbs
	Perennial	Drought deciduous	Often palatable but hairy	N-fixing	N-fixing forbs
				Non-N-fixing	Perennial non-N-fixers
Geophytes	Anti-seasonal	Water storage bulb	Toxic		Geophytes

Frost, P. Menaut, J.-C., Walker, B., Medina, E., Solbrig, O. T. & Swift, M. (1986) Responses of savannas to stress and disturbance: a proposal for a collaborative programme of research. *Biology International*, Special Issue 10.

Gillison, A. N. (1970) Towards a functional vegetation classification. In Leeper, G. W. (ed.) *The Australian Environment*, Melbourne: CSIRO/Melbourne University Press.

Grime, J. P. (1979) *Plant Strategies and Vegetation Processes*. Chichester: John Wiley.

Huntley, B. H. (1982) Southern African savannas. In Huntley, B. J. & Walker, B. H. (eds), *Ecology of tropical savannas*, pp. 101–19. Ecological Studies, No. 42. Berlin: Springer-Verlag.

Knight, H. K. & Loukes, O. L. (1968) A quantitative analysis of Wisconsin forest vegetation on the basis of plant function and gross morphology. *Ecology*, **50**, 219–34.

Menaut, J.-C. & Noble, I. R. (1988) A functional classification of savanna plants. In Walker, B. H. & Menaut, J.-C. (eds), *Responses of savannas to stress and disturbance: Research Procedure and Experimental Design for Savanna Ecology and Management*, pp. 8–11. Report 2. Paris: International Union of Biological Sciences.

Noble, I. R. & Slatyer, R. O. (1980) The use of vital attributes to predict successional changes in plant communities subject to recurrent disturbances. *Vegetatio*, **43**, 5–21.

Pickett, G. A., (1994) *A functional classification of a range of southern African savanna types*. MSc thesis, University of the Witwatersrand, Johannesburg.

Raunkiaer, C. (1934) The life forms of plants and statistical plant geography. Oxford: Clarendon Press.

Rutherford, M. C. & Westfall, R. H. (1986) Biomes of southern Africa: an objective classification. *Memoirs of the Botanical Survey of Southern Africa*, **54**, 94 pp.

Part four

14 Using plant functional types in a global vegetation model

W. Cramer

Introduction

Global biogeography has long been based on correlation analysis of the spatial patterns of vegetation and climate. This approach has proved valuable for investigation of the biosphere–atmosphere interactions of the earth system as a whole, and has been used to interpret climatic patterns in a way that is more meaningful for an understanding of climatic constraints to human land use than are the statistics derived from meteorological station networks. It has also been used to map climate systems where meteorological networks were sparse or unavailable (Holdridge 1947; Cramer & Leemans 1993).

The coincidence between the distributional limits of species or ecosystem types and major climatic boundaries, such as temperature isolines, is striking, and has fostered the notion that climate might be the single most important habitat variable for the survival and performance of living organisms. Ecophysiological studies have shown that many correlations between climate and plant performance are in fact based on essential requirements of plants carrying out their life cycle (Walter & Breckle 1983a; Woodward 1987). Many field observations, however, also show the limitations of this idea; the availability of nutrients or the presence of toxic substances in the soil are only two out of many important habitat factors that are seen to have a strong influence on the performance of plants.

The complexity of the problem is further increased by between-plant interactions: the presence of one species can either inhibit or enhance the growth and development of others. This is commonly due to the modifying effect that some plants have on the microscale climate as it is experienced by their competitors, e.g. by shading or owing to enhanced moisture availability. Between-species interactions not only add noise to the correlation between species distributions and climate, but are also likely to reduce the predictive capabilities of any given model. Climate change may lead to new combinations of habitat factors, and therefore the competitive relationships may change under the new climatic boundary conditions. Consequently, species composition and/or relative abundances may change, and the ecosystem may change into a different state.

The threat of rapid climate change within the next few decades has caused increasing interest in the development of simulation tools for those aspects of structural vegetation dynamics that are primarily driven by climate. However, there is also a requirement for scenarios about other changes in vegetation, for example those that are influenced by land use, as well as their impact on the atmospheric systems. For both issues, the nature of climate–vegetation interactions needs to be understood in a more mechanistic way.

Plant functional types (PFTs) have an important role in the development of more appropriate models of climate–vegetation interactions. This chapter is concerned with the broad-scale pattern of ecosystems, distinguished by major structural features. Its purpose is (1) to briefly review some earlier attempts to build global vegetation models, (2) to describe one specific global vegetation model, focusing on the way it uses the plant functional type concept, and (3) to show the implementation of the PFT-based model for a climate-change impact assessment study in Central Europe.

This chapter deals explicitly with a 'top-down' oriented derivation of the list of plant functional types. It therefore differs from 'bottom-up' aggregations of similar species into a common class (see for example Gitay & Noble (Chapter 1, this volume), and Grime *et al.* (Chapter 7, this volume)). Ideally, these two approaches should eventually merge, but this is not a realistic short-term prospect. Instead, they will complement each other, whenever appropriate, or they will be used separately. For the global model, a breakdown of the world's vegetation can only be done based on major physiognomic or otherwise easily recognizable features, because the screening of all the world's species for certain functional characteristics is impossible.

Matching climate and vegetation on a global scale

Climate classifications

Many coincidences between climatic and ecosystem gradients across a continent are obvious (Köppen 1884, 1936). These coincidences may be used to find the essential climatic requirements that need to be fulfilled for certain species or ecosystems to occur. Often these correlations have been investigated using visual map comparisons, and they have proven to be useful in many biogeographical studies of the world (e.g. Sjörs 1963).

Climate, as it is measured by meteorological stations, is more likely to be the dominant determining factor of vegetation pattern at the global scale rather than at the local scale. This statement is true as long as vegetation is treated at a fairly coarse level of resolution, e.g. by major vegetation types (biomes). Biomes contain a variety of local climate and soil systems, and the description of an entire biome usually summarizes the types of local habitat that may be found in a given region. The range of soil texture and nutrient classes, as well as the variety of topographic features, usually differ more *within* a biome than *between* them. It is therefore fair to assume that the major differences between biomes are the latitudinally constrained radiation regimes, and the broad-scale features of atmospheric circulation. Biomes, being summaries of the vegetation features at roughly similar spatial scales, are therefore well predictable from climatic summaries' data.

Köppen's studies on classifying climate on an objective basis (Köppen 1936) are the classic example for climate–vegetation models, although the terminology used is that of a classification system (Cramer & Leemans 1993). Broad climatic zones, which are defined on the basis of certain climatological indices, were given the names of the major biome occurring in each of them. Inspection of Köppen's maps today gives reasons to believe that some kind of an 'implicit model' has actually been used to fill the large gaps in the meteorological station network, for example in mountainous regions, to achieve a better global coverage of the climate map.

The work of L. R. Holdridge (Holdridge 1947) on the *life zone* concept is complementary to this. The goal of this concept is to estimate climatic conditions based on the observation of vegetation. It also includes a more schematic formulation of the ranges in climate space in which the various ecosystems are predicted to occur. Holdridge's model has been used in a now classic study to give the first prediction of large-scale biome shifts following climate change (Emanuel *et al.* 1985a,b).

Although improvements to both models have been suggested and are in use (Guetter & Kutzbach (1990) for Köppen, and Prentice (1990) for the Holdridge scheme), the line of development for this type of model has probably reached its end. Based as they are on long-term means of climatological variables, and on aggregations of ecosystems that cannot be related to organisms or types of organisms, they are not likely to reflect essential limiting factors of ecosystem behaviour any more in the future than they are doing today.

Individualistic models of ecosystem structure

Ecosystems do not interact with their environment as unchangeable entities. Instead, their components (the organisms) respond to habitat factors as individualistic units *sensu* Gleason (Gleason 1926), as has been shown by several

Table 14.1 *The plant functional types defined by E. O. Box*

Tropical rainforest trees
Tropical montane rainforest trees
Tropical evergreen sclerophyll trees
Tropical evergreen microphyll trees
Warm-temperate broad-leaved evergreen trees
Mediterranean broad-leaved evergreen trees
Temperate broad-leaved evergreen trees
Monsoon broad-leaved raingreen trees
Montane broad-leaved raingreen trees
Xeric raingreen trees
Summergreen broad-leaved trees
Boreal broad-leaved summergreen trees
Tropical linear-leaved trees
Tropical xeric needle-trees
Heliophilic long-needled trees
Sub-mediterranean needle-trees
Temperate rainforest needle-trees
Temperate needle-trees
Boreal/montane short-needled trees
Swamp summergreen needle-trees
Boreal summergreen needle-trees
Tropical broad-leaved evergreen dwarf-trees
Tropical broad-leaved evergreen small trees
Tropical cloud-forest dwarf-trees
Broad-leaved evergreen small trees
Subpolar broad-leaved evergreen small trees
Broad-leaved raingreen small trees
Broad-leaved summergreen small trees
Dwarf-needle small trees
Palmiform tuft-trees
Palmiform tuft-treelets
Tree ferns
Tropical alpine tuft-treelets
Xeric evergreen tuft-treelets
Evergreen giant-scrub
Raingreen thorn-scrub

Summergreen giant-scrub
Leafless xeromorphic large-scrub
Needle-leaved treeline krummholz
Tropical broad-leaved evergreen shrubs
Mediterranean evergreen shrubs
Broad-leaved ericoid evergreen shrubs
Temperate broad-leaved evergreen shrubs
Hot desert evergreen shrubs
Leaf-succulent evergreen shrubs
Cold-winter xeromorphic shrubs
Broad-leaved summergreen mesic shrubs
Xeric summergreen shrubs
Needle-leaved evergreen shrubs
Mediterranean dwarf-shrubs
Temperate evergreen dwarf-shrubs
Maritime heath dwarf-shrubs
Summergreen tundra dwarf-shrubs
Xeric dwarf-shrubs
Palmiform mesic rosette-shrubs
Xeric rosette-shrubs
Mesic evergreen cushion-shrubs
Xeric cushion-shrubs
Arborescent stem-succulents
Typical stem-succulents
Bush stem-succulents
Arborescent grasses
Tall cane-graminoids
Tall grasses
Short sward-grasses
Short bunch-grasses
Tall tussock-grasses
Short tussock-grasses
Sclerophyllous grasses
Desert-grasses
Tropical evergreen forbs
Temperate evergreen forbs
Raingreen forbs
Summergreen forbs
Succulent forbs

Table 14.1. *Continued*

Xeric cushion-herbs	Tropical broad-leaved evergreen
Ephemeral desert herbs	epiphytes
Seasonal cold-desert herbs	Narrow-leaved epiphytes
Raingreen cold-desert herbs	Broad-leaved wintergreen epiphytes
Tropical broad-leaved evergreen lianas	Evergreen ferns
Broad-leaved evergreen vines	Summergreen ferns
Broad-leaved raingreen vines	Mat-forming thallophytes
Broad-leaved summergreen vines	Xeric thallophytes

Source: Box (1981).

recent studies (Huntley & Webb 1988; Prentice 1992; Huntley 1996). The ecological behaviour of organisms is somewhat flexible, but it is largely constrained by the organisms' taxonomic status at a species or ecotype level. This notion represents a manageable challenge for model developers dealing with relatively species-poor, but well-studied, ecosystems. It should also be noted, however, that in most of these cases only the dominating vascular plants are part of the model (Shugart *et al.* 1992). For broad-scale, biome-level modelling, the challenge of describing every species in relation to its environment clearly becomes insurmountable.

This limit is one of several main reasons for the development of vegetation models based on plant functional types (Steffen *et al.* 1992). Based on previous ideas about plant life forms by Raunkiaer (Raunkiaer 1907), E. O. Box formulated a fairly detailed system of 90 plant functional types as fundamental components of the world's ecosystems (Table 14.1; Box 1981). In Box's model, the presence or absence of every plant type is predicted by an 'environmental envelope', consisting of eight bioclimatic indices (Table 14.2), derived from meteorological data. For each index, an upper and lower limit are derived, based on an extensive literature survey. Competition between different plant forms is treated indirectly by the application of a dominance hierarchy, which effectively excludes certain types of plants from a site, based on the presence of others, rather than the plants being excluded by climate. Although this method represents only a rather crude description of the complex processes involved in between-species competition, it nevertheless mimics the individualistic behaviour of the ecosystem as a whole. Changes in climate, for example, can allow the appearance of previously absent species, which may then suppress others although they may still be inside their climatic envelope, but further down the dominance hierarchy.

Table 14.2 *Bioclimatic indices as they are used by the Box model*

T_{max}	Mean temperature of the warmest month (°C)
T_{min}	Mean temperature of the coldest month (°C)
D_T	Range between T_{min} and T_{max} (°C)
P	Mean total annual precipitation (mm)
MI	Moisture index defined as the ratio between P and annual potential evapotranspiration (PET, estimated according to Thornthwaite & Mather (1957))
P_{max}	Mean total precipitation of the wettest month (mm)
P_{min}	Mean total precipitation of the driest month (mm)
P_{Tmax}	Mean total precipitation of the warmest month (mm)

Source: Box (1981).

Box's approach has been reviewed in more detail and re-implemented into a global database of the major climatic driving variables (Cramer & Leemans 1993). Some parts of it have also been used in a climate-change impact assessment for the Canadian boreal forest (Sargent 1988). A basic problem is the calibration and validation of the individual plant functional types' response to climate. Although the initial presentation of the model is based on a huge number of field observations from sites all over the world, it is not precisely clear how the basic ecophysiological observations made in one area translate into ecosystems of other continents. It appears that the number of types that are distinguished is too low to cover the world's structural vegetation diversity satisfactorily, and at the same time too large to be parameterized in a way that would be valid for all areas with a given climatic profile. The work is pioneering, however, in its conceptual contribution to the description of both environmental constraints and competitive relationships within a global vegetation model. Biomes essentially 'emerge' as communities through the combination of plant functional types and thereby mimic an important aspect of global vegetation composition.

A plant functional type model for global vegetation

General considerations

Given the essentially correlative nature of the previous approaches, a step towards a more mechanistic model of global climate–vegetation relationships should involve a set of ecophysiological arguments for the selected set of plant functional types. It must also include the definition of appropriate response functions between the habitat variables and the performance of plant types. This requires the aggregation of many different species' behaviour into very few major classes and therefore necessarily ignores a wide range of processes that *also* affect performance of vegetation. The criteria for the selection of attributes for the plant functional types are primarily determined by the spatial scale of the model (Noble & Slatyer 1980). If the aim is to distinguish the major global classes of ecosystems, assuming that they are in a steady-state equilibrium with their habitat, then a relatively small number of types and a limited number of parameters should be sufficient to model this distribution. Any further requirement, such as realistic transient behaviour over time, or greater specificity in terms of the plant types considered, must be compensated for either by increased data requirements or by the development of response functions that are currently not available, or both.

Data requirements for global vegetation models

An important constraint for the development of a global vegetation model is the availability of data. Observations of the variables involved are primarily needed for the following purposes:

1. model parameterization and calibration (i.e. parameters for the functional model for broad-scale ecophysiological processes);
2. validation (i.e. comparison of predicted ecosystem structure with observations of 'reality');
3. model application in a global, spatially explicit context (i.e. climatic and other habitat variables for every site where the model is to be run).

Given only one planet, no two sites with strictly replicable conditions occur, so there are severe limitations for the databases available to any one of these components. *Parameterization/calibration* based on empirical comparison of spatial patterns between climate and vegetation structure is not a desirable procedure for global applications, even if detailed data for every point on a dense grid were available. This is because (1) a model based on actual process measurements from controlled experiments (rather than derived from correlation analysis) would be more likely to have global validity, and (2) the process would leave no data for validation, because all data points would already have been used in the calibration process. The typical situation, on the other hand, is one of data scarcity, rather than overflow. There is no general solution to this dilemma. A good strategy, however, will aim for the smallest number of parameters, because there is a higher chance that these can be derived from experimental studies in the scientific literature.

Model *validation* is data-limited, too. First, few databases exist about the distribution of vegetation on the earth. Secondly, those that do exist are either descriptions of the existing vegetation (i.e. the global land cover as it had been modified by human land use until the time of data collection), or implicit interpretations of the potential vegetation by individual scientists (i.e. they effectively imply another vegetation model). To a certain extent, the validation data set that is already chosen involves a decision about certain features of the model. The global map of actual ecosystem complexes (Olson *et al.* 1983) appears to be one of the most suitable compilations of actual land-cover data. For comparison with a model of the potentially natural vegetation, however, only those regions that are classified by Olson as having little modified vegetative cover can be used.

The availability of climatic and other habitat data for *application* of a global vegetation model is rapidly improving, but there are still significant gaps, even for a coarse spatial grid. Bioclimatic characterizations of sites should involve indices for the major resources required for, or constraints to, the growth of plants. The definition of a particular plant functional type scheme is closely connected to this problem. Distinguishing two different types of plant based on their climatic requirements only makes sense for the global model if the pattern of the climatic variables involved is known in sufficient detail.

The spatial pattern of most climatic variables is only partly captured by the meteorological station network, because it is practically impossible for a network to cover all climatic gradients at all scales. The particular short-comings of the global station network have been reviewed elsewhere (Legates & Willmott 1990; Leemans & Cramer 1991; Willmott & Legates 1991). They mostly concern less developed countries, sparsely populated regions and high-altitude areas. Coastal regions and the world's major airports are over-represented. Some of these problems can be reduced by involving topography as a major determinant of spatial pattern in climatic characteristics. Most climatic variables follow strong gradients with increasing elevation (e.g. temperature lapse rates), and these can be derived empirically from the weather station data.

Updating an existing global climate database (Leemans & Cramer 1991), we have used a trivariate thin-plate splining technique, developed by M. Hutchinson (Hutchinson & Bischof 1983), to generate climatic surfaces for the terrestrial surface of the earth, excluding Antarctica. The three independent variables are longitude, latitude and elevation, and the fit of these surfaces is appropriate to generate interpolated climatic data for grid-points at 10' spatial resolution. The grid-points are obtained from a digital topography database.

Because the defined aim of the model is to describe a steady-state situation, long-term means of the climatic variables are appropriate. The seasonal response of vegetation to climate, however, is crucial for the model, because it distinguishes major biomes such as the temperate deciduous and the boreal evergreen forest, or the raingreen and the evergreen tropical forests. Therefore the climatic values are chosen as long-term *monthly*, rather than annual, means.

Bioclimatic indices

Plants usually avoid frost risk, and they have requirements for heat and moisture. Many temperate woody plants also require a winter chilling period. F. I. Woodward has provided detailed ecophysiological arguments for minimum temperature tolerances of major woody plant forms (Woodward 1987). The major temperature-related indices are straightforward to approximate from monthly means of temperature. The temperature of the coldest month, for example, is closely correlated with the absolute minimum temperature (Prentice *et al.* 1992). Heat sums for the long-term average climate at a given site can be calculated from linear or spline interpolations of the monthly temperature records.

Moisture availability or drought risk, however, are only inadequately described by precipitation values alone (Thornthwaite & Mather 1957), because the amount of water available to plants is controlled by both precipitation and temperature. A wide range of indices that couple precipitation and temperature data into one value reflecting available moisture have been proposed. It is possible to develop such an index by a simple soil water balance model experiencing quasi-daily values of radiation, precipitation and temperature (Cramer & Prentice 1988; Prentice *et al.* 1993). The main limitation of this approach, however, is that the seasonal pattern of moisture availability is not accounted for. Also, the long-term means hide the interannual variability, which is likely to be particularly important in many moisture-limited ecosystems (Walter & Breckle 1983b). Table 14.3 shows how the bioclimatic indices used by the BIOME model are derived from interpolated climatic surfaces (for details see Prentice *et al.* 1992).

Plant functional types

The functioning of plants is constrained by the availability of resources to the individual specimen, which is, in turn, a function of habitat. A plant's efficiency in exploiting the local habitat conditions varies, not only owing to the climatic conditions at the site, but also owing to the phase of its life-cycle and to competition with other plants. Killing frosts are particularly dangerous during the seedling stage and during budburst, but a plant

Table 14.3 *Derivation of bioclimatic indices from interpolated climatic data in the BIOME model*

Tolerance/ requirement	Ecophysiological mechanism	Bioclimatic index	Climatic variable used (monthly means)
Cold tolerance	killing temperature during coldest period of the year	T_{min}: temperature of the coldest month (lower limit)	temperature
Chilling requirement	winter chilling period required for budburst of woody plants	T_{min}: temperature of the coldest month (upper limit)	temperature
Heat requirement	annual growth respiration requirement	GDD: growing degree days above 0 °C and 5 °C	temperature
Moisture require-ment/drought tolerance	soil moisture availability	AET/PET: annual actual evapo-transpiration/annual potential evapotranspiration	temperature, precipitation, cloudiness

Source: Prentice *et al.* (1992).

under a closed canopy of other species may be less sensitive to the climatic risk. It is therefore desirable that the functional-type classification reflects these sensitivities as much as possible.

The development of a list of plant functional types for a global, climate-driven model is constrained by the very limited information available about the functional relationships between the world's plants and their climatic habitat. The goal for the development of the BIOME model was therefore set to *find the simplest possible model with the smallest number of plant functional types, constraints and driving variables* that could still simulate the broad features of present vegetation as indicated in the data of Olson *et al.* (1983). The result of this work gave not only a rather small number of PFTs (fourteen), but also relatively few climatic constraints (Table 14.4). Additional constraints were only defined where this was necessary for the simulation of a particular distribution limit. Further details of the reasoning for the individual types can be found in Prentice *et al.* (1992).

Assembling 'vegetation' from plant types

If plants respond to habitat in an individualistic way, then different assemblages (communities) of plants are possible. Habitat changes, such as those driven by climate change or by land use, may be experienced differently by different plants and new communities can form. The model described here attempts to mimic this process in the simplest possible form. First, it is implicit in the climatic constraints table that distributions of plant types can overlap. Secondly, a simple set of rules was defined that prescribes 'competitive' relationships between the plant types in such a way that certain plant types exclude the presence of certain others within their distributional area. Because the effect of trees on the microclimate experienced by grasses cannot be modelled on a globally generalized scale, for example, it was assumed that trees 'exclude' the occurrence of grasses. This dominance hierarchy is clearly a very simple approximation and will be refined by more mechanistic relationships between plant types in later global vegetation models.

In combination, the overlapping distributional areas of all plant types, and the limits for co-occurrence due to the dominance hierarchy, result in a finite list of vegetation types, or biomes. It should be noted that this list is a *result*, rather than a component, of the model, and that it would change if new plant functional types were added, or if the response function parameters of individual types were changed. The list of biomes can, however, be *interpreted* in a way that is similar to the description of global potential vegetation, comprising major types of plants in different combinations (Table 14.5).

Table 14.4 *Environmental constraints for each plant functional type in the BIOME model*

Plant functional type	No.	T_{min}		GDD_0	GDD_5	T_{max}	AET/PET	
		min	max	min	min	min	min	max
Trees								
tropical evergreen	1	15.5	—	—	—	—	0.80	—
tropical raingreen	2	15.5	—	—	—	—	0.45	0.95
warm–temperate evergreen	3	5.0	—	—	—	—	0.65	—
temperate summergreen	4	−15.0	15.5	—	1200	—	0.65	—
cool-temperate conifer	5	−19.0	5.0	—	900	—	0.65	—
boreal evergreen conifer	6	−35.0	−2.0	—	350	—	0.75	—
boreal summergreen	7	—	5.0	—	350	—	0.65	—
Non-trees								
sclerophyll/succulent	8	5.0	—	—	—	—	0.28	—
warm grass/shrub	9	—	—	—	—	22.0	0.18	—
cool grass/shrub	10	—	—	—	500	—	0.33	—
cold grass/shrub	11	—	—	100	—	—	0.33	—
hot desert shrub	12	—	—	—	—	22.0	—	—
cold desert shrub	13	—	—	100	—	—	—	—
No plants (dummy type)	14	—	—	—	—	—	—	—

Source: Prentice *et al.* (1992).

Table 14.5 *Combinations of plant functional types in biomes, resulting from the BIOME model*

Biome type	PFT no. 1	2	3	4	5	6	7	8	9	10	11	12	13	14
Tropical rainforest	x													
Tropical seasonal forest	x	x												
Tropical dry forest/savanna		x												
Broad-leaved evergreen forest			x											
Temperate deciduous forest				x	x		x							
Cool mixed forest				x	x	x	x							
Cool conifer forest					x	x	x							
Boreal forest (taiga)						x	x							
Cold mixed forest					x		x							
Cold deciduous forest							x							
Xerophytic woods/shrub								x						
Warm grass/shrub									x					
Cold grass/shrub										x	x			
Tundra											x			
Hot desert												x		
Semidesert													x	
Ice/polar desert														x

Source: Prentice *et al.* (1992).

Example: European potential natural vegetation

Vegetation distribution under baseline climate conditions

The resulting map of biome distribution, produced at a spatial resolution of 0.5° × 0.5° latitude/longitude, has been published by Prentice *et al.* (1992). It has been shown to correspond well to an aggregation of the terrestrial vegetation types in the database of Olson *et al.* (1983) and it can be used to assess the likely impact of climate change on terrestrial ecosystem distribution and carbon storage (Solomon *et al.* 1993). Here, output from the same model is shown at a higher spatial resolution (10' × 10' latitude/longitude) using a database with improved climatic variables, reflecting more detailed pattern, for example in mountainous areas. Plate 1 shows the distribution of potential natural vegetation in a large section of Europe. Although the distribution of *potential* natural vegetation cannot be validated against an observation of *actual* vegetation on a continent with major human modifications of terrestrial vegetation, the map clearly reflects the major ecosystem types that have been described by biogeographers. It should be noted that many altitudinal boundaries in the mountains of Europe appear roughly where they occur in reality, without any further modifications of the global model.

Vegetation redistribution under a changed climate

Ecosystem redistribution under a future climate can currently not be *predicted*, for several reasons, including the following: (1) there is still considerable uncertainty about predictions of future climate itself, and (2) there is no current global model of ecosystem distribution that accounts for dynamic factors such as migration or other aspects of succession. However, climate-change impact assessments using the current type of model, such as BIOME, are of interest because (1) they can given an indication of the sensitivity of terrestrial ecosystems to climate change of the magnitude and pattern suggested by current general circulation model (GCM) scenarios, and (2) they can help to generate more mechanistic land surface parameterization schemes for the GCMs, and hence potentially improve the predictions of future climate itself.

Plates 2 and 3 illustrate two potential impact scenarios for Europe. Both maps were generated from a comparison between a 'control run' with normal CO_2 concentration in the atmosphere and a 'scenario run' with higher CO_2 concentration. This difference was calculated for the monthly means of temperature (absolute difference) and precipitation (relative difference) at the spatial resolution of the GCM involved. The resulting anomalies were interpolated to the 10' grid of the baseline climate database and added to or multiplied by the current climate data held there. Output from this process was fed through the BIOME model; this yielded new distributions of plant functional types, and hence biomes.

Plate 2 shows the new biome distribution that would result from the $2 \times CO_2$ scenario of the GISS model (Hansen *et al.* 1988). The high sensitivity of natural vegetation in Europe to a climate change of this magnitude is clearly visible: the temperate deciduous forests of Western and Central Europe are almost entirely replaced by broad-leaved evergreen forests. The treeline in most alpine areas moves up, and the boreal vegetation types move northwards.

Plate 3 illustrates the sensitivity of the same biome distribution to a more recent analysis made with the Max Planck Institute for Meteorology ECHAM T42 model (Cubasch *et al.* 1992). This is a coupled ocean–atmosphere model, and the scenario was generated using a $3 \times CO_2$ scenario. The displacement of northern biome types with more warmth-demanding forests is less dramatic than in the GISS scenario, but there is a considerable expansion of drier biome types in the continental parts of Europe.

Perspectives and limitations

It must be stressed that none of the scenarios shown in Plates 2 and 3 is capable of indicating a *forecast* of specific ecosystem changes from current conditions to a future map of Europe, for two main reasons. First, the baseline scenario corresponds to current conditions only in so far as it reflects the major trends of current *potential* rather than actual distribution. Secondly, the current scenarios of future climatic conditions, particularly concerning precipitation, are relatively uncertain in both their magnitude and their spatial pattern. The main usefulness of the model output shown here is to show the importance of spatial pattern in terrestrial topography for the distribution of ecosystem types. This opens the possibility of improved schemes for land surface parameterizations in general circulation models. If we can generate a plausible pattern of high-resolution ecosystem distribution from climate and topography, then phenomena at scales finer than current general circulation models can be simulated. These simulations would yield both an improved understanding of the likely qualitative changes that the earth's surface might undergo under changed climate conditions, and also a means to describe those changes quantitatively, for example through a fractional area linkage within a grid cell of the circulation model.

Plant functional types are an essential component of both developments. They allow the addition of more realism to the ecophysiological underpinning of a climate–vegetation model (and, in turn, to the feedback mechanisms that can be parameterized between the land surface and the atmosphere). Plant functional types also improve the potential of the mapping of changes in spatial pattern at the scales which are relevant for terrestrial vegetation. These are essential for a subgrid-scale analysis of terrestrial biosphere sensitivity to climate change.

Acknowledgements

This work is based on the results of an international cooperative project ('Sustainable Development of the Biosphere') sponsored by the International Institute for Applied Systems Analysis (IIASA), in Laxenburg, Austria, while Allen M. Solomon was leader of the project. The BIOME model itself was developed by I. Colin Prentice (Lund, Sweden), Sandy P. Harrison (Lund, Sweden), Rik Leemans (Bilthoven, The Netherlands), Robert A. Monserud (Moscow, ID, USA) and Allen M. Solomon (Corvallis, OR, USA). Support for the development of the global climatic surface is acknowledged from Michael Hutchinson (Canberra, Australia) and Brian Huntley (Durham, UK).

References

Box, E. O. (1981) *Macroclimate and Plant Forms: An Introduction to Predictive Modeling in Phytogeography*. The Hague: Dr W. Junk Publishers.

Cramer, W. & Leemans, R. (1993) Assessing impacts of climate change on vegetation using climate classification systems. In Solomon, A. M. & Shugart, H. H. (eds), *Vegetation Dynamics Modeling and Global Change*, pp. 190–217. New York: Chapman and Hall.

Cramer, W. & Prentice, I. C. (1988) Simulation of soil moisture deficits on a European scale. *Norsk Geografisk Tidskrift*, **42**, 149–51.

Cubasch, U., Santer, B. D., Hellbach, A., Hegerl, G., Höck, H., Maier-Reimer, E., Mikolajewicz, U., Stössel, A. & Voss, R. (1992) *Monte Carlo climate change forecasts with a global coupled ocean-atmosphere model*. Report, Max-Planck-Institut für Meteorologie, Hamburg, Germany.

Emanuel, W. R., Shugart, H. H. & Stevenson, M. P. (1985a). Climatic change and the broad-scale distribution of terrestrial ecosystems complexes. *Climate Change*, **7**, 29–43.

Emanuel, W. R., Shugart, H. H. & Stevenson, M. P. (1985b) Response to comment: Climatic change and the broad-scale distribution of terrestrial ecosystems complexes. *Climate Change*, **7**, 457–60.

Gleason, H. A. (1926) The individualistic concept of the plant association. *Bulletin of the Torrey Botanical Club*, **57**, 7–26.

Guetter, P. J. & Kutzbach, J. E. (1990) A modified Köppen classification applied to model simulations of glacial and interglacial climates. *Climate Change*, **16**, 193–215.

Hansen, J. E., Fung, I. Y., Lacis, A. A., Rind, D., Lebedeff, S., Ruedy, R., Russell, G. & Stone, P. (1988) Global climate changes as forecast by the Goddard Institute for Space Studies three dimensional model. *Journal of Geophysical Research*, **93**(D8), 9541–64.

Holdridge, L. R. (1947) *Life Zone Ecology*. San José, Costa Rica: Tropical Science Center.

Huntley, B. (1996) The responses of vegetation to past and future climate changes. In Holten J. & Oechel, W. (eds), *Global*

Change and Arctic Terrestrial Ecosystems. Trondheim: Norwegian Institute of Nature Research. (In press.)

Huntley, B. & Webb, T. III (eds) (1988) *Vegetation History.* Dordrecht: Kluwer Academic Publishers.

Hutchinson, M. F. & Bischof, R. J. (1983) A new method for estimating the spatial distribution of mean seasonal and annual rainfall applied to the Hunter Valley, New South Wales. *Australian Meteorological Magazine,* **31**, 179–84.

Köppen, W. (1884) Die Wärmezonen der Erde, nach der Dauer der heissen, gemässsigten und kalten Zeit und nach der Wirkung der Wärme auf die organische Welt betrachtet. *Meteorologische Zeitschrift,* **1**, 215–26 (+ map).

Köppen, W. (1936) Das geographische System der Klimate. In Köppen. W. & Geiger, R. (eds), *Handbuch der Klimatologie,* pp. 1–46. Berlin: Gebrüder Bornträger.

Leemans, R. & Cramer, W. (1991) *The IIASA database for mean monthly values of temperature, precipitation and cloudiness on a global terrestrial grid.* Laxenburg: International Institute for Applied Systems Analysis. Research Report RR-91-18.

Legates, D. R. & Willmott, C. J. (1990) Mean seasonal and spatial variability in global surface air temperature. *Theoretical and Applied Climatology,* **41**, 11–21.

Noble, I. R. & Slatyer, R. O. (1980) The use of vital attributes to predict successional changes in plant communities subject to recurrent disturbances. *Vegetatio,* **43**, 5–21.

Olson, J., Watts, J. A. & Allison, L. J. (1983) *Carbon in Live Vegetation of Major World Ecosystems.* Oak Ridge National Laboratory (ORNL-5862).

Prentice, I. C. (1992) Climate change and long-term vegetation dynamics. In Glenn-Lewin, D. C., Peet, R. K. & Veblen, T. T. (eds), *Plant Succession: Theory and Prediction,* pp. 293–339. London: Chapman & Hall.

Prentice, I. C., Cramer, W., Harrison, S. P., Leemans, R., Monserud, R. A. & Solomon, A. M. (1992) A global biome model based on plant physiology and dominance, soil properties and climate. *Journal of Biogeography,* **19**, 117–34.

Prentice, I. C., Sykes, M. T. & Cramer, W. (1993) A simulation model for the transient effects of climate change on forest landscapes. *Ecological Modelling,* **65**, 51–70.

Prentice, K. C. (1990) Bioclimatic distribution of vegetation for general circulation model studies. *Journal of Geophysical Research,* **95**(D8), 11811–30.

Raunkiaer, C. (1907) *Planterigets Livsformer.* Copenhagen/Kristiania: Gyldendalske Boghandel & Nordisk Forlag.

Sargent, N. E. (1988) Redistribution of the Canadian boreal forest under a warmed climate. *Climatological Bulletin,* **22**, 23–34.

Shugart, H. H., Leemans, R. & Bonan, G. B. (eds), (1992) *A Systems Analysis of the Global Boreal Forest.* Cambridge University Press.

Sjörs, H. (1963) Amphi-atlantic zonation, nemoral to arctic. In Löve, A. & Löve, D. (eds), *North Atlantic Biota and their History,* pp. 109–25. Oxford: Pergamon Press.

Solomon, A. M., Prentice, I. C., Leemans, R. & Cramer, W. (1993) The interaction of climate and land use in future terrestrial carbon storage and release. *Water, Air and Soil Pollution,* **70**(1–4), 595–614.

Steffen, W. L., Walker, B. H., Ingram, J. S. & Koch, G. W. (1992) *Global Change and Terrestrial Ecosystems: The Operational Plan.* International Geosphere–Biosphere Programme. Global Change Report (21).

Thornthwaite, C. W. & Mather, J. R. (1957) *Instructions and Tables for Computing Potential Evapotranspiration and the Water Balance.* Drexel Institute of Technology, Laboratory of Climatology. (*Publications in Climatology* Vol. X, No. 3).

Walter, H. & Breckle, S.-W. (1983a) *Ökologie der Erde,* vol. 1: *Ökologische Grundlagen in globaler Sicht.* Stuttgart: Gustav Fischer Verlag.

Walter, H. & Breckle, S.-W. (1983b) *Ökologie der Erde,* vol. 2: *Spezielle Ökologie der Tropischen und Subtropischen Zonen.* Stuttgart: Gustav Fischer Verlag.

Willmott, C. J. & Legates, D. R. (1991) Rising

estimates of terrestrial and global precipitation. *Climate Research*, **1**, 179–80.

Woodward, F. I. (1987) *Climate and Plant
Distribution*. Cambridge University Press.

15 The use of plant functional type classifications to model global land cover and simulate the interactions between the terrestrial biosphere and the atmosphere

R. Leemans

Introduction

Changing atmospheric composition, land-cover patterns and climate will have large impacts on the functioning of the terrestrial biosphere at global, regional and local levels. These impacts have recently been recognized as one of the current major environmental concerns (Bolin *et al.* 1986). The International Geosphere–Biosphere Programme (IGBP; Anon. 1990) was created to address these concerns by developing a global change research programme spanning many scientific disciplines and fields. Meanwhile national governments and international organizations have felt the need to implement policies to mitigate and/or adapt the negative effects of these impacts and changes, which are unprecedented in human history. They have set up an Intergovernmental Panel for Climate Change (IPCC) to thoroughly review and address our knowledge and understanding of global change issues (Bernthal 1990; Houghton *et al.* 1990; Izrael *et al.* 1990; Houghton *et al.* 1992). All these activities led to the Global Climate Convention, which was signed by 152 countries at the 1992 UNCED conference in Rio de Janeiro. This Convention provides a framework for discussions on policies to deal with global climate change. These events could well become a turning point in international policy and have already led to an increased recognized need for comprehensive understanding and modelling of global change issues (e.g. Turner *et al.* 1993).

The work presented here is an effort to enhance the understanding of global change issues scientifically by using and developing state-of-the-art integrated models. It follows on from earlier work on land-cover modelling at the International Institute for Applied Systems Analysis (IIASA) for specific biomes (e.g. Shugart *et al.* 1992) and the whole globe (e.g. Prentice *et al.* 1989, 1992; Solomon & Shugart 1993). These undertakings were mainly based on a top-down modelling approach, in which vegetation patterns are described in terms of their environmental requirements. The approach was embedded in a global Geographic Information System (GIS) with a resolution of 0.5° longitude and latitude (Leemans 1992). One of the central databases in this grid was the IIASA climate database (Leemans & Cramer 1991), which has been used to implement and review global vegetation–climate classifications (e.g. Leemans 1989; Smith *et al.* 1992; Cramer & Leemans 1993; Tchebakova *et al.* 1993a; Cramer, Chapter 14 of this volume). These studies clearly displayed the performance, advantages and disadvantages of the different types of models (Table 15.1). In summary, the more superior models all included a realistic water balance and/or seasonality. The best performance was shown by the BIOME model (Prentice *et al.* 1992; Table 15.1). This was due not only to the choice of adequate climatic variables, but also to the model's structure (see Cramer, Chapter 14, this volume). The model uses plant functional types (PFTs) as its basic units. This approach is derived from that of Box (1981; cf. Cramer & Leemans, 1993), but a globally more adequate selection of PFTs was selected. The PFTs of BIOME were selected as a minimum set that should represent the most important facets of global plant physiognomy and physiology. Such a model should be capable of successfully simulating past, present and future vegetation patterns. The poorest-performing models (e.g. Holdridge 1947; Table 15.1) are less adequate for this purpose, although they are still frequently used (e.g. Emanuel *et al.* 1985a,b; Guetter & Kutzbach 1990; Cramer & Leemans, 1993; Tchebakova *et al.* 1993b). All of these models use biomes (e.g. boreal forest, tropical savanna, tropical rainforest) as basic units and define climatic limits to delimit the borders of each biome. It is implicitly assumed that under changed climatic conditions biomes react as entities. Studies of historic vegetation patterns have shown that this is not a very valid assumption (e.g. Huntley & Webb 1988). These models could therefore give only a coarse indication of the potential change. The unit of vegetation change should not be an individual biome, but should involve species. Species respond in an individualistic way to climatic change, so that no-analogue species assemblages (biomes) could emerge under changed conditions.

PFT-based models are more appropriate for simulating vegetation response under different environmental conditions. An adequately selected set of PFTs could mimic the wide range of species responses (Smith & Huston 1989). PFTs are a powerful concept for aggregating the large number of species which occur, and are therefore a very suitable approach for realistic and comprehensive land-cover modelling.

PFTs can be defined at many different levels. The BIOME model (Prentice *et al.* 1992) uses only 14 PFTs, but this number can easily be expanded to cope with specific questions and/or regional applications. The models

Table 15.1 *Performance of global vegetation–climate classifications using the Kappa statistic (Monserud & Leemans 1992)*

The digitized maps to which each vegetation–climate classification is compared, actual vegetation cover (Olson *et al.* 1985) and natural vegetation (A. W. Küchler (in Espenshade & Morrison 1991)). The classes of the maps and climate classifications are aggregated in order to create comparable units.

Model and/or references	Region	Actual vegetation	Natural vegetation
Woodward & Williams (1987)	global	very poor	poor
Thornthwaite (1948)	global	poor	poor
Holdridge (1947, as in Monserud & Leemans 1992)	global	poor	poor
Köppen (1936, as in Guetter & Kutzbach 1990)	global	poor	fair
Box (1981)	global	fair	fair
Budyko (1986, as in Tchebakova *et al.* 1993a)	global	fair	fair
BIOME (Prentice *et al.* 1992)	global	fair	good

can be used as a vegetation interface for Atmospheric General Circulation Models (AGCMs; e.g. Henderson-Sellers 1991; Claussen & Esch 1994). Simulated past, current and future vegetation patterns can be used to drive geographically explicit carbon cycle models. Such a carbon model was first developed with the Holdridge Life Zone Classification for different climatic conditions by Prentice & Fung (1990) and their analysis was further improved by Smith *et al.* (1992). The BIOME model has also been used for determining carbon dynamics since the last glacial maximum (LGM) (Prentice *et al.* 1994). All the simulations showed that since the LGM, the carbon content in the terrestrial biosphere has increased and that this increase could be enhanced under climatic warming. These conclusions are similar to results obtained by other methods (cf. Sundquist 1993).

Recently, BIOME has been coupled to a simple climatic envelope, which could delimit current and future agricultural regions (Cramer & Solomon 1993). This approach was linked to future climate-change scenarios and a large expansion of potential agricultural areas was demonstrated. Such human-induced land-cover change should have a much larger impact on the sink-functioning of the terrestrial biosphere than changing climatic patterns alone. However, the simple climatic-envelope approach is not completely adequate. Agricultural patterns are driven not only by climatic suitability but also, more strongly, by individual crop requirements and distributions; furthermore, socioeconomic driving forces such as fertilization, irrigation and mechanization could also influence productivity. Here adequately selected crop PFTs could also improve determination of global agricultural potential and subsequently give an improved assessment of changed land use and cover. This is of utmost importance for a more comprehensive analysis of all consequences of global change.

In this chapter I present a preliminary version of a recent development in integrated environmental modelling. The model, IMAGE 2 (*I*ntegrated *M*odel for *A*ssessment of the *G*reenhouse *E*ffect; Alcamo 1994), is designed to link socioeconomic origins of global change with the environmental consequences. Although greenhouse gas emissions resulting from energy and industrial production and their uptake into the atmosphere and oceans are taken into account, I limit this presentation to the terrestrial biosphere component. I illustrate how ecological and agricultural PFTs could be defined and used to simulate changes in land cover and the influences on the carbon cycle of the terrestrial biosphere. Finally, I emphasize the linkages between the different compartments of the global system and the potential for quantifying feedbacks between and within these compartments.

The IMAGE 2 model

The IMAGE 2 model evolved from earlier versions of the IMAGE model (Rotmans 1990; Rotmans *et al.* 1990) and from the ESCAPE framework (*E*valuation of *S*trategies for *C*limate change by *A*dapting to and *P*reventing *E*missions; CRU and ERL 1992; Rotmans 1992). IMAGE 1 consisted of a globally averaged model for climate-change issues which combined (1) an energy model for greenhouse gas emissions, (2) a global carbon cycle model, and (3) highly parameterized mathematical expressions of global radiative forcing, atmospheric temperature response, and sea-level rise (Rotmans 1990). The global-average calculations of IMAGE 1 were useful for evaluating policies at both national (Dutch) and international level (e.g. Bernthal 1990). Under a consortium organized by Environmental Resources Limited and financed by the Commission of the European Communities, the developers of IMAGE 1 then contributed to the development of the ESCAPE framework, which combined parameterized global-average climate calculations with grid-based impact calculations for Europe (CRU and ERL 1992). These geographically explicit impact calculations were derived from approaches developed by Parry *et al.* (1990, agriculture) and Leemans (1992, ecosystems). As part of the ESCAPE framework, an innovative approach was adopted to estimate emissions from energy (CRU and ERL 1992) and land use (Bouwman *et al.* 1992) for several world regions. The

IMAGE 2 model builds especially on these geographically and regionally explicit components of the ESCAPE framework.

The IMAGE 2 model is a multidisciplinary integrated model of global change (Fig. 15.1). The model is designed to provide a scientifically based overview of global change issues to support the national and international evaluation of policies. It consists of three fully linked components: energy–industry, terrestrial environment, and the ocean–atmosphere system (Fig. 15.1). Dynamics calculations are performed for a 100 year time horizon. Some calculations are geographically explicit (grid-based with a resolution of 0.5° longitude and latitude); others are regionally explicit (13 regions are currently defined; see Table 15.2). These two very different dimensions depend on the purpose and structure of each specific submodel. In general, all socioeconomic submodels are regionally explicit, whereas all environmental and physical submodels are geographically explicit. Recently, several global compilations of geographical databases have been created (e.g. EPA/NOAA 1991; Leemans & Cramer 1991). These databases are fundamental in order to run most of our submodels. Regional (county, country, continent) data are readily available from national and international census bureaux. The most innovative accomplishment of IMAGE 2 is the

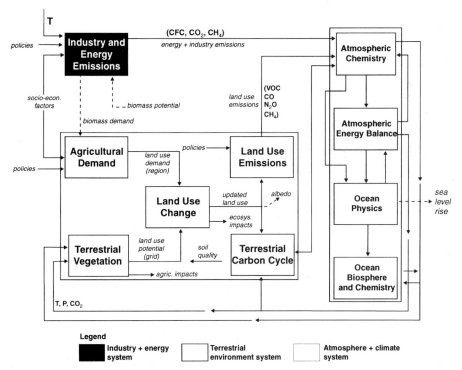

Figure 15.1 *A flow diagram for the IMAGE 2 model, illustrating the different components and their linkages.*

linkage between these two, structurally different dimensions. The whole model will be embedded in a global dedicated GIS with advanced user support systems to assist the (non-)expert user in selecting comprehensive sets of model parameters and scenario options.

The objective of the energy–industry set of models is to compute the emissions of greenhouse gases in each region as a function of energy consumption and industrial production. The models are designed especially for investigating the effects of improved energy efficiency and technological development on future emission in each region, and can be used to assess the consequences of different policies and socioeconomic trends on future emissions.

The purpose of the atmosphere–ocean set of models is to compute the build-up of greenhouse gases in the atmosphere and the resulting change in global temperature and precipitation patterns. Emissions from the energy–industry and terrestrial environment components are combined and used to determine the uptake of C (both physically and biologically) by the oceans, and the atmospheric composition. Temporal trends of global-average carbon dioxide, methane and other greenhouse gases are computed. The atmospheric composition model consists of a simple chemical scheme describing global-average tropospheric and stratospheric chemistry (Prather 1989). Finally an atmospheric energy balance model uses the computed atmospheric levels of greenhouse gases, together with other data, to compute the earth's energy balance. Surface air temperature and precipitation changes on $10°$ latitudinal bands covering the whole globe are derived. The heat sink of the ocean is accounted for in this model. The climatic changes can be linked to AGCMs to obtain a longitudinally explicit climate-change scenario.

The Terrestrial Environment Component

The objectives of the Terrestrial Environment Component (TEC) are to simulate global land use and land cover dynamically through time and employ these changes to determine greenhouse gas emissions from the terrestrial biosphere to the atmosphere. This set of models can be used to evaluate the effectiveness of land-use policies for controlling the build-up of greenhouse gases (e.g. changed agricultural practices or sequestration of carbon in forest plantations). They can also be used to evaluate the impact of climate change and changing atmospheric composition on global ecosystems and agriculture.

The core of TEC is the Land Cover Change model (Fig. 15.1), which reconciles (1) the demand for land stemming from economic and demographic trends, and (2) the potential of land as indicated by potential vegetation and plant productivity. These two factors are reconciled by heuristic rules, contained in the Land Cover Change model, which reflect key factors in land-use changes such as proximity to infrastructure or population, or government settlement policy.

Land-use demand is computed by the Agricultural Demand model, which uses socioeconomic variables as input, and calculates the demands for agricultural products (including crops, lumber and livestock) that require land in each region of the world. Land-use potential is computed by the Terrestrial Vegetation model in the form of potential vegetation types and crop productivity on the terrestrial grid; the model uses climate and soil characteristics as input.

The computed land-cover patterns serve as a basis for calculating methane, nitrous oxide and other greenhouse gas emissions from land-related activities, such as livestock raising, fertilizer use, etc. These calculations are performed by the Land Use Emissions model. The conversions of land cover between years are used by the Terrestrial Carbon model to compute C fluxes between the biosphere and the atmosphere.

TEC is linked to the energy–industry component via the demand for fuelwood, and the land requirements for alcohol and other biofuels. TEC and the atmosphere–ocean component are at present linked only through greenhouse gas emissions, but this elementary linkage will be improved in future IMAGE versions by taking into account changes in potential evapotranspiration, soil moisture and albedo that stem from land-cover changes and could influence the global radiation balance. The atmosphere–ocean component is linked with TEC through the changed climate. The whole IMAGE 2 framework thus provides an integrated simulation of the earth's system, where changes in one compartment directly influences processes in other compartments. This is an important improvement over earlier models, such as IMAGE 1 (Rotmans 1990) and ESCAPE (CRU and ERL 1992), where impacts did not feed back into the causes.

The use of plant functional types in IMAGE 2

Two different sets of PFTs are used in the Terrestrial Vegetation model of TEC. First, the global distributions of PFTs of the BIOME model (Prentice *et al.* 1992; Cramer, Chapter 14, this volume) are used to define potential global vegetation patterns, and second, a set of crop PFTs is defined to compute the potential distribution and productivity of agriculture. This model is based on the Crop Suitability approach of the Food and Agriculture Organization of the United Nations (FAO; Anon. 1978; Higgins *et al.* 1987; Leemans & Solomon 1993).

Potential vegetation patterns

No changes are made to the original BIOME approach. This approach is based on a comprehensive set of climatic indices that represent the seasonal and spatial patterns of temperature and moisture availability. Moisture availability is based on the water balance model developed by Prentice *et al.* (1993). This water balance model was developed especially for use at the landscape scale and it adequately captures regional differences in soil water content and actual evapotranspiration. The model utilizes monthly mean temperatures, precipitation and cloudiness values, and some soil characteristics to determine the annual pattern of water availability. The only adaptation that we allowed to the BIOME model was a more aggregated classification of the final biomes (Table 15.2). Two classes (agriculture and wetlands) were added to the original BIOME classification. These classes are important for the Land Cover Change model and the Terrestrial Emissions and Carbon models. All classes are compatible with the global land-cover database of Olson *et al.* (1985). We have created a similar aggregation of this database and assumed that it is representative of global land-cover patterns (both annual and agricultural vegetation) in 1970, the starting point of most IMAGE 2 simulations.

An important advantage of incorporating the BIOME model in IMAGE 2 is that changes in land cover due to shifting vegetation zones (caused by changes in climate and atmospheric composition) can be taken into account directly. The climatic linkages between the atmosphere–ocean component and TEC immediately determine the potential changes. Within the C cycle module the fluxes are computed dynamically and a transient response is simulated. The implementation of this response is very similar to the approach proposed by Smith & Shugart (1993). Another globally important feedback is the changing water use efficiency (WUE) due to increased atmospheric CO_2 levels. Gifford (1979), for example, found that under extreme xeric conditions plants could grow at high atmospheric CO_2 concentrations, whereas they were unable to grow at lower CO_2. This has far-reaching consequences for setting the soil water requirements of each PFT in BIOME. Plants tolerate a more severe drought stress, and thus have lower moisture requirements. We (Vloedbeld & Leemans 1993) have implemented this by assuming that WUE increases by 30% with a doubling of atmospheric CO_2 concentration (Körner 1993).

Potential agricultural patterns

The Crop Suitability approach was adopted, because it involves one of the few systems that can determine crop distribution and productivity comprehensively on a global grid. Most other approaches are either too

Table 15.2 *Regional aggregation for the socioeconomic compartments of IMAGE 2*

This aggregation is designed so that the socioeconomic driving factors for global change can be assumed to be homogeneous within a region.

1	Canada	
2	USA	(including Bermuda, St Pierre and Miquelon and the US minor outlying islands)
3	Latin America	(South America, Central America, Virgin Islands and Falklands)
4	Africa	(including Comoros and Brit. Indian Ocean Territories, Mauritius and São Tomé)
5	OECD Europe	(EU, EFTA and Andorra, Faeroe Islands, Gibraltar, Greenland, Liechtenstein, Monaco, San Marino, Svalbard, Jan Mayen and Vatican City State)
6	Eastern Europe	(Albania, Bulgaria, Czech Republic, Slovakia, Hungary, Poland, Romania and former Yugoslavia)
7	CIS	(Former USSR)
8	Middle East	(Afghanistan, Bahrain, Cyprus, Islamic Republic of Iran, Iraq, Israel, Jordan, Kuwait, Lebanon, Neutral Zone, Oman, Qatar, Saudi Arabia, Syrian Arab Republic, Turkey, United Arab Emirates, Yemen and Democratic Yemen)
9	India	(including Bangladesh, Bhutan, Maldives, Myanmar, Nepal, Pakistan, Sikkim, Sri Lanka)
10	China	(including Hong Kong, Democratic Kampuchea, People's Republic of China, Laos, Macau, Mongolia, Taiwan, Vietnam)
11	East Asia	(Indonesia, Malaysia, Republic of Korea, Papua New Guinea, Philippines, Singapore, Thailand and East Timor)
12	Pacifica	(Australia, New Zealand, Polynesia)
13	Japan	

data-demanding or too regionally specific (Leemans & Solomon 1993). The FAO approach was initially developed for tropical and subtropical regions, but we modified the methodology for cooler regions as well. The central concept of the approach is that of the growing period, i.e. that period during the year when warmth and soil moisture are adequate for growth (including germination, growth and maturation). We implemented this definition computationally by effectively using the same water balance model as the BIOME model (Cramer & Prentice 1988; Prentice *et al.* 1993) and some similar climatic indices (Leemans & Solomon 1993).

The growing period for crops (Fig. 15.2) is characterized by the annual pattern of daily temperature, precipitation and soil moisture values. By

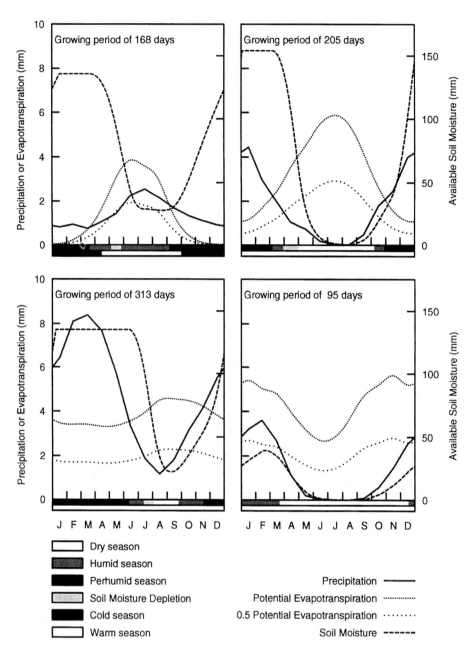

Figure 15.2   *Annual curves that determine the characteristics of the growing period. The upper left graph presents a typical temperate climate with a distinct winter period and no pronounced dry period. The upper right graph presents a subtropical climate without a cold period, but with a distinct dry period during the summer period. The lower left graph presents a tropical climate with a distinct seasonality in rainfall, resulting in a short dry period. The lower right graph presents a hot desert climate with little precipitation and no growing period. The definitions of the growing period follow those given in Anon (1978).*

definition, growth occurs only above 5 °C. The definition of the moisture season is more complex and follows the rationale given in Anon. (1978). In regions with a distinct dry season (inadequate moisture availability), the growing period starts arbitrarily when precipitation (P) equals half the potential evapotranspiration (PET). The rationale is that the first rains that fall on the surface of a soil with a large moisture deficit can be utilized immediately for seed germination and early growth. The soil moisture therefore does not need to be replenished completely (Anon. 1978). A successful crop life cycle further requires a period when the evapotranspiration demands of a crop under a full canopy cover can be met.

A successful crop harvest requires success in all life-cycle stages: germination, growth, flowering, and seed maturation. The phenology of these events is influenced by different climatic constraints. The model defines a minimum period to grow each crop to mimic its time-dependent phenology (Table 15.3). In some crops a cold or dry season is required to obtain harvestable yields. For winter varieties like winter wheat, the model assumes that germination takes place at the end of the previous growing period and that development is completed in the current growing period. The required length of the growing period is adjusted accordingly. However, a complete set of crop requirements is more complex than the model can handle and the length of the growing period alone does not define all constraints.

Chilling requirements for temperate crops are simulated using regressions between the mean temperature of the coldest month and the absolute minimum temperature. Specific warmth requirements during the growing period are defined using growing degree days above 5 °C. Drought requirements during the maturation periods are simulated by maximum values of the Priestley-Taylor index (α, which is estimated by AET : PET; see Prentice et al. 1992 and Cramer, Chapter 14, this volume). The ranges of the variables that define the climatic constraints for the several common crop varieties we simulated are given in Table 15.3.

If a crop could be successfully grown on the basis of climatic constraints, we used the model of de Wit (1965) to determine its productivity. The model was adapted to calculate the net biomass production and yield of crops from their climatic and photosynthetic properties (Anon. 1978; Leemans & Solomon 1993). Total yield of a crop (B_y) is determined by both photosynthesis and respiration, given by:

$$B_y = H_i (B_g - R) \tag{15.1}$$

where B_g is the gross biomass production, R is respiration loss and H_i the harvest index, an index that converts biomass production into the fraction of each crop representing economically useful yield.

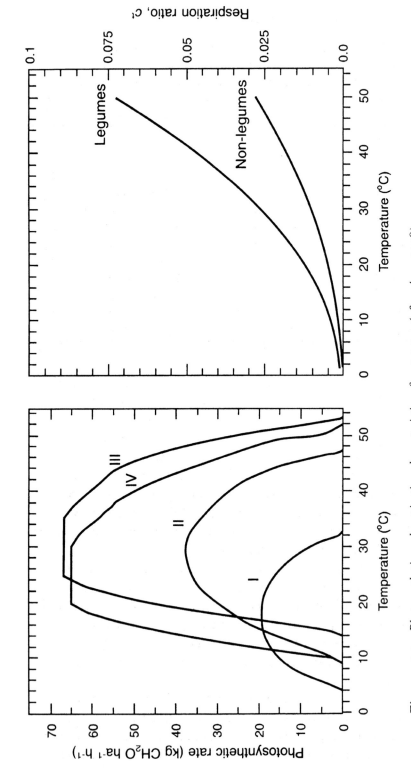

Figure 15.3 *Photosynthetic and respiration characteristics of crop types (after Anon 1978).*

Table 15.3 *Climatic crop requirements for ten major crops*

GPL, length of the growing period; MTR, temperature of the coldest month; GGD, growing degree days (°C); MR, α-moisture index (AET : PET); CT, crop type (see Fig 15.3); H_i, harvest index. The rank specifies combinations of crops.

Rank	Crop variety	GPL	MTR	GDD	MR	CT	H_i
1	Spring wheat	≥80	≥5.0	≥1000	none	I	0.4
1	Winter wheat	≥170	<10.0	≥1250	none	I	0.3
2	Rice	≥135	≥7.5	≥2250	≥0.95	II	0.3
3	Temperate maize	≥130	−20.0−15.0	≥1500	none	IV	0.4
3	Tropical maize	≥175	≥5.0	≥3000	none	III	0.3
4	Millet	≥80	≥25.0	≥1500	<0.95	IV	0.2
5	Potatoes	≥90	<15.0	≥900	none	I	0.5
5	Cassava	≥180	≥10.0	≥4500	none	II	0.7
6	Beans	≥110	≥20.0	≥1000	none	I	0.4
7	Sugar cane	≥240	≥5.0	≥4500	≥0.95	III	1.0
8	Soybeans	≥125	≥20.0	≥2000	none	II	0.4
8	Oil palm	≥330	>5.0	≥3500	none	II	0.2

Photosynthesis is limited by the amount of photosynthetically active radiation (PAR), which depends on location and cloudiness. Maximum gross biomass production is further strongly temperature dependent. This temperature dependence is crop-specific. All crops are classified into four crop PFTs with respect to temperature response of photosynthesis and photosynthetic pathway (Table 15.3 and Fig. 15.3). Crop types I and II incorporate the so-called C3 photosynthetic pathway and types III and IV the C4 pathway (Anon. 1978; for a description of these pathways see Bazzaz & Fajer 1992). A minimum, optimal and maximum temperature can be established for each crop type (Fig. 15.3).

Productivity (t ha^{-1}) is corrected for differences in soil productivity by using the soil classes of a digitized version of the FAO soil database (Zobler 1986). The outcome of this submodel is a geographically explicit potential crop distribution and productivity for each crop PFT. Productivity is further regionally adjusted for specific technology input levels (yield is adjusted downwards and upwards in the poorest regions and the industrialized regions, respectively).

The agricultural model simulates the global distribution of each crop realistically (Fig. 15.4). It also gives a reasonable indication of productivity (Leemans & Solomon 1993). The major discrepancy is in those agricultural regions where agriculture depends not upon precipitation but rather on

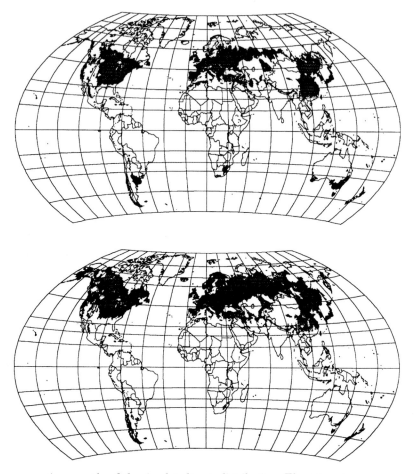

Figure 15.4 *An example of the simulated crop distribution. The top map shows the distribution for spring wheat under current climate; the bottom map shows a plausible distribution for future climate.*

support by the water supply of rivers, deep groundwater and irrigation. These agricultural systems are not taken into account in the current version of TEC, but this could easily be done by adding an irrigation overlay to the simulated crop distributions. This would be especially important for the Middle East (Table 15.2), where most agriculture depends on irrigation.

The simulation of actual land-cover patterns

The Land Cover Change model combines the potential vegetation and agricultural distributions and crop productivities with the demand for agricultural land. This demand is generated by the Agricultural Demand model and is merely a socioeconomic function of regionally explicit (human) food (cereals, vegetables and meat), animal fodder (mainly pasture and grass-lands, but also competing for cereals), energy crops (e.g. oils and alcohol) and wood (both timber and fuelwood) requirements. Historic trends are used as an empirical basis to parameterize the main mechanism and trends in each region (Table 15.2). Both demographic and developmental trends are incorporated. Different assumptions on the future projections of those socioeconomic demands can be included for each region to generate specific scenarios and analyse their global implications.

The demand for arable lands (needed to produce food, fodder and energy crops) must be satisfied by the potential crop distribution and productivity. This is done for each crop separately, but where combinations of crops are grown in each region, crops are ranked according to impor-tance (Table 15.3). The demand for the highest-ranking crops is satisfied first, followed by the others. If several of the required crops can be grown in a single grid cell, the ratios between the actual (1970) productivities of each crop for that region are used to fulfil the demands. The actual productivities are based on census data of the FAO (FAO 1991). If the productivity of current arable land (class 1, Table 15.4) does not meet the demands, arable land will expand at the expense of surrounding natural vegetation. Heuristic rules based on productivity and/or distance criteria are applied to satisfy such additional demand. The Land Cover Change model thus simulates a geographically explicit expansion (or contraction) of arable land. This is a major advance over earlier approaches (e.g. Rotmans 1990, 1992), which were often based on simple empirical transition matrices, and where deforestation always occurred in tropical forests. Here we do not discriminate when expanding agricultural areas.

A somewhat different approach is taken to fulfil the agricultural demands for grazing areas and pastures. This is mainly driven by the human demand for meat and dairy products. Pastures are not part of agricultural lands (Table 15.4) but are spread among the other classes (especially grasslands and savannas). This is in agreement with the definitions of the Olson *et al.* (1985) database. This also explains the smaller extent of agricultural land, when compared with other land-cover databases (e.g. Matthews 1985). If current grasslands and pastures do not satisfy the demand, other natural vegetation will be converted into grasslands within the same ecoclimatolo-gical zone of BIOME or Olson *et al.* (1985) (see Table 15.4).

The demand for forestry products (fuelwood, pulp and timber) is also a regionally specific function of demographic and socioeconomic driving factors. Wood products are closely linked to the C cycle (see next section), where net primary production (NPP) defines the large-scale sustainable

Table 15.4 *Aggregation of plant functional types into IMAGE 2 land-cover types*

No.	Land-cover type	Combinations of PFTs
1	Agricultural lands	None
2	Ice	Polar ice
3	Cool semidesert	Cool desert/shrub
4	Hot desert	Hot desert/shrub
5	Tundra	Cold grass/shrub
6	Cool grass and shrubs	Cold grass/shrub and cool grass/shrub
7	Warm grass and shrubs	Warm grass/shrub
8	Xerophytic woods	Sclerophyll/succulent shrub
9	Taiga	Boreal summergreen trees and boreal evergreen trees
10	Cool conifer forest	Boreal summergreen trees and cool-temperate evergreen trees
11	Cool mixed forest	Boreal summergreen trees, boreal evergreen trees, cool-temperate evergreen trees and temperate summergreen trees
12	Temperate deciduous forest	Boreal evergreen trees, cool-temperate evergreen trees and temperate summergreen trees
13	Warm mixed forest	Warm-temperate evergreen trees
14	Tropical dry savanna	Tropical raingreen trees
15	Tropical seasonal forest	Tropical rainforest trees and tropical evergreen trees
16	Tropical rainforest	Tropical evergreen trees
17	Wetlands	None

limits for forest productivity. The annual biomass increment (see section on C cycle) is used to satisfy the demand. If this increment is insufficient, standing biomass is used, which leads automatically to an increased deforestation rate. The forestry product module is calibrated for known years against forestry data (e.g. FAO 1983).

Additional requirements for pastures and forestry products are geographically satisfied by expanding current pasture areas or edges from existing forested regions, respectively. Only part of the biomass released by pasture-related deforestation is used to satisfy all forestry product demands. Fuel-wood is given a higher priority here.

All agricultural and forestry demands lead to geographically explicit changes in patterns of land cover. Shifts in vegetation patterns, caused by changing atmospheric CO_2 levels and climatic change, are taken into account directly (Vloedbeld & Leemans 1993); changes in land use, including those caused by different agricultural possibilities under changed

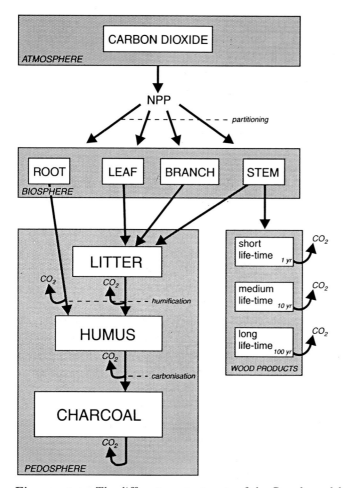

Figure 15.5 *The different compartments of the C cycle model.*

conditions, give additional shifts in global land-cover patterns. These global patterns (Fig. 15.7 and Plate 4, Table 15.4) are used to drive the calculations of greenhouse gas emissions from the terrestrial biosphere.

Emissions from changes in land use

The emissions of specific land-cover patterns are based on two different submodels (Fig. 15.1). The first model deals explicitly with the terrestrial C cycle, while the other accounts for all other greenhouse gases. The latter model is based on the land-use model of Bouwman *et al.* (1992), and is a mixture of geographically and regionally explicit parameterization of the most important fluxes of greenhouse gases. Some of the fluxes are climate- and soil-dependent (modelled geographically explicitly) whereas others are directly driven by human activities (modelled regionally explicitly).

The C cycle model is derived from the structure of Goudriaan &

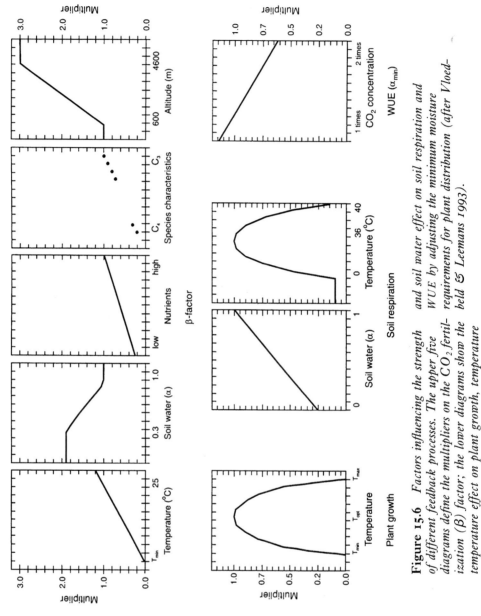

Figure 15.6 Factors influencing the strength of different feedback processes. The upper five diagrams define the multipliers on the CO_2 fertilization (β) factor; the lower diagrams show the temperature effect on plant growth, temperature and soil water effect on soil respiration and WUE by adjusting the minimum moisture requirements for plant distribution (after Vloedbeld & Leemans 1993).

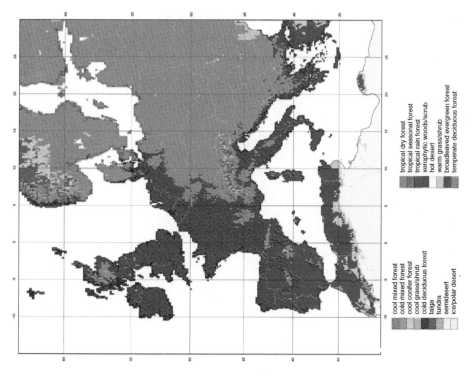

Plate 1 *Map of the distribution of major types of potential natural vegetation in Europe under 'current' climate, based on the BIOME model (Prentice et al. 1992) and a detailed database of climatic variables at 10′ latitude/longitude resolution.*

cool mixed forest
cold mixed forest
cool conifer forest
cool grass/shrub
cold deciduous forest
taiga
tundra
semidesert
ice/polar desert

tropical dry forest
tropical seasonal forest
tropical rain forest
xerophytic woods/scrub
hot desert
warm grass/shrub
broadleaved evergreen forest
temperate deciduous forest

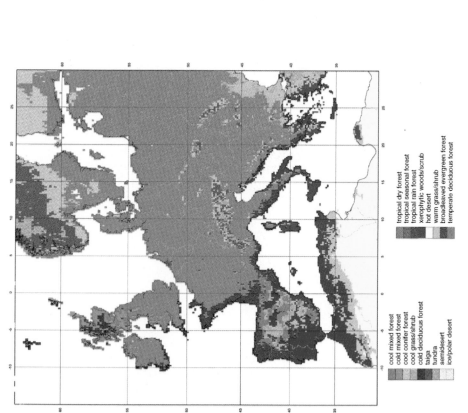

Plate 2 *Map of the distribution of major types of potential natural vegetation in Europe under a climate scenario generated from current baseline climate and climatic anomalies for temperature and precipitation, derived from the GISS model run at $2 \times CO_2$ concentration (Hansen et al. 1988).*

cool mixed forest
cold mixed forest
cool conifer forest
cool grass/shrub
cold deciduous forest
taiga
tundra
semidesert
ice/polar desert

tropical dry forest
tropical seasonal forest
tropical rain forest
xerophytic woods/scrub
hot desert
warm grass/shrub
broadleaved evergreen forest
temperate deciduous forest

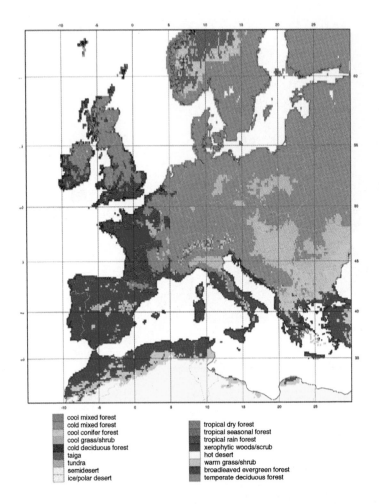

cool mixed forest	tropical dry forest
cold mixed forest	tropical seasonal forest
cool conifer forest	tropical rain forest
cool grass/shrub	xerophytic woods/scrub
cold deciduous forest	hot desert
taiga	warm grass/shrub
tundra	broadleaved evergreen forest
semidesert	temperate deciduous forest
ice/polar desert	

Plate 3 *Map of the distribution of major types of potential natural vegetation in Europe under a climate scenario generated from current baseline climate and climatic anomalies for temperature and precipitation, derived from the MPI T42 model run at 3×CO₂ concentration (Cubasch et al. 1992).*

Plate 4 (OPPOSITE) *The distribution of potential global vegetation patterns under current climate (top), future climate (centre) and future climate with a WUE feedback (bottom).*

Current climate

Future climate

Future climate and WUE feedback

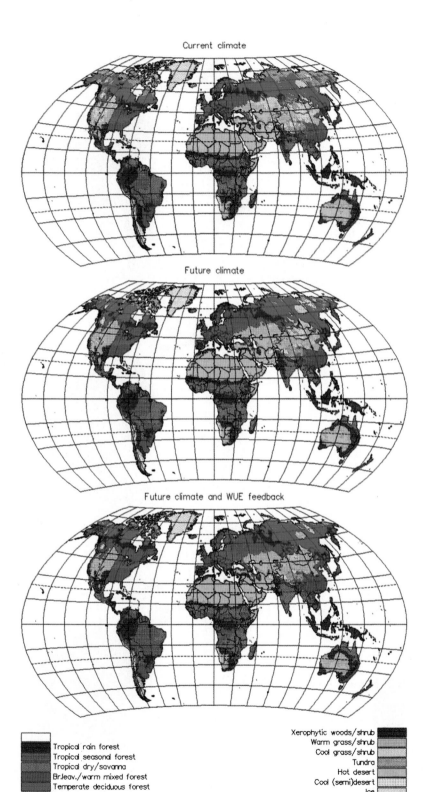

Tropical rain forest
Tropical seasonal forest
Tropical dry/savanna
Br.leav./warm mixed forest
Temperate deciduous forest
Cool mixed forest
Cool conifer forest
Taiga

Xerophytic woods/shrub
Warm grass/shrub
Cool grass/shrub
Tundra
Hot desert
Cool (semi)desert
Ice

1970

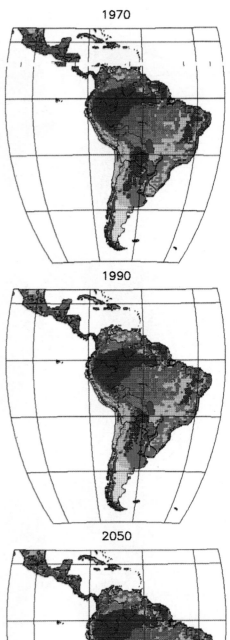

Plate 5 *Land-cover map
for Latin America as simulated
with the terrestrial environment
system of IMAGE 2, showing
initial (1970), current (1990)
and future patterns (2050).*

1990

2050

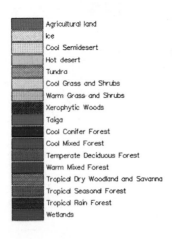

Agricultural land
Ice
Cool Semidesert
Hot desert
Tundra
Cool Grass and Shrubs
Warm Grass and Shrubs
Xerophytic Woods
Taiga
Cool Conifer Forest
Cool Mixed Forest
Temperate Deciduous Forest
Warm Mixed Forest
Tropical Dry Woodland and Savanna
Tropical Seasonal Forest
Tropical Rain Forest
Wetlands

Ketner's (1984) model. This was initially developed for a limited number of global ecosystems worldwide, but we adapted its structure in order to make it geographically explicit and fully compatible with the IMAGE 2 set of land-cover types (Table 15.4). This updated C cycle model can take into account C fluxes from changes in land cover.

Each land-cover type is characterized by an NPP and a series of compartments (leaves, branches, stems, roots, litter humus and charcoal; Fig. 15.5). NPP is partitioned between these compartments, each compartment having its own specific C turnover time (Table 15.5). The increase in biomass and NPP of undisturbed ecosystems through time is modelled with a simple logistic function (Cooper 1983). The C fluxes between different compartments and the atmosphere are modelled so that they mimic the time-dependent dynamics between them (Klein Goldewijk & Vloedbeld 1995). NPP is adjusted for environmental conditions, such as atmospheric CO_2 concentrations, nutrient availability, topography and climate. Feedbacks that influence the C fluxes from the terrestrial biosphere, such as enhanced soil respiration at increasing temperatures, C fertilization, changing WUE, and combined temperature response of photosynthesis and respiration, are modelled explicitly. The growth rate, soil respiration rate and CO_2 fertilization factor (β) are adjusted according to local conditions (Vloedbeld & Leemans 1993; Fig. 15.6).

The geographically explicit C dynamics can easily be aggregated, so that regional and global C budgets can be presented and compared with other analyses. This aggregation consists of summing all (area-corrected) fluxes of separate grid cells. The C stock values for each compartment are stored; this allows for a detailed analysis of the transient C dynamics.

Applications of the Terrestrial Environment Component of IMAGE 2

IMAGE 2 is still under development; in particular, parts of the energy–industry and atmosphere–ocean components are not yet completed. Within TEC the socioeconomic demand functions are still rudimentary and could be improved for most regions. Only the demand functions for Latin America are implemented in a satisfactory way. This region is further chosen because land-cover changes are well documented here. Potential vegetation and crop distribution patterns, however, are already determined for the whole globe. I therefore used a standard socioeconomic and climatic scenario for Latin America as an illustration of the potential of land-use change modelling in IMAGE 2. Completely linked runs with IMAGE 2 (energy, climate–ocean and terrestrial biosphere) should become available during the latter half of 1993 (see Note added in proof, p. 313).

Table 15.5 *Characteristics of the terrestrial carbon cycle model for each land-cover type*

The life time of charcoal in all land-cover types is set to 1000 years.

Land-cover type	Area (Mha)	Init. NPP (t C/ha^{-1})	Partitioning (unitless)				Lifetime (yr)					
			leaf	branch	stem	root	leaf	branch	stem	root	litter	humus
Agricultural lands	2509.6	4.0	0.8	0.0	0.0	0.2	1	1	1	1	1	20
Ice	278.5	0.0	0.5	0.1	0.1	0.3	1	10	50	3	5	100
Cool semidesert	200.8	0.5	0.5	0.1	0.1	0.3	1	10	50	3	5	100
Hot desert	1622.8	0.5	0.5	0.1	0.1	0.3	1	10	50	3	1	50
Tundra	1111.9	1.0	0.5	0.1	0.1	0.3	1	10	50	10	5	100
Cool grass and shrubs	481.9	3.0	0.6	0.0	0.0	0.4	1	10	50	3	1	60
Warm grass and shrubs	1736.0	4.0	0.6	0.0	0.0	0.4	1	10	50	3	1	20
Xerophytic woods	850.9	4.0	0.3	0.2	0.3	0.2	1	10	50	10	1	50
Taiga	1146.1	5.0	0.3	0.2	0.3	0.2	3	20	50	10	5	60
Cool conifer forests	309.4	5.0	0.3	0.2	0.3	0.2	3	20	50	10	3	50
Cool mixed forests	153.5	6.0	0.3	0.2	0.3	0.2	1	20	50	10	1	40
Temperate deciduous forests	78.3	6.0	0.3	0.2	0.3	0.2	1	20	50	10	1	40
Warm mixed forests	431.5	6.0	0.3	0.2	0.3	0.2	1	20	50	10	1	40
Tropical dry savanna	1151.1	4.0	0.3	0.2	0.3	0.2	2	10	30	5	1	20
Tropical seasonal forests	621.0	8.0	0.3	0.2	0.3	0.2	2	10	30	8	1	20
Tropical rainforests	427.1	9.0	0.3	0.2	0.3	0.2	2	10	30	8	1	20
Nat. wetlands	370.7	5.0	0.3	0.2	0.3	0.2	2	10	30	8	1	50

The socioeconomic scenario was based on the IPCC business-as-usual scenario (Houghton *et al.* 1992), with an increase in world population of 93% at 2050 (82% for South America). The accompanying change in Gross National Product (GNP) is 134% (242% for South America). I used a climate scenario that is based on the General Fluid Dynamics Laboratory Atmospheric General Circulation Model (GFDL-AGCM) (Manabe & Wetherald 1987), which shows a large temperature increase in high-latitude regions during winter. The creation of climate-change scenarios using AGCMs and existing databases is described by Leemans (1992). The model was initialized with the global land-cover database of Olson *et al.* (1983) and climate database of Leemans & Cramer (1991). The simulation with the TEC started at 1970 and continued up to 2050, which resulted in more than a doubling of CO_2-equivalents in greenhouse gases (686 ppm vs. 340 ppm in 1970) and almost a doubling of the CO_2 concentration (570 ppm vs. 320 ppm in 1970).

The simulation of potential vegetation patterns (Plate 4) was in good agreement with the actual vegetation patterns (Table 15.1: BIOME). Although there was good agreement in global patterns, a significant number of terrestrial grid cells (*ca.* 20%) did not match with those of the Olson *et al.* (1985) database. Those cells occurred mainly along the edges of land-cover classes, but also in those areas where vegetation patterns are not strictly dominated by climate, but rather by other factors, such as soil, topography and anthropogenic influences. (This contradiction probably also reflects the very different origins of the land-cover databases and demonstrates the need for the development of a globally comprehensive land-cover database.) The contradiction between land-cover databases, however, provided a serious problem in the smooth transition of vegetation between current (1970, i.e. Olson *et al.*) patterns to future (i.e. BIOME) patterns. Major changes occur during the early simulated years. We have met this inconsistency by assuming that a cell only changes vegetation if there is a shift in potential vegetation (i.e. current and future BIOME patterns). This is a reasonable compromise and results in a more even transition through time.

Plate 4 also illustrates the potential of including feedback mechanisms in the determination of the vegetation patterns. Earlier studies have emphasized the redistribution of vegetation under a changing climate (e.g. Emanuel *et al.* 1985a,b; Leemans 1992; Cramer & Leemans 1993). Such feedback is also included in IMAGE 2 and is a direct result of the linkage between the atmosphere–ocean component (here the result of a changed climate by an AGCM) and TEC. Climate change immediately feeds forward into the potential vegetation patterns and agricultural productivity (Plates 4 and 5), which determine land cover and thus the carbon cycle. A different

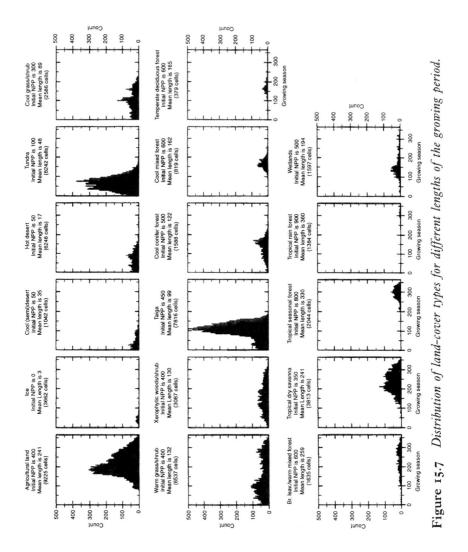

Figure 15.7 *Distribution of land-cover types for different lengths of the growing period.*

and more complex feedback involves a changed WUE under increasing atmospheric CO_2 concentrations. This should both enhance growth and influence the distribution of PFTs into more xeric conditions. This feedback is implemented in two different parts of the TEC. Growth enhancement directly influences the C fertilization factor, while the distributions are determined through changes in the climatic moisture requirements (Vloedbeld & Leemans 1993). Fig. 15.7 and Plate 4 illustrate the consequences of the latter feedback. An expansion of grasslands in desert can be observed, as well as a significant expansion of forested ecosystems. This could have a major impact on the carbon cycle and is not yet considered in most other studies.

Figure 15.4 illustrates the simulation of the distribution of wheat under current (1970) and future (2050) climate. Other crops are presented and discussed by Leemans & Solomon (1993). The distribution is the combined outcome of two simulated varieties, winter wheat and spring wheat (Table 15.3), but it is typical for a whole range of temperate cereal crops, including barley and oats. The aggregation is made because the data supporting the Agricultural Demand model do not support all crop varieties separately. For the land-cover simulation, all equally ranked crops are aggregated (Table 15.3).

Plate 5 presents a preliminary assessment of changes in land cover for Latin America. Small shifts can be observed for the period from 1970 to 1990 and these are caused mainly by changes in land-use patterns. Agricultural land in Argentina and Paraguay expands, mainly along the coast into Brazil, while some forested regions in southern Brazil are converted into grassland and pastures. This pattern is a direct result of the simple set of heuristic rules weighting productivity and distance. The observed pattern of deforestation, pasture and agricultural expansion resembled the simulated pattern. This is a surprise, because many publications (e.g. Turner *et al.* 1990) present much more complex socioeconomic causes of these changes than can be grasped with our heuristic model. These initial results give us at least some confidence that human behaviour on large scales can be modelled adequately with simple rule-based systems.

The simulation further develops increasing deforestation and pasture formation in the southern parts of Brazil, increasing the extent of agricultural regions along the coast and in Argentina. In 2050 most parts of Latin America are strongly influenced by human activities. This has a much stronger impact on the C cycle than changes in climate or the biochemical feedbacks, such as CO_2 fertilization. The latter process increases the sink function of the terrestrial biosphere, whereas changed land use generates a much larger source. A large difference between this model and the traditional C cycle models should be mentioned here, however. TEC spreads out land use over all different land-cover classes. There is no discrimination for any class, but the emerging patterns are based only on the agricultural potential and historic patterns. The traditional C cycle models that use a transition approach all decrease the total amount of tropical forests rapidly. Some of them show scenarios where these forests have completely disappeared by 2025 or somewhat later (Rotmans *et al.* 1990). IMAGE 2 gives a more optimistic scenario. A significant amount of tropical forests remains, leaving possibilities for conservation of biodiversity.

Concluding remarks

The IMAGE 2 approach indicates the usefulness of plant functional types (PFTs). PFTs can be implemented so that most important environmental constraints that limit their distribution and productivity are accounted for. Different sets of PFTs can be set up for different tasks. In the IMAGE model there are sets for agricultural crops and natural vegetation. Linkages between those sets provide possibilities for generating modelling frameworks that mimic the behaviour of relatively complex systems. The use of sets of PFTs also increases the flexibility of the modelling approach. The modularity allows for an exchange of unique sets. In future applications this could become advantageous, when issues different from global change, such as sustainable agriculture and forestry, and biodiversity, have to be addressed.

The above examples of simulations with IMAGE show the significance of land-use changes when projecting causes and impacts of global change. Without an adequate assessment of land use and the consequent changes and modifications of land cover, future projections of greenhouse gas emissions cannot be achieved. Furthermore, feedbacks, environmental constraints and regional differences also have to be taken into account. The IMAGE 2 model presents a framework for starting to address these important scientific, policy and social issues. The model will not resolve all scientific problems or develop the most effective policies, but it will assist in focusing on the most important issues and in evaluating policy scenarios. In this respect it has already been used to confine some of the socioeconomic driving forces of the IGBP-HDP core project on 'Land Use and Land Cover'.

A new field of integrated modelling that links scientific theory, models, Geographic Information Systems, environmental databases, and socioeconomic scenarios is clearly emerging, and many core projects of IGBP and IPCC are facilitating and enhancing this process. In the coming decade many approaches will be developed, improving our understanding of the basic functioning of the global system in all its facets.

Acknowledgements

The foregoing work owes its development and implementation to many people. Wolfgang Cramer and Allen M. Solomon provided useful ideas and discussions for the vegetation modelling, and Robert Brinkmann (FAO) and Gunter Fischer (IIASA) provided considerable assistance in agricultural modelling and obtaining crop data. Finally, without the IMAGE team headed by Joe Alcamo, the work would not have been possible. The work was supported by the Ministry of Housing, Physical Planning and the Environment of The Netherlands (MAP 481507 and MAP 482507 to RIVM) and the Dutch National Research Programme 'Global Air Pollution and Climate Change' under agreements MAP 482510 and MAP 482509. The research also contributes to IGBP–GCTE core research.

Note added in proof

Since the preparation of this paper in January 1993, the IMAGE 2 model has been further developed and used to evaluate the policy influence on future trends in the atmospheric concentrations of several greenhouse gases. The model and some applications are fully documented in Alcamo (1994).

References

Alcamo, J. (ed.) (1994) *IMAGE 2.0: Integrated Modeling of Global Climate Change.* Dordrecht: Kluwer Academic Publishers, 318 pp.

Anon. (1978) *Report on the Agro-ecological Zones Project. Vol 3. Methodology and Results for South and Central America.* Rome: Food and Agriculture Organization of the United Nations. World Soil Resources Report 48/3.

Anon. (1990) *The International Geosphere-Biosphere Programme: A Study of Global Change. The Initial Core Projects.* Stockholm: International Geosphere-Biosphere Programme. IGBP-Report No. 12.

Bazzaz, F. A. & Fajer, E. D. (1992) Plant life in a CO_2-rich world. *Scientific American,* **1992,** 18–21.

Bernthal, F. M. (ed.) (1990) *Climate Change: The IPCC Response Strategies.* Geneva: World Meteorological Organization/United Nations Environment Program.

Bolin, B., Döös, B. R., Jäger, J. & Warrick, R. A. (eds) (1986) *The Greenhouse Effect, Climate Change, and Ecosystems: A Synthesis of Present Knowledge.* Chichester: John Wiley & Sons.

Bouwman, A. F., Van Staalduinen, L. & Swart, R. J. (1992) *The IMAGE land use model to analyze trends in land-use related emissions.* Bilthoven: National Institute of Public Health and Environmental Protection, RIVM. Report nr. 222901009.

Box, E. O. (1981) *Macroclimate and Plant Forms: an Introduction to Predictive Modeling in Phytogeography.* The Hague: Dr W. Junk Publishers.

Budyko, M. I. (1986) *The Evolution of the Biosphere.* Dordrecht: D. Reidel Publishing Company.

Claussen, M. & Esch, M. (1994) Biomes computed from simulated climatologies. *Climate Dynamics,* **9,** 235–43.

Cooper, C. F. (1983) Carbon storage in managed forests. *Canadian Journal of Forest Research,* **13,** 155–66.

Cramer, W. & Leemans, R. (1993) Assessing impacts of climate change on vegetation using climate classification systems. In Solomon, A. M. & Shugart, H. H. (eds), *Vegetation Dynamics Modeling and Global Change,* pp. 190–217. New York: Chapman and Hall.

Cramer, W. & Prentice, I. C. (1988) Simulation of regional soil moisture deficits on a European scale. *Norsk Geografisk Tidskrift,* **42,** 149–51.

Cramer, W. & Solomon, A. M. (1993) Climatic classification and future global redistribution of agricultural land. *Climate Research,* **3,** 97–110.

CRU and ERL (1992) *Development of a framework for the evaluation of policy options to deal with the Greenhouse Effect. A scientific description of the ESCAPE model version 1.1.* Norwich and London: Climatic Research Unit and Environmental Resources Limited.

Emanuel, W. R., Shugart, H. H. & Stevenson, M. P. (1985a) Climatic change and the broadscale distribution of terrestrial ecosystem complexes. *Climatic Change*, 7, 29–43.

Emanuel, W. R., Shugart, H. H. & Stevenson, M. P. (1985b) Response to comment: Climatic change and the broad-scale distribution of terrestrial ecosystem complexes. *Climatic Change*, 7, 457–60.

EPA/NOAA (1991) *Global ecosystems database, Version 0.1 (Beta-test).* Boulder, CO: EPA Global Climate Research Program and NOAA/NDGC Global Change Database Program. Geophysical Records Documentation No. 25.

Espenshade, E. B. Jr & Morrison, J. L. (eds) (1991) *Goode's World Atlas.* Chicago: Rand McNally & Company.

FAO (1983) *FAO Production Handbook.* FAO Database series 1. Rome: Food and Agriculture Organization of the United Nations.

FAO (1991) *Agrostat PC.* Rome: Food and Agriculture Organization of the United Nations.

Gifford, R. M. (1979) Growth and yield of CO_2-enriched wheat under water-limited conditions. *Australian Journal of Plant Physiology*, 6, 367–78.

Goudriaan, J. & Ketner, P. (1984) A simulation study for the global carbon cycle, including man's impact on the biosphere. *Climatic Change*, 6, 167–92.

Guetter, P. J. & Kutzbach, J. E. (1990) A modified Köppen classification applied to model simulations of glacial and interglacial climate. *Climatic Change* 6, 193–215.

Henderson-Sellers, A. (1991) Developing an interactive biosphere for global climate models. *Vegetatio*, 91, 149–66.

Higgins, G. M., Kassam, A. H., van Velthuizen, H. T. & Purnell, M. F. (1987) Methods used by FAO to estimate environmental resources, potential outputs of crops, and population-supporting capacities in developing nations. In Bunting, A. H. (ed.), *Agricultural Environments: Characterization, Classification and Mapping*, pp. 171–83. Wallingford, UK: CAB International.

Holdridge, L. R. (1947) Determination of world plant formations from simple climatic data. *Science*, 105, 367–8.

Houghton, J. T., Callander, B. A. & Varney, S. K. (eds) (1992) *Climate Change 1992. The Supplementary Report to the IPCC Scientific Assessment.* Cambridge University Press.

Houghton, J. T., Jenkins, G. J. & Ephraums, J. J. (eds) (1990) *Climate Change: The IPCC Scientific Assessment.* Cambridge University Press.

Huntley, B. & Webb, T. III (eds) (1988) *Vegetation History.* Dordrecht: Kluwer Academic Publishers.

Izrael, Y. A., Hashimoto, M. & Tegart, W. J. M. (eds) (1990) *Climate Change: The IPCC Impact Assessment.* Canberra: Australian Government Publishing Service.

Klein Goldewijk, K. & Vloedbeld, M. (1995) The exchange of carbon between the atmosphere and the terrestrial biosphere in Latin America. In Lal, R., Kimble, J., Levine E. & Stewart, B. A. (eds), *Soils and Global Change*, pp. 395–413. Boca Raton, Florida: Lewis.

Köppen, W. (1936) Das geographische System der Klimate. In Köppen, W. & Geiger, R. (eds), *Handbuch der Klimatologie*, pp. 1–46. Berlin: Gebrüder Borntraeger.

Körner, C. (1993) CO_2 fertilization: The great uncertainty in future vegetation development. In Solomon, A. M. & Shugart, H. H. (eds), *Vegetation Dynamics Modeling and Global Change*, pp. 53–70. New York: Chapman and Hall.

Leemans, R. (1989) Possible changes in natural vegetation patterns due to a global warming. In Hackl, A. (ed.), *Der Treibhauseffeckt: das Problem – Mögliche Folgen – Erforderliche Maßnahmen*, pp. 105–22.

Laxenburg, Austria: Akademie für Umwelt und Energie.

Leemans, R. (1992) Modelling ecological and agricultural impacts of global change on a global scale. *Journal of Scientific and Industrial Research*, **51**, 709–24.

Leemans, R. & Cramer, W. (1991) *The IIASA database for mean monthly values of temperature, precipitation and cloudiness on a global terrestrial grid*. Laxenburg: International Institute for Applied Systems Analysis. Research Report RR-91-18.

Leemans, R. & Solomon, A. M. (1993) The potential reponse and redistribution of crops under a doubled CO_2 climate. *Climate Research*, **3**, 79–96.

Manabe, S. & Wetherald, R. T. (1987) Large-scale changes in soil wetness induced by an increase in carbon dioxide. *Journal of Atmospheric Science*, **44**, 1211–35.

Matthews, E. (1985) *Atlas of archived vegetation, land-use and seasonal albedo data sets*. New York: NASA. Technical Memorandum 86199.

Monserud, R. A. & Leemans, R. (1992) The comparison of global vegetation maps. *Ecological Modelling*, **62**, 275–93.

Olson, J., Watts, J. A. & Allison, L. J. (1983) *Carbon in live vegetation of major world ecosystems*. Oak Ridge, Tennessee: Oak Ridge National Laboratory. ORNL-5862.

Olson, J., Watts, J. A. & Allison, L. J. (1985) *Major world ecosystem complexes ranked by carbon in live vegetation: a database*. Oak Ridge, Tennessee: Oak Ridge National Laboratory. Report NDP-017.

Parry, M. L., Porter, J. H. & Carter, T. R. (1990) Agriculture: climate change and its implications. *Trends in Ecology and Evolution*, **5**, 318–22.

Prather, M. (ed.) (1989) *An assessment model for atmospheric composition: report of a workshop*. New York: NASA Goddard Institute.

Prentice, I. C., Cramer, W., Harrison, S. P., Leemans, R., Monserud, R. A. & Solomon, A. M. (1992) A global biome model based on plant physiology and dominance, soil properties and climate. *Journal of Biogeography*, **19**, 117–34.

Prentice, I. C., Sykes, M. T. & Cramer, W.

(1993) A simulation model for the transient effects of climate change on forest landscapes. *Ecological Modelling*, **65**, 51–70.

Prentice, I. C., Sykes, M. T., Lautenschlager, M., Harrison, S. P., Denissenko, O. & Bartlein, P. J. (1994) Modelling global vegetation patterns and terrestrial carbon storage at the last glacial maximum. *Global Ecological and Biogeographical Letters*, **3**, 67–76.

Prentice, I. C., Webb, R. S., Ter-Mikhaelian, M. T., Solomon, A. M., Smith, T. M., Pitovranov, S. E., Nikolov, N. T., Minin, A. A., Leemans, R., Lavorel, S., Korzukhin, M. D., Helmisaari, H. O., Hrabovsky, J. P., Harrison, S. P., Emanuel, R. W. & Bonan, G. B. (1989) *Developing a Global Vegetation Dynamics Model: Results of an IIASA Summer Workshop*. Laxenburg, Austria: International Institute of Applied Systems Analysis. IIASA Research Report RR-89-7.

Prentice, K. C. & Fung, I. Y. (1990) The sensitivity of terrestrial carbon storage to climate change. *Nature*, **346**, 48–51.

Rotmans, J. (1990) *IMAGE: An Integrated Model to Assess the Greenhouse Effect*. Dordrecht: Kluwer Academic Publishers.

Rotmans, J. (1992) ESCAPE: an integrated climate model for the EC. *Change*, **11**, 1-4.

Rotmans, J., de Boois, H. & Swart, R. J. (1990) An integrated model for the assessment of the greenhouse effect: the Dutch approach. *Climatic Change*, **16**, 331–56.

Shugart, H. H., Leemans, R. & Bonan, G. B. (eds) (1992) *A Systems Analysis of the Global Boreal Forest*. Cambridge University Press.

Smith, T. & Huston, M. (1989) A theory of the spatial and temporal dynamics of plant communities. *Vegetatio*, **83**, 49–69.

Smith, T. M., Leemans, R. & Shugart, H. H. (1992) Sensitivity of terrestrial carbon storage to CO_2 induced climate change: Comparison of four scenarios based on general circulation models. *Climatic Change*, **21**, 367–84.

Smith, T. M. & Shugart, H. H. (1993). The potential response of global terrestrial carbon storage to a climate change. *Water, Air and Soil Pollution*, **70**, 629–42.

Solomon, A. M. & Shugart, H. H. (eds) (1993) *Vegetation Dynamics Modeling and Global Change*. New York: Chapman and Hall.

Sundquist, E. T. (1993) The global carbon dioxide budget. *Science*, **259**, 934–41.

Tchebakova, N., Monserud, R. A., Leemans, R. & Golovanov, S. (1993a) A global vegetation model based on the climatological approach of Budyko. *Journal of Biogeography*, **20**, 219–44.

Tchebakova, N. M., Monserud, R. A. & Leemans, R. (1993b) Global vegetation change predicted by the modified Budyko model. *Climatic Change*, **25**, 59–83.

Thornthwaite, C. W. (1948) An approach toward a rational classification of climate. *Geographical Reviews*, **38**, 55–94.

Turner, B. L., II, Clark, W. C., Kates, R. W., Richards, J. F., Mathews, J. T. & Meyer, W. B. (eds) (1990) *The Earth as Transformed by Human Action: Global and Regional Changes in the Biosphere over the Past 300 Years*. Cambridge University Press.

Turner, B. L., Moss, R. H. & Skole, D. L. (1993) *Relating Land Use and Global Change: A Proposal for an IGBP-HDP Core Project*. Stockholm: International Geosphere–Biosphere Programme and the Human Dimensions of Global Environmental Change Programme. IGBP Report No. 24 and HDP Report No. 5.

Vloedbeld, M. & Leemans, R. (1993) Quantifying feedback processes in the response of the terrestrial carbon cycle to global change – the modeling approach of IMAGE-2. *Water, Air and Soil Pollution*, **70**, 615–28.

Wit, C. T. de (1965) *Photosynthesis of leaf canopies*. Wageningen: Centre for Agricultural Publication and Documentation. Agricultural Research Report 663.

Woodward, F. I. & Williams, B. G. (1987) Climate and plant distribution at global and local scales. *Vegetatio*, **69**, 189–97.

Zobler, L. (1986) *A world soil file for global climate modeling*. New York: NASA. Technical Memorandum.

Part five

Examining the consequences of classifying species into functional types: a simulation model analysis

T. M. Smith

Introduction

The approach of categorizing species into functional types based on adaptations for specific environments has formed the basis of a number of plant classification systems. Approaches include *r*, *K*, and adversity strategies (MacArthur & Wilson 1967; Southwood 1977; Greenslade 1983); early- and late-successional species (Budowski 1965, 1970; Whittaker 1975; Bazzaz 1979; Finegan 1984); exploitative and conservative (Bormann & Likens 1981); ruderal, stress-tolerant, and competitive (Grime 1977, 1979); gap and non-gap species (Hartshorn 1978; Brokaw 1985a,b); structural classifications (Raunkiaer 1934; Webb *et al.* 1970; Hallé 1974; Hallé & Oldeman 1975; Walker *et al.* 1981); and vital attributes (Noble & Slatyer 1980). Many of these schemes for classifying species into functional categories have been used within a modelling framework to examine community and ecosystem dynamics; however, little if any focus has been placed on examining the consequences of aggregation on the predicted patterns. Such comparisons become more important when examining the response of ecosystems to changing environmental conditions.

The previous chapters have been devoted to defining the concept of plant functional types and exploring the development and application of functional classifications of plants in different ecosystems. If the task is to simplify the description of vegetation for the purposes of examining ecosystem dynamics, the critical question becomes 'What are the consequences of aggregating species into functional types for the purposes of predicting the spatial and temporal dynamics of ecosystems?' In the context of global change issues, this question must be evaluated not only for a given environment, but also with respect to changing (temporal and spatial) environmental conditions.

The objectives of this chapter are: (1) to examine the consequences of using an aggregated description of vegetation (i.e. functional types) to

predict ecosystem dynamics under changing environmental conditions, and
(2) in the light of the model results, to evaluate the basis for developing
aggregation schemes (i.e. functional classifications). To meet the first objective, one must examine the predicted dynamics of an ecosystem based on a
description of the vegetation at the species level, as well as using some
aggregated description where species are categorized into functional groups.
Rather than examining any given ecosystem, I will examine the consequences of aggregating species in hypothetical communities, where the
communities are defined by an array of species with given life-history
characteristics and environmental response functions. I will use the individual-based forest simulation model ZELIG (Smith & Urban 1988). This
model has been used in a variety of studies examining the consequences of
patterns of species life history on community and ecosystem dynamics (e.g.
Huston & Smith 1987; Smith & Huston 1989; Smith & Urban 1988; Urban
& Smith 1989).

Model description

ZELIG is an individual-based forest stand simulator derived from the
JABOWA (Botkin *et al.* 1972) and FORET (Shugart & West 1977) models.
Each individual plant is modelled as a unique entity with respect to the
processes of establishment, growth and mortality. This allows the model to
track species- and size-specific demographic behaviours. The model structure includes two features important to a dynamic description of vegetation:
(a) the response of the individual to the prevailing environmental
conditions, and (b) how the individual modifies those environmental
conditions (i.e. feedback between vegetation structure/composition and the
environment).

Response of the individual to the environment

The growth of an individual is calculated by using a function that is
species-specific and predicts, under optimal conditions, the expected growth
increment given an individual's current size (Figure 16.1). This optimal
growth increment is then modified by a set of environmental response functions, and the realized increment is added to the individual.

Environmental responses are modelled via a 'constrained potential' paradigm. In this, an individual has a maximum potential behaviour (i.e.
maximum diameter increment, survivorship or establishment rate) under
optimal conditions. This optimum is then reduced according to the environmental conditions of the plot (e.g. shading, drought), to yield the realized

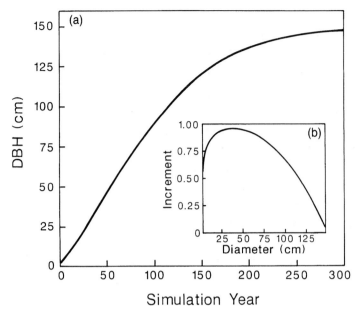

Figure 16.1   *Optimal growth curve describing (a) growth (diameter at breast height) as a function of time, and (b) annual diameter increment as a function of diameter.*

behaviour under the ambient conditions. The functions used to describe species' response to environmental factors are scaled between 0.0 and 1.0 to reflect proportion of optimal growth (see Figure 16.2 and later discussion for an example). These values are then multiplied by the optimal growth rate as defined in Figure 16.1.

The death of individuals is modelled as a stochastic process. Two components of mortality are simulated: age-related and stress-induced. The age-related component applies equally to all individuals of a species and is a function of the maximum expected longevity for the species. Stress is defined with respect to a minimum growth increment; individuals failing to meet this minimal condition are subjected to an elevated mortality rate.

Effect of the individual on the environment

The vertical structure of the canopy is modelled explicitly. The sizes of individuals (height and leaf area) are used to construct a leaf area profile. Using a light extinction equation, the vertical profile of available light is then calculated so that the light environment for each individual can be defined.

Plant influence on other features of the environment (e.g. temperature, nutrient and water availability) has been included in other applications of the model; however, for the purposes of this exercise only plant feedback

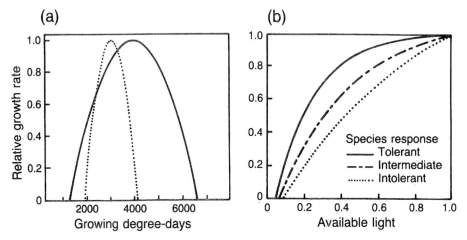

Figure 16.2 *Environmental response function relating proportion of optimal growth (see Fig. 16.1) as a function of environmental conditions on the simulated plot. Examples shown are (a) temperature expressed as growing degree-days, and (b) available light expressed as proportion of full sun.*

on the light environment will be considered. This class of models has been used to examine patterns of vegetation dynamics for a variety of forest ecosystems (see Shugart 1984), including examining the potential impacts of climate change, both past (Solomon *et al.* 1980, 1981; Solomon & Shugart 1984; Solomon & Webb 1985; Bonan & Hayden 1990) and future (Solomon *et al.* 1984; Solomon 1986; Pastor & Post 1988; Urban & Shugart 1989; Bonan *et al.* 1990; Overpeck *et al.* 1990; Smith *et al.* 1992b).

Consequences of aggregating species into functional types

Approach

For the purposes of the following model experiments, species will differ only in their environmental response functions. Optimal growth functions (see Fig. 16.1) and all other life-history characteristics are assumed to be identical (e.g. seedling establishment, maximum size, etc.). This simplification of the problem is intended to aid in interpretation of the results. The consequences of these simplifying assumptions will be discussed later, after the presentation of results.

Let us assume a simple environmental gradient (Fig. 16.2) where species growth response to the gradient is in the form of a parabola. Two parameters define the species response: the minimum (a_{min}) and maximum (a_{max}) values that define the upper and lower tolerances beyond which growth is not possible. The value of a along the gradient at which growth is optimal (a_{opt}) is defined as the mid point between a_{min} and a_{max}. Examples of environmental gradients along which species exhibit this type of response are temperature and pH (see Larcher (1980) for a general discussion).

To explore the consequences of aggregating species into functional types based on their response to the environmental gradient, I will focus on total stand biomass. A suite of species will be defined and the temporal dynamics of biomass will be evaluated at a given point along the environmental gradient. The value along the gradient will then be changed over a period of time to mimic changing environmental conditions and the dynamics of stand biomass will be examined. Species will then be aggregated according to various criteria and the procedure repeated with the new functional plant type(s).

Model experiments

Single functional type

The simplest case to evaluate is that of a single functional type. The first example to be evaluated is the case where all species that have been classi-fied together as a functional type have identical environmental response functions (ERF). Five species were defined, each having identical parameters (Table 16.1). The dynamics of biomass were simulated for 500 years (iterations) at a fixed value a^1 along the environmental gradient (Figure 16.3). Simulations were started from bare ground (i.e. no plants present) with a constant recruitment rate. Between years 500 and 600 the value for the environmental gradient was shifted from a^1 to a^2 using a linear interpolation. From year 600 to year 1000 the value remained constant at a^2. Results from the simulations are shown in Fig. 16.4. As expected, there is no difference in the predicted dynamics for the two simulations. In both cases the standing biomass and net primary productivity (defined here as the annual biomass increment) decline as the environmental conditions change from optimal (a^1) to suboptimal (a^2).

Table 16.1 *List of parameters used to describe the species characteristics*
These parameters are used to calculate the optimal growth curve and allometry (see Shugart 1984). All species and functional types were characterized using these parameters. Species and functional types differed only in their environ-mental response functions (see Figs. 16.5 and 16.10).

Maximum size (diameter at breast height, cm)	100.0
Maximum height (m)	30.0
Maximum longevity (yr)	300

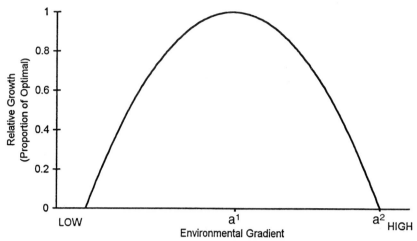

Figure 16.3   *Environmental response function (ERF) used to characterize species for model experiments. Points a¹ and a² along the x-axis represent the environmental conditions corresponding to those used in the model simulations. See text for explanation.*

The more interesting case is when there is variation among the species in their ERFs. Let us now examine five species with different but overlapping ERFs (Fig. 16.5a). Two procedures were used to define the ERF for the aggregated functional type. In the first procedure, the average growth rate for the five species was used to define the value of growth for the functional type at each point along the gradient (Fig. 16.5b). The plant functional type defined using this procedure will be referred to as FT_{avg}. In the second procedure, the maximum value was used (Fig. 16.5c). The plant functional type defined using this procedure will be referred to as FT_{max}.

The simulated biomass dynamics for (a) the five-species community, (b) FT_{avg}, and (c) FT_{max}, at the point a¹ along the gradient are shown in Fig. 16.6. Although the resulting patterns from all three simulations are similar, there are some important differences. The three simulations differ in the initial rate of biomass accumulation over the first 150 years of stand development. The simulation using the single functional type FT_{max} has the highest initial rate of biomass accumulation. The functional type FT_{avg} has the lowest initial rate of biomass accumulation and maintained a lower value of stand biomass over the remainder of the simulation. The five species community has an intermediate rate of initial biomass accumulation and maintained comparable values to those observed for FT_{max} after year 200.

The differences in the biomass dynamics for the simulated stands can be directly interpreted from the ERFs. In the simulation using the FT_{max}, all individuals have a potential for maximum growth rate (i.e. growth multiplier of 1.0; see Fig. 16.5c). Growth rates will be reduced below optimum only under conditions of shading. In the FT_{avg} simulation, the growth

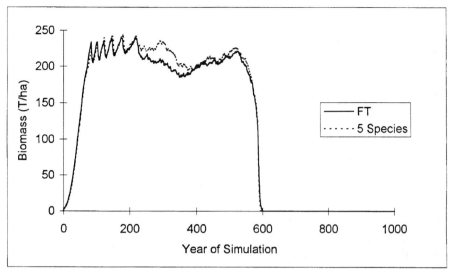

Figure 16.4 *Simulated dynamics of stand biomass (T/ha) under both fixed environmental conditions (a¹) and changing environmental conditions. Results include both the five species community and the single functional type. Changing environmental conditions were simulated by linearly interpolating between points a¹ and a² along the environmental gradient (see Fig. 16.3) between years 500 and 600 of the simulation. Environmental conditions remain constant at a² from year 600 to 1000 of the simulation.*

multiplier at a^1 was 0.76 (see Fig. 16.5b), reducing both maximum potential growth rates and sustainable biomass. In the simulation using the five species, the community is composed of individuals exhibiting a range of potential maximum growth rates for the prevailing environmental conditions (i.e. point a^1 along the environmental gradient). The growth multipliers for the five species at a^1 ranged from 1.0 to 0.36 (see Fig. 16.5a). Although the FT_{avg} represents the average of these values, to include the five species with the range of growth responses is very different from defining the response of each individual by the average value (i.e. FT_{avg}). The reason for this difference is that each of the five species is not equally represented in the community (Fig. 16.7). Those species having the highest growth rates at point a^1 on the gradient dominate, resulting in an overall higher net primary productivity (annual biomass increment) for the community than would be predicted based on the average growth response.

The simulations of biomass dynamics under changing environmental conditions are shown in Fig. 16.8. Despite the similarities in the overall biomass dynamics among the three simulations under fixed environmental conditions (i.e. a^1), there is a divergence in the predictions under changed environmental conditions. The simulation using FT_{max} does not differ from the predictions under fixed environmental conditions (i.e. a^1). This result is

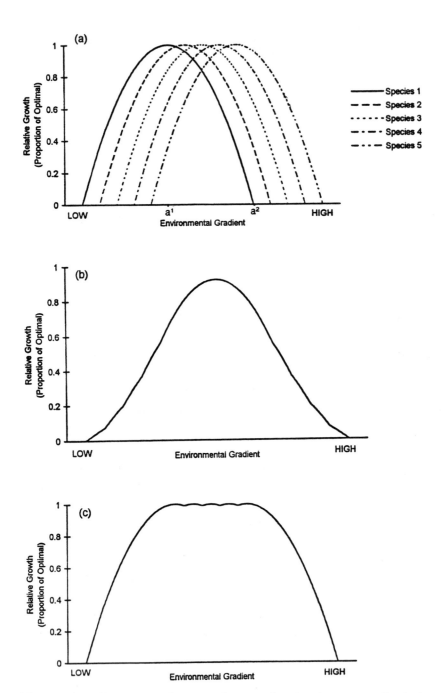

Figure 16.5   *Environmental response functions for: (a) five species, (b) single functional type representing the average response (FT$_{avg}$) of the five species, and (c) single functional type representing the maximum response (FT$_{max}$) of the five species at any point along the gradient.*

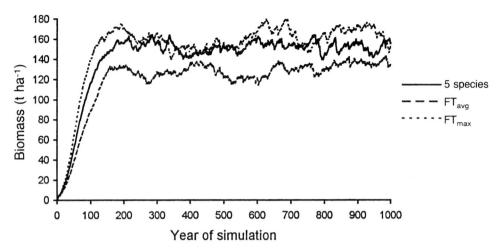

Figure 16.6 *Simulated dynamics of stand biomass for the five-species community, FT$_{avg}$, and FT$_{max}$ under fixed environmental conditions (point a^1 along the gradient; see Fig. 16.3).*

to be expected, since the change in environmental conditions has little effect on the optimal growth rate of individual plants (see Fig. 16.5c). In the simulation using FT$_{avg}$ there is a decline in biomass for a period of approximately 250 years following the onset of changing environmental conditions (years 500–750), after which a 'steady state' is achieved with the new environmental conditions (i.e. a^2). The decline in biomass and net primary productivity is a result of the decline in growth rates with the change in environmental conditions (see Fig. 16.5b). Note the initial increase in biomass beginning at year 500 followed by a gradual decline. This pattern is a direct function of the ERF characterizing FT$_{avg}$. Initially, as conditions change from a^1 to a^2, the potential growth rate increases; however, as environmental conditions continue to change, growth rates decline.

In the simulation using all five species from which the functional types were defined, the response of the community to changed environmental conditions is markedly different from the patterns seen in the two simulations which use the plant functional types. The changes in environmental conditions between years 500 and 600 result in a 75% decline in biomass from approximately 170 t ha^{-1} to 40 t ha^{-1}. This decline is a result of the dieback of species 1, which dominated the community (Fig. 16.9). The dieback is a result of the environmental conditions exceeding the tolerance of the species as defined by the ERF (see Fig. 16.5a). As environmental conditions change, the species composition changes, with species 5 replacing species 1 and 2 as the dominant. This is to be predicted since species 5 has the highest potential growth rate under the new environmental conditions.

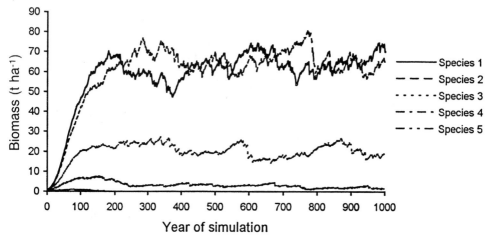

Figure 16.7   *Simulated dynamics of stand biomass for the five species shown in Fig. 16.5(a) under fixed environmental conditions (a¹).*

The above example shows some of the potential limitations in aggregating species into functional categories. The differences in the rates of biomass accumulation and sustainable biomass for the three simulations are a direct function of the manner in which the ERFs (i.e. functional types) are defined. When the ERF of the functional types is defined by the average of the species, the growth rate (and subsequently biomass and net productivity) is underestimated. This difference is due to the fact that in the five species (community) simulation, the species having the higher growth rates under a given set of environmental conditions come to dominate. In the case where the ERF of the functional type is defined as the maximum of the five species being aggregated, the rate of biomass accumulation is overestimated. This overestimation occurs because, although the species with the highest growth rate dominates (when all five species are considered), the community is composed of a mix of species, some of which exhibit suboptimal growth.

The differences among the three simulations in their response to changing environmental conditions are also a direct function of the manner in which the functional type is characterized; specifically, the loss of information regarding differences in species' environmental tolerances. In the case where all species exhibit the same environmental limits to growth (Fig. 16.3), aggregating species into a single functional type described by a single ERF is of no consequence. However, if the species differ in their environment limits to growth (i.e. tolerances; see Fig. 16.5), then any aggregation into a single response function will result in error in the predicted dynamics. In the example presented above, the change in environmental conditions resulted in exceeding the environmental tolerance of the

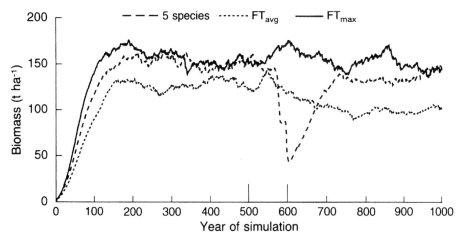

Figure 16.8 *Simulated dynamics of stand biomass for the five-species community, FT_{avg}, and FT_{max} under changing environmental conditions. The simulation is divided into three periods: between years 0 and 500 of the simulation, environmental conditions are set at a^1 (see Fig. 16.5); between years 500 and 600 of the simulation, environmental conditions are linearly interpolating between points a^1 and a^2 (see Fig. 16.5); environmental conditions then remain constant at a^2 from years 600 to 1000 of the simulation.*

dominant species. This dynamic was not captured by defining a single functional type based on either the average or maximum response of the five species to the environmental gradient.

Multiple functional types

The example presented above focuses on the consequences of aggregating a group of similar species into a single functional type. In most cases, the species present in a region (or ecosystem) will not be categorized into a single functional type, but rather into a number of functional types based on similarity in their life history characteristics and/or response to environmental factors. We can expand the previous analyses using a single functional type to include multiple functional types.

As in the last example, let us define a set of species all having the same life-history characters (i.e. parameters; see Table 16.1) but differing in their ERFs. The ERFs of ten species are shown in Fig. 16.10a. Let us assume that the ten species can be grouped into two categories based on the differences in their tolerance to low values along the environmental gradient. Group A species (1–5) are able to tolerate lower values along the environmental gradient than group B species (6–10), whereas group B species are able to continue growth at higher values along the environmental gradient. This pattern of species response might be expected along a temperature gradient, where group B species grow in warmer environments (e.g. subtropical) and are unable to tolerate freezing temperatures, whereas group

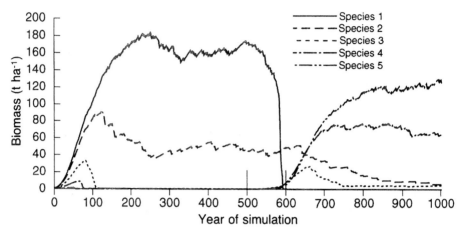

Figure 16.9 *Simulated dynamics of stand biomass for the five species shown in Fig. 16.5(a) under changing environmental conditions. The simulation is divided into three periods as described for Fig. 16.8.*

A species are more indicative of a temperate climate and can tolerate much lower minimum temperatures (see discussion by Woodward & Kelly, Chapter 3, this volume). If we categorize species into these two groups (A and B) we can define two functional types by using the procedures defined above (average and maximum; Fig. 16.10b,c) and contrast the predicted patterns of dynamics to those simulated by using all ten species.

Predicted patterns of biomass for (a) the ten-species community, (b) FT_{avg1} and FT_{avg2}, and (c) FT_{max1} and FT_{max2} at point a^1 along the environmental gradient are presented in Fig. 16.11. As in the case of the analyses using the single functional type, the patterns of biomass dynamics are similar, but a number of differences can be seen. As was the case in the five species community using a single functional type, the simulations differ in the rate of biomass accumulation in the period of initial stand development. The simulation using $FT_{max1,2}$ has the highest initial rate, $FT_{avg1,2}$ the lowest, and the ten-species community an intermediate rate of biomass accumulation. Likewise, the simulation using $FT_{avg1,2}$ has a lower stand biomass throughout the simulation.

Three environmental change scenarios were examined. The three scenarios were chosen to characterize the possible qualitative differences in the response of the ecosystem when characterized by either individual species or the functional types. The three scenarios represent changes in environmental conditions that either: (1) do not exceed the tolerances of either the species or functional types; (2) exceed the environmental tolerance of the dominant (or codominant) species but not the functional types; and (3) exceed the tolerances of both the dominant (or codominant) species and the

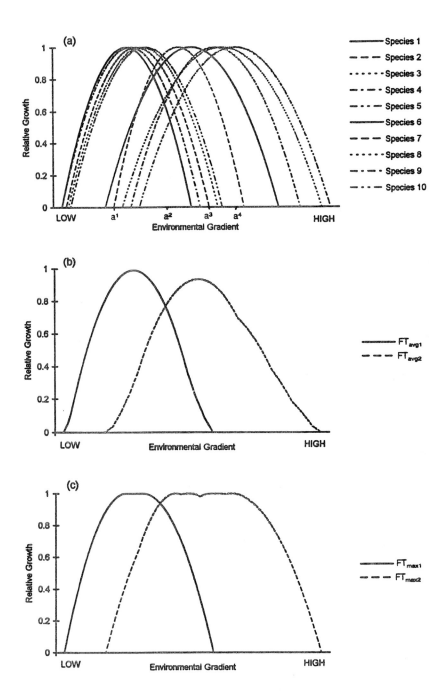

Figure 16.10 *Environmental response functions for: (a) ten species making up groups A (species 1–5) and B (species 6–10); (b) two functional types representing the average response of group A (FT$_{avg1}$) and group B (FT$_{avg2}$) species; and (c) two functional types representing the maximum response of group A (FT$_{max1}$) and group B (FT$_{max2}$) species at any point along the gradient.*

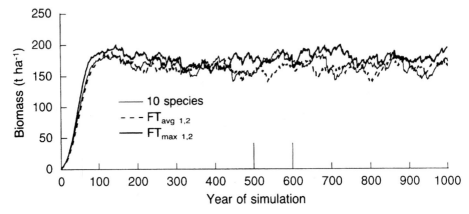

Figure 16.11   *Simulated dynamics of stand biomass for the ten-species community, FT$_{avg1,2}$, and FT$_{max1,2}$ under fixed environmental conditions (point a^1 along the gradient; see Fig. 16.10).*

dominant functional type. The equilibrium environmental values for these three scenarios (a^2, a^3, and a^4) are shown in Fig. 16.10.

In the first scenario, environmental conditions change from a^1 to a^2 over the period from 500 to 600 years in the simulation (Fig. 16.12). This change in environmental conditions results in a shift in species dominance from species in group A (1, 2 and 4) to group B species (5, 6 and 7). This shift is a result of the change in relative growth rates. The shift in species dominance reflects different competitive abilities at the regeneration stage, primarily following gap formation (death of a canopy dominant tree). The change in environmental conditions does not result in a dieback of the dominant species or any significant increase in the rate of mortality. The gradual transition in species dominance results in no significant change in the dynamics of standing biomass or productivity of the stand.

Both FT$_{avg}$ and FT$_{max}$ simulations produce similar results to those observed for the ten-species simulation. In both cases a gradual transition in dominance occurs as a result of changing relative growth rates and subsequent competitive abilities. As with the simulation using ten species, there is no detectable change in the dynamics of standing biomass or productivity associated with the change in environmental conditions (Fig. 16.12).

In the second scenario, environmental conditions change from a^1 to a^3 over the period from 500 to 600 years in the simulation (Fig. 16.10). This change in environmental conditions results in a shift in dominance from group A to group B species. In contrast to the first scenario, this change in environmental conditions exceeds the maximum tolerance (a$_{max}$) of the dominant species (species a; see Fig. 16.10a). The resulting shift in dominance is associated with a marked dieback of the dominant species, with an

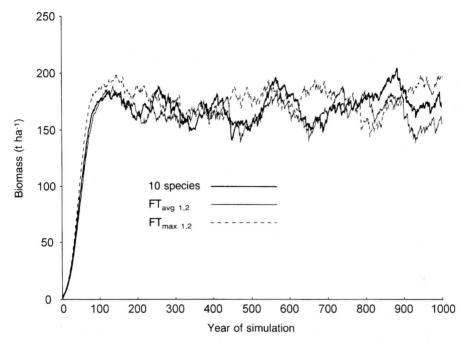

Figure 16.12 *Simulated dynamics of stand biomass under scenario one for the ten-species community, $FT_{avg1,2}$, and $FT_{max1,2}$. The simulation is divided into three periods: between years 0 and 500 of the simulation, environmental conditions are set at a^1 (see Fig. 16.5); between years 500 and 600 of the simulation, environmental conditions are linearly interpolating between points a^1 and a^2 (see Fig. 16.10); environmental conditions then remain constant at a^2 from years 600 to 1000 of the simulation.*

associated 60% drop in standing biomass from 190 t ha^{-1} to 75 t ha^{-1} between years 550 and 610 of the simulation (Fig. 16.13).

In contrast, there is a less marked decline of 20% from 175 t ha^{-1} to 130 t ha^{-1} in the $FT_{avg1,2}$ simulations associated with the transition in relative dominance between FT_{avg1} and FT_{avg2} (Fig. 16.13). In the $FT_{max1,2}$ simulation associated there is a shift in dominance between FT_{max1} and FT_{max2} (Fig. 16.12); however, no decline in biomass is associated with the change in environmental conditions. In all three simulations, by year 700, standing biomass recovers to values observed prior to changes in environmental conditions.

In the third scenario, environmental conditions change from a^1 to a^4 (Fig. 16.10). Under this scenario, all three simulations show a marked decline (90%) in standing biomass following the onset of changing environmental conditions (Fig. 16.14). The rate of recovery differed among the three scenarios. The simulation using $FT_{max1,2}$ recovered the most rapidly; the $FT_{avg1,2}$ simulation was the slowest to recover. In addition, the stand biomass of the $FT_{avg1,2}$ community is reduced to half that observed prior to the onset of the

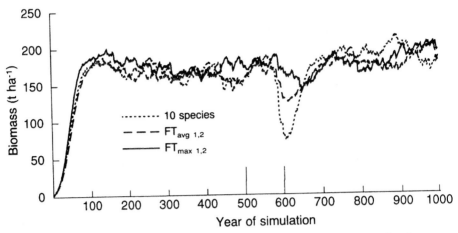

Figure 16.13 *Simulated dynamics of stand biomass under scenario two for the ten-species community, $FT_{avg1,2}$ and $FT_{max1,2}$. The simulation is divided into three periods: between years 0 and 500 of the simulation, environmental conditions are set at a^1 (see Fig. 16.5); between years 500 and 600 of the simulation, environmental conditions are linearly interpolating between points a^1 and a^3 (see Fig. 16.10); environmental conditions then remain constant at a^3 from years 600 to 1000 of the simulation.*

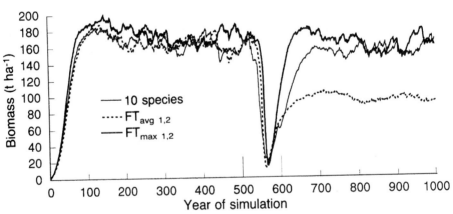

Figure 16.14 *Simulated dynamics of stand biomass under scenario three for the ten-species community, $FT_{avg1,2}$ and $FT_{max1,2}$. The simulation is divided into three periods: between years 0 and 500 of the simulation, environmental conditions are set at a^1 (see Fig. 16.5); between years 500 and 600 of the simulation, environmental conditions are linearly interpolating between points a^1 and a^4 (see Fig. 16.10); environmental conditions then remain constant at a^4 from year 600 to 1000 of the simulation.*

change in environmental conditions. The simulation using all ten species showed an intermediate rate of recovery following the change in environmental conditions.

This divergence in the predictions of biomass dynamics for the three environmental change scenarios is a result of representing the environmental tolerances of the five species in each functional type (Fig. 16.10) with a single statistic (i.e. average or maximum). In the first scenario, a gradual transition in dominance was observed in both the species and functional type simulations. In both cases, the transition resulted from a shift in competitive ability, primarily in the regeneration stage. The environmental change did not exceed the environmental tolerance of any of the species. Therefore, the functional categorizations did not diverge from the general patterns observed when all species were included.

In contrast, the decline in stand biomass in the ten-species simulation in the second scenario is a result of the environmental conditions exceeding the tolerance of the dominant species in group A (species 1). Following the dieback of species 1, the stand is dominated by a mix of species from group A and B (i.e. species 5, 6 and 7). By aggregating the five species into a single functional type categorized by a single ERF, the biomass dynamics associated with the shift in species dominance are largely missed.

In contrast to the second scenario, the patterns of biomass dynamics for the three simulations under the third scenario are similar. In all three simulations the stand biomass declined dramatically during the change in environmental conditions. The general agreement among the three simulations is due to the fact that the environmental change exceeded the upper tolerance limits of all five species in group A. Therefore, the environmental change also exceeded the tolerance limits for both FT_{avg1} and FT_{max1}.

The reason for differences among the three simulations in the rate of recovery following environmental change is identical to that discussed above for initial stand development. In comparison to the ten species community, the higher growth rates and resulting net primary productivity for $FT_{max1,2}$ results in a faster rate of biomass accumulation. In contrast, the lower growth rates for $FT_{avg1,2}$ result in a slower recovery and a lower sustainable biomass under the new environmental conditions.

General comments on the development of functional classifications

The simulations presented above provide some insight into the limitations of aggregating species into functional categories for the purpose of predicting the response of communities or ecosystems to changing environmental conditions.

In the examples presented, defining the aggregated response of a functional type along an environmental gradient by taking the average response of the component species consistently underestimates productivity and sustainable biomass. In contrast, defining the aggregated response by taking the maximum value among the species overestimates the productivity and standing biomass. Despite these differences, under a given set of environmental conditions there is a general agreement between the simulations that use a full species complement, and those that classify the species into functional categories or types. However, as environmental conditions change, the predictions of the simulations diverge, both quantitatively and qualitatively. This divergence in predicted dynamics is directly attributable to the loss of information concerning differences in species tolerances when using an aggregated parameterization (a single statistic to describe all species).

The divergence of results for the three environmental change scenarios using multiple functional types point to some of the inherent problems associated with characterizing the environmental tolerances of a set of species using a single statistical representation. In the scenario where the environmental change did not exceed the tolerances of any of the component species (scenario one), the functional classification provided an adequate description of the environmental responses of the component species. Likewise, when the change in environmental conditions exceeded the range of tolerance for both the component species and the corresponding functional type (scenario three) the predicted dynamics are similar. However, when the environmental change exceeds the tolerance of one or more of the component species yet is not sufficient to exceed the tolerance defined for the functional type (scenario two), the predicted patterns diverge. The occurrence of this result will be related to the degree of variation in the environmental tolerances exhibited by the group of species that are being aggregated to form the functional type. Unfortunately, there is a general lack of data relating to the environmental constraints on important features of species' life history relating to reproduction, establishment and growth. The aggregation of species into functional types implies a similarity in response to environmental factors, yet very little research has been reported in this area of study. Certain approaches for categorizing species into functional types relating to their response to environmental conditions have proven very successful, such as the dichotomy of shade-tolerant and shade-intolerant species (see Larcher (1980) for a general discussion). However, similar classifications have not been investigated for other critical environmental factors (e.g. temperature, moisture, nutrient limitations), although Woodward & Kelly (Chapter 3, this volume) and Smith *et al.* (1992c) have attempted a first step at identifying key features necessary for such a classification.

The characterization of environmental tolerances is a problem common to all classification systems used in modelling the dynamics of terrestrial ecosystems. In biogeographical models such as Holdridge (1967), BIOME (Prentice *et al.* 1992) and MAPSS (Neilson 1995), vegetation is characterized into groups relating to major biomes or vegetation types. The potential distribution of these units is limited by defined tolerances to environmental variables relating to temperature and precipitation. A common environmental tolerance for the species comprising the classification unit is implicit in these approaches. In these classifications (i.e. biogeographical models), as the classification unit (e.g. biome or vegetation type) becomes more inclusive (includes an increasing number of species), the range of environmental conditions used to define the tolerances or distributional limits of the unit become broader. As a result, the change in environmental conditions required to influence vegetation distribution becomes greater (Smith *et al.* 1992a). This is the same phenomenon observed in the current analyses, where aggregating species into plant functional types reduces the overall responsiveness (i.e. biomass dynamics) to environmental change.

The problems inherent in aggregating species into functional groupings is not limited to biogeographical models or the individual-based model used in the current exercise; most models that examine ecosystem dynamics (e.g. TEM) (Melillo *et al.* 1993) define vegetation in terms of biome or vegetation types similar to those used by the biogeographical models discussed above. These vegetation types are then assumed to share common attributes relating to biochemistry, structure, physiology and environmental response.

The results presented suggest that caution should be used in the development and application of functional classifications of species for the purpose of exploring community or ecosystem dynamics under changing environmental conditions. Particularly troublesome are the potential problems associated with model validation. As the current exercise shows, the ability of the classification to be used within a dynamic modelling framework to predict currently observed vegetation (ecosystem) patterns does not imply that the classification will 'realistically' predict the response of vegetation (ecosystem) to changing environmental conditions.

The observed divergence among the simulations under changing environmental conditions is due solely to the use of a single statistic to describe the ERF for the functional types. If there are differences among the species being aggregated in their environmental tolerance (reproduction, growth and survival), the use of a single parameter to describe tolerance will produce errors as the environmental conditions exceed the tolerance limits of the component species. The degree of error will depend on the magnitude of difference among the species relative to the environmental change, and the patterns of relative species dominance.

In the examples presented above, all species were identical with the exception of their ERFs. This is an unrealistic assumption (or expectation). Even species having similar environmental responses to basic resources such as light, water and nitrogen show a diversity of life-history characteristics. The greater the variation among the species in key characteristics and processes (e.g. fecundity, conditions for establishment, growth rate, environmental tolerances, longevity, maximum size) the greater will be the errors associated with aggregating species into functional types described by a single parameter set.

In conclusion, the use of functional type classifications should not be viewed as a simple alternative to understanding the characteristics of vegetation at the species (taxonomic) level. Species are the basic demographic unit, and any attempt at a functional aggregation for purposes of evaluating the dynamics of plant distribution must address the environmental responses of the component species relating to dispersal, establishment, growth, reproduction and mortality.

References

Bazzaz, F. A. (1979) The physiological ecology of plant succession. *Annual Review of Ecology and Systematics*, **10**, 425–86.

Bonan, G. B., Shugart, H. H. & Urban, D. L. (1990) The sensitivity of some high-latitude boreal forests to climatic parameters. *Climatic Change*, **16**, 9–29.

Bonan, G. B. & Hayden, B. P. (1990) Using a forest stand simulation model to examine the ecological and climatic significance of the late-Quaternary pine-spruce pollen zone in eastern Virginia, U. S. A. *Quaternary Research*, **33**, 204–18.

Bormann, F. H. & Likens, G. E. (1981) *Pattern and Process in a Forest Ecosystem*. New York: Springer-Verlag.

Botkin, D. B., Janak, J. F. & Wallis, J. R. (1972). Some ecological consequences of a computer model of forest growth. *Journal of Ecology*, **60**, 849–73.

Brokaw, N. V. L. (1985a) Gap-phase regeneration in a tropical forest. *Ecology*, **66**, 682–7.

Brokaw, N. V. L. (1985b) Treefalls, regrowth, and community structure in tropical forests. In Pickett, S. T. A. & White, P. S. (eds), *The Ecology of Natural Disturbance and Patch Dynamics*, pp. 101–8. New York: Academic Press.

Budowski, G. (1965) Distribution of tropical American trees in the light of the successional process. *Turrialba*, **15**, 40–2.

Budowski, G. (1970) The distinction between old secondary and climax species in tropical Central American lowland forests. *Tropical Ecology*, **11**, 44–8.

Finegan, B. (1984) Forest succession. *Nature*, **312**, 109–14.

Greenslade, P. J. M. (1983) Adversity selection and the habitat templet. *American Naturalist*, **122**, 352–65.

Grime, J. P. (1977) Evidence for the existence of three primary strategies in plants and its relevance to ecological and evolutionary theory. *American Naturalist*, **111**, 1169–94.

Grime, J. P. (1979) *Plant Strategies and Vegetation Processes*. New York: John Wiley.

Hallé, F. (1974) Architecture of trees in the rainforest of Morobe District, New Guinea. *Biotropica*, **6**, 43–50.

Hallé, F. & Oldeman, R. A. A. (1975) *Essay on the Architecture and Dynamics of Growth of Tropical Trees*. Kuala Lumpur, Malaysia: Penerbit University.

Hartshorn, G. S. (1978) Tree falls and tropical forest dynamics. In Tomlinson, P. B. & Zimmermann, M. H. (eds), *Tropical Trees as Living Systems*, pp. 617–38. Cambridge University Press.

Holdridge, L. R. (1967) *Life Zone Ecology*. San Jose, Costa Rica: Tropical Science Center.

Huston, M. & Smith, T. M. (1987) Plant succession: life history and competition. *American Naturalist*, **130**, 168–98.

Larcher, W. (1980) *Physiological Plant Ecology*, 2nd edn. Berlin: Springer-Verlag.

MacArthur, R. H. & Wilson, E. O. (1967) *The Theory of Island Biogeography*. New Jersey: Princeton University Press.

Melillo, J. M., McGuire, A. D., Kicklighter, D. W., Moore, B. III, Vorosmarty, C. J. & Schloss, A. L. (1993) Global climate change and net primary production. *Nature*, **363**, 234–40.

Neilson, R. P. (1995) A model for predicting continental-scale vegetation distribution and water balance. *Ecological Applications* **5**, 362–85.

Noble, I. R. & Slatyer, R. O. (1980) The use of vital attributes to predict successional changes in plant communities subject to recurrent disturbances. *Vegetatio*, **43**, 5–21.

Overpeck, J. T., Rind, D. & Goldberg, R. (1990) Climate-induced changes in forest disturbance and vegetation. *Nature*, **343**, 51–3.

Pastor, J. & Post, W. M. (1988) Response of northern forests to CO_2-induced climate change. *Nature*, **334**, 55–8.

Prentice, I. C., Cramer, W., Harrison, S. P., Leemans, R., Monserud, R. A. & Solomon, A. M. (1992) A global biome model based on plant physiology and dominance, soil properties and climate. *Journal of Biogeography*, **19**, 117–34.

Raunkiaer, C. (1934) *The Life-forms of Plants and Statistical Plant Geography*. Oxford University Press.

Shugart, H. H. (1984) *A Theory of Forest Dynamics*. New York: Springer-Verlag.

Shugart, H. H. & West, D. C. (1977) Development of an Appalachian deciduous forest succession model and its application to the assessment of the impact of the chestnut blight. *Journal of Environmental Management*, **5**, 161–79.

Smith, T. M. & Huston, M. (1989) A theory of spatial and temporal dynamics in plant communities. *Vegetatio*, **83**, 49–69.

Smith, T. M., Leemans, R. & Shugart, H. H. (1992a) Sensitivity of terrestrial carbon storage to CO_2 induced climate change: Comparison of five scenarios based on general circulation models. *Climatic Change*, **21**, 367–84.

Smith, T. M., Shugart, H. H., Bonan, G. B. & Smith, J. B. (1992b) Modeling the potential response of vegetation to global climate change. *Advances in Ecological Research*, **20**, 93–113.

Smith, T. M., Shugart, H. H., Woodward, F. I. & Burton, P. J. (1992c) Plant functional types. In Solomon, A. M. & Shugart, H. H. (eds), *Vegetation Dynamics and*

Global Change, pp. 272–92. New York: Chapman & Hall.

Smith, T. M. & Urban, D. L. (1988) Scale and resolution of forest structural pattern. *Vegetatio*, **74**, 143–50.

Solomon, A. M. (1986) Transient responses of forests to CO_2-induced climate change: Simulation modeling experiments in eastern North America. *Oecologia*, **68**, 567–9.

Solomon, A. M., Delcourt, H. R., West, D. C. & Blasings, T. J. (1980) Testing a simulation model for reconstruction of prehistoric forest-stand dynamics. *Quaternary Research*, **14**, 275–93.

Solomon, A. M. & Shugart, H. H. (1984) Integrating forest-stand simulations with paleoecological records to examine long-term forest dynamics. In Agren, G. I. (ed.), *State and Change of Forest Ecosystems: Indicators in Current Research, Report Number 13*. pp. 333–57. Uppsala, Sweden: Swedish University of Agricultural Science.

Solomon, A. M., Tharp, M. L., West, D. C., Taylor, G. E., Webb, J. M. & Trimble, J. L. (1984) *Response of unmanaged forests to CO_2-induced climate change: Available information, initial tests and data requirements*. Tech. Report TR009, US DOE Carbon Dioxide Research Division, Washington, D.C.

Solomon, A. M. & Webb, T. III (1985) Computer-aided reconstruction of late-quaternary landscape dynamics. *Annual Review of Ecology and Systematics*, **16**, 63–84.

Solomon, A. M., West, D. C. & Solomon, J. A. (1981) Simulating the role of climate change and species immigration in forest succession. In West, D. C., Shugart, H. H. & Botkin, D. B. (eds), *Forest Succession*, pp. 154–77. New York: Springer-Verlag.

Southwood, T. R. E. (1977) Habitat, the templet for ecological strategies. *Journal of Animal Ecology*, **46**, 337–65.

Urban, D. L. & Shugart, H. H. (1989) Forest response to climate change: a simulation study for Southeastern forests. In Smith, J. & Tirpak, D. (eds), *The Potential Effects of Global Climate Change on the United States*, pp. 31–345. EPA-230-05-89-054. Washington, D.C.: US Environmental Protection Agency.

Urban, D. L. & Smith, T. M. (1989) Microhabitat pattern and the structure of forest bird communities. *American Naturalist*, **133**, 811–29.

Walker, B. H., Ludwig, D., Holling, C. S. & Peterman, R. M. (1981) Stability of semi-arid savanna grazing systems. *Journal of Ecology*, **69**, 473–98.

Webb, L. J., Tracey, J. G., Williams, W. T. & Lance, G. N. (1970) Studies in the numerical analysis of complex rain-forest communities. V. A comparison of the properties of floristic and physiognomic-structural data. *Journal of Ecology*, **58**, 203–32.

Whittaker, R. H. (1975) *Communities and Ecosystems*. New York: MacMillan.

17 Ecosystem function of biodiversity: the basis of the viewpoint

H. A. Mooney

Introduction

We are witnessing a dramatic depauperation of the earth's biotic capital due to the activities of humans. This loss is of great concern for a number of reasons that have been well established, including the loss of irreplaceable products of potential use to humans. Another concern has been the resultant losses of 'ecosystem services' provided to humans by these species. It is this latter concern on which I focus here. Phrased in other ways, the questions have been asked, 'What is the ecosystem role of biodiversity?', 'Do species count in the functioning of ecosystems?' and, 'Is the welfare of humans at risk as the losses of species mount'? It has been only recently that these latter questions have been posed and added to the general concerns related to the losses of biodiversity. As is evident from how the above questions are posed, the principal thrust of my discussion will focus on species as the basic unit of biodiversity upon which lower or higher levels of integration can be added.

The biodiversity–function question is of considerable importance since, except in the case of preserves, decisions on the management of land and seascapes will increasingly be made on a functional basis rather than on the basis of conservation of the numbers, or kinds, of species *per se*. We will need to know how landscape and ecosystem processes are controlled by species functions and interactions.

A little history

Systematists and population biologists have, for the most part, been the leaders in alerting society to the extinctions that are occurring and of the losses of potentially useful food and drug products. On the other hand it has been the ecological community that has focused on the functional consequences of these losses. It is important to note this since there is a fundamental divergence in viewpoint between these two groups. It can be stated somewhat simplistically that systematists and population biologists in general focus on biotic differences and ecologists on biotic similarities. The former viewpoint is of course fundamental to classification, where the least common denominator separating groups of individuals is sought. On the other hand the fundamental tenet of population and evolutionary biology is that no two species can coexist in exactly the same habitat since one would outcompete the other for resources. The conceptual outcome of this tenet (Gause 1934) is that every species has a unique resource requirement and hence use pattern.

The history of the development of ecology explains why ecologists have concentrated on the functional similarities among species rather than on their differences. The early ecological plant geographers such as Warming (1909) classified, or lumped, the biota into such functional groups as xerophytes, mesophytes, and so forth. Warming noted that large numbers of species of quite different taxonomic relationships shared ecological characteristics, in this particular example, evolutionary response to habitat water abundance. Another example is the classification by Raunkiaer (1934) of all of the world's plants into a few categories related to leaf size and the height of their perennating buds. More elaborate functional group classifications were developed that took into account multiple characteristics of plants, but they too had as a principal objective the lumping of taxonomic diversity into small numbers of functional categories (see recent review by Smith *et al.* (1993)).

This functional grouping of organisms by ecologists has continued to modern times even as the general viewpoint of the discipline has broadened and become more complex. Odum's (1963) powerful paradigm of the functional construction of ecosystems lumped all organisms into a very few functional categories: producers, consumers and decomposers. This categorization is certainly at the opposite pole from that of systematists, who have divided the biotic world into millions of specific entities, and that of population biologists, who would make the divisions even finer when viewing ecologically distinct populations even within specific categories. The resurgence of interest in the consequences of convergent evolution in the 1970s (Mooney & Dunn 1970; Mooney 1977; Orians & Solbrig 1977) led to further consolidation of the concepts of the limitations on the numbers of 'strategy types' that were possible in a given region. More recently, those concerned with the metabolism of the biosphere have divided the vegetation of the world into a very few large groupings that encompass information on roughness characteristics and very simple physiological traits, such as the abundances of C_3 vs. C_4 plants (Sellers *et al.* 1992). Thus, it is evident that various types of biologists view biotic diversity through different 'glasses':

one emphasizing differences and the other similarities. It is important to note, as these various constituencies debate the priorities for research and policy related to biodiversity, that they have different central tendencies and are asking different questions. In the context of the issue of ecosystem function, we need to examine more carefully what we mean by ecosystem function and relate these functions to the roles that various species play within an ecosystem.

Functional properties of ecosystems

A simplistic and organism-centred view of a habitat is that of a pool of resources that is available to support growth and reproduction. These resources are light (energy), carbon, nutrients and water. It is the uptake, fixation and transfer of these resources by a collective group of organisms, and their exchange with the physical components of the habitat, that define basic ecosystem processes. These exchanges encompass system processes that are termed productivity, decomposition and nutrient and water cycling. Imbalances between rates of fixation and rates of loss, through litterfall or decomposition and mineralization, result in accumulation, biomass production or soil formation (Table 17.1). A more human-centred view of ecosystem function properties has been developed through the years by Paul Ehrlich in the context of ecosystem goods and services (Table 17.2).

These two viewpoints are similar, although the latter indicates a role for ecosystems not only in maintaining suitable conditions for all organisms but also in repairing damage done to them by the action of humans as well as providing services that are human-specific. Further, the latter viewpoint indicates the feedback role of ecosystems in maintaining atmospheric properties, including climate, a role which has only recently become more fully appreciated, at least at regional and global scales.

There seems to be little question that ecosystems play an important role in accumulating and processing resources in a way that is life-maintaining. The question, though, which I posed earlier, is what is the role of individual species in the maintenance of these functions in a particular habitat?

Productivity

There is a considerable literature on the role of diversity in controlling the rates of production of communities. The statement has been made that to a large degree 'productivity is independent of species numbers, at least for a given growth form' (Mooney & Gulmon 1983). Observations from a variety of sources, including explicit experiments, support this statement. Mooney

Table 17.1 *Ecosystem functions and processes related to the transfer of energy, nutrients, water and genetic information*

Function/process	Action
Nutrient capture	Nutrient cycling
Decomposition and soil formation	Nutrient cycling
Mycorrhizal activity	Enhanced nutrient uptake
Photosynthesis	Energy capture and productivity
Herbivory	Energy and nutrient capture
Pollination	Genetic information transfer
Species interactions (mutualism, symbiosis, predation, parasitism, competition)	Energy, water and nutrient transfer
Water uptake and loss	Water transfer

Source: From Hobbs (1992a).

Table 17.2 *Ecosystem goods and services*

Ecosystem 'goods'	
	Food
	Construction materials
	Medicinal plants
	Wild genes for domestic plants and animals
Ecosystem services	
	Maintaining hydrological cycles
	Regulating climate
	Cleansing water and air
	Maintaining the gaseous composition of the atmosphere
	Pollinating crops and other important plants
	Generating and maintaining soils
	Storing and cycling essential nutrients
	Absorbing and detoxifying pollutants
	Providing sites for inspiration, tourism, recreation and research

Source: From Lubchenco *et al.* (1993) after Ehrlich & Ehrlich (1991) and McNeeley *et al.* (1990).

& Gulmon (1983) cite evidence from classical successional studies that show that above-ground production increases with time but this is unrelated to the changes in species numbers, which generally decline (Fig. 17.1). The increases in productivity with time are attributed to the introduction of a new growth form, for example a shrub, into a herbaceous early-successional stage; this new growth form is able to tap an unutilized nutrient or water resource of the habitat by virtue of its deeper roots. This phenomenon leads to the idea that not all species are equal in relation to the coarse measures of ecosystem function.

There is similarly evidence that increases in productivity per unit of land may increase with multicropping, but only to a limited extent (Swift & Anderson 1993), although other ecosytem properties relating to nutrient cycling may continue to be enhanced with the further addition of species (Fig. 17.2).

Accumulation

There is similar evidence for a limited change in ecosystem function with species additions in relation to the rate of soil development, as indicated by the accumulation of nitrogen and organic matter in the soil (Fig. 17.3) (Vitousek & Hooper 1993).

These observations are based on relatively short-term explicit experiments of Ewel *et al.* (1991) that examine the relationships between species

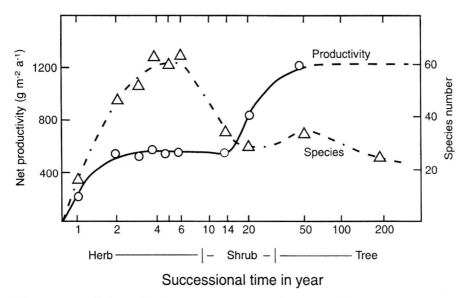

Figure 17.1 *Relationship between community productivity and species numbers in a successional series for an oak–pine forest in New York (from Mooney & Gulmon (1983), after Whittaker (1975) and Woodwell (1974)).*

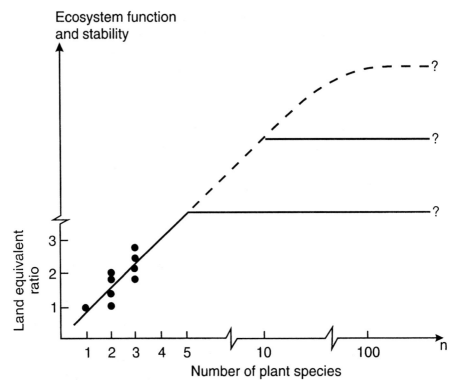

Figure 17.2   *Productivity of monoculture and mixed cropping in cassava. Swift & Anderson (1993) propose that productivity increases may not continue after three species; however, there may be an increase in other functions of the systems owing to greater nutrient retention.*

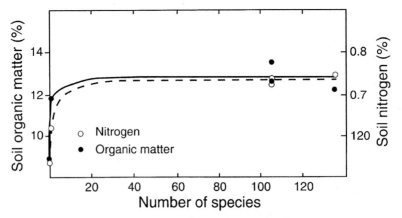

Figure 17.3   *Relationship between species numbers in model communities and soil nitrogen and organic matter. (From Vitousek & Hooper (1993), after Ewel et al. (1991).)*

numbers and ecosystem function. Thus in the model 'possibilities of species versus function' proposed by Vitousek & Hooper (1993) (Fig. 17.4) soil development in these experimental communities would be of type 3, with little relationship between numbers and function, *in the short term*.

Decomposition

Springett (1976), in his study of varied-aged plantations of *Pinus pinaster* in Australia, found that the rate of decomposition of leaf litter increased in relation to the number of microarthropod species found in the soil, which increased with stand age (Fig. 17.5).

All species are not equal: keystone species

Although it appears that there may be considerable redundancy in ecosystems in relation to ecosystem function, this does not mean that all species are equal and substitutable. There is increasing evidence, some experimental, that some species count to an inordinate degree in controlling community structure. These species, termed keystone species by Paine (1966), control the stability of communities. Their removal can result in a large alteration of community structure and function. Bond (1993), in a recent analysis of the role of keystone species, noted that there are many types of keystone species; they include predators (otters), herbivores (rabbits), competitors (dominant trees), mutualists (certain pollinators), earth movers (pocket gophers), and certain species that play a particularly important role in ecosystem processes, such as nitrogen fixers. As described by Bond, keystone species may be 'rare, or common, generalists or dietary specialists. They need not be important energy transformers and so would

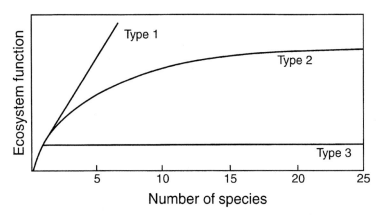

Figure 17.4 *Theoretical relationship between species function and species numbers (from Vitousek & Hooper 1993).*

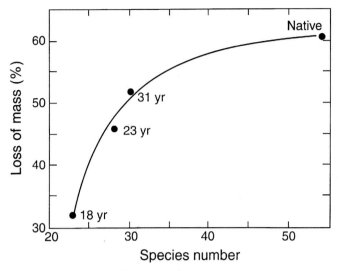

Figure 17.5   *Litter-bag mass loss in relation to soil microarthropod species numbers in pine stands of differing ages and a native woodland vegetation (data of Springett (1976) as presented by Hobbs (1992b)).*

not necessarily be identified by the ecological accounting of system ecologists'. Food web analysis (Pimm 1993) can illustrate keystone roles of species in both simple and complex webs, as shown in Fig. 17.6.

The keystone concept is, then, an important addition to our basic thinking about species and function. Not only may a few species perform the bulk of the gross ecosystem functions, but it also appears that not all species are equal in the role that they play in ecosystem integration. There is a need for refining keystone concepts and to quantify strengths of species interactions (Mills *et al.* 1993) in order to utilize this concept in ecosystem analysis and management.

The functional type question

The keystone species concept offers a very important guideline for targeting species with particularly important roles in the functioning of ecosystems. What about functional types, or guilds: groupings of species with analogous functional traits in a ecosystem? If these could be identified, *a priori*, such knowledge would tell us how much redundancy there is in an ecosystem, if any, and hence give us important information about the management of that system. Smith *et al.* (1993) have noted that 'clearly the number of functional types should be much smaller than the number of plant species (indeed, smaller than even the number of species of dominant plants) to have significant applicability and utility in generic or regionally specific

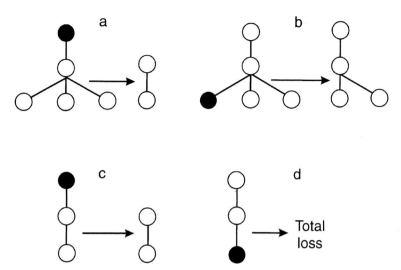

Figure 17.6   *Food web stability. In case (a), the complex web is greatly simplified by the loss of the top predator that in turn increases the amount and impact of the herbivore. The predator in this case would be considered a keystone species. In a simple food chain, as in (d), the total food chain collapses with the loss of the only source of food for the herbivore and hence the predator. In the complex web the removal of a herbivore has little effect since there are multiple food sources for the higher levels of the food chain. Similarly, in the simple food chain shown in (c), the loss of the predator does not result in the loss of the herbivore because the herbivore cannot eliminate its only food source (from Pimm 1993).*

models'. The concept is thus scale-dependent, in addition to being time- and question-dependent.

In theory, one could imagine that there would be more functional groups than there are species, since each species has a number of different traits, and most species have ecologically differentiated races. Thus within a community the ecologically important entities, or trait combinations, could exceed the number of species. Because functional group classifications are designed around traits – morphological, physiological, developmental and reproductive – a single species could belong to more than one functional group category. Thus, for example, a given species may belong to a number of the different exemplary categories as shown in Table 17.2, and classified as a member of one functional group by one trait and of another by a different trait. Chapin (1993) and Mooney & Godron (1983) have argued that this is not the case since many traits are interrelated and thus represent syndromes of response to a given resource base. In the example given by Chapin, he notes that plants with a high relative growth rate have high rates of photosynthesis, transpiration, herbivory and decomposition. Thus, any one trait can be utilized as a surrogate for another and, more to the point, there will be a limited number of combinations of traits among a pool of co-occurring species. Mooney & Godron similarly described the genetic, demographic and physiological traits that characterize the syndrome of plants of disturbed habitats. Grime (1979, and Chapter 7, this volume) has fully developed these ideas in relation to his plant strategy construct.

There is no doubt that the functional type approach can be very useful in reducing the biotic complexity of a given system in order to answer specific questions about ecosystem organization and function. However, it does not totally help in answering the species redundancy question because of the multiple-function possibilities of species, and the possibilities that these functions change with changing environmental conditions and biotic assemblages (see Gitay & Noble, Chapter 1, this volume, for a full discussion of these issues). It is likely that as the functional type concept receives more attention it will be seen that any functional type classification will be time-, space- and system-dependent and will be question-driven (see Hobbs and Walker, Chapters 4 and 5, this volume).

Stability and diversity

The widely held view that ecosystem complexity promoted stability, which dominated the ecological literature of the 1950s to 1970s (McNaughton 1993), has been challenged by the keystone concept (Bond 1993). McNaughton (1993) notes, though, that experimental approaches to the stability–diversity paradigm have led to new insights when the details of particular systems are examined and the mechanisms of change are discerned. He shows, for example, how greater diversity can lead to greater resilience to grazing in an African grassland (Fig. 17.7), owing to the presence of a pool of species capable of regrowth during early showers after the drought. Thus in this case the increased resilience with greater diversity was due to the presence of a specialized 'strategy', or functional group. McNaughton suggests that the question should be 'How and why does biodiversity affect ecosystem function?' rather than 'Does biodiversity affect ecosystem function?'.

Explicit experiments

It is unfortunate that we have so little direct information on the role of species in ecosystem function. Although experimental approaches have become more common in community ecology (Hairston 1989) they have rarely been employed at the ecosystem level (Mooney *et al.* 1991) and those that have been directed towards learning of whole-system responses to perturbations have not generally examined species properties or community

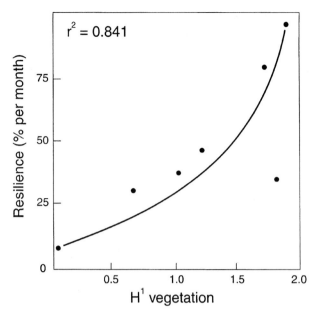

Figure 17.7 *Resilience of grassland communities to grazing as related to community diversity (from McNaughton 1993).*

properties in relation to ecosystem function. An important exception is the experiments of Schindler (1991) on whole lakes. He found that species were more sensitive to stress than were such ecosystem functions as primary production, respiration and nutrient cycling, apparently owing to redundancy in those groups that carry out key ecosystem functions. He notes, however, that there are exceptions to this generality and that 'the conclusion that species changes are more sensitive than changes in function must be a cautious and tentative one'. He further notes that the present increase in methane in the atmosphere may be the result of a relatively minor change in a component of a biogeochemical cycle.

The lack of an experimental base from which to address more explicitly the question of the ecosystem function of biodiversity is clearly hindering progress in this area. The programme described below is designed to provide new information in this area.

A new programme for the study of ecosystem function

SCOPE, the Scientific Committee on Problems of the Environment, has embarked on a three-year programme to synthesize information on the ecosystem function of biodiversity. The questions that are being asked are:

1. Does biodiversity 'count' in system processes (e.g. nutrient retention, decomposition, production, etc.), including atmospheric feedbacks,

over short and long-term time spans, and in the face of global change (climate change, land-use change, invasions)?

2. How is system stability and resistance affected by species diversity, and how will global change affect these relationships?

The programme is taking a global view of the problem and is focusing on a number of ecosystem types, terrestrial, freshwater and marine, including temperate and tropical forests, savannas, arid zones, upwelling systems and coral reefs.

For each system there will be an analysis of the natural diversity of the system, and of the impact of change on diversity including climate change, land-use change, and biological invasions. The impact of the ecosystem role of species will be evaluated by noting the consequences of additions (invasion analogues) and subtractions (harvesting and disease effects) as well as those of disturbance and habitat fragmentation. Most importantly, an analysis will be made of the kinds of experiments and long-term observations that could lead to a better understanding of the role of species in the operation of ecosystems. The intent is that this information would provide input to the emerging research effort of the International Geosphere–Biosphere Programme.

Conclusions

As ecosystems become more simplified through the impact of human activities, will they lose some of their major functional properties? Preliminary evidence indicates that they may not, unless radically simplified. However, this conclusion does not take into account issues of ecosystem stability in a changing climate, for example, nor the more subtle ecosystem function properties related to the results of species interactions. Intensified direct experimental research in this area is necessary to bring us beyond the few examples from which we are now generalizing. Further, work on understanding the role and extent of keystone species and functional types will help us to develop the knowledge we need to manage ecosystems in a changing world.

Acknowledgements

I thank Richard Hobbs, Osvaldo Sala and Detlef Schulze for stimulating discussions on this topic.

Note added in proof

See UNEP (1995), *Global Biodiversity Assessment*, Sections 5 and 6, Cambridge University Press, for an update on the issues discussed here.

References

Bond, W. J. (1993) Keystone species. In Schulze, E. D. & Mooney, H. A. (eds), *Biodiversity and Ecosystem Function*, pp. 237–53. Berlin: Springer-Verlag.

Chapin, F. S. III. (1993) The functional role of growth forms in ecosystem and global processes. In Ehleringer, J. & Field, C. (eds), *Scaling Physiological Processes: Leaf to Globe*, pp. 287–312. San Diego: Academic Press.

Ehrlich, P. R. & Ehrlich, A. H. (1991) *Healing the Planet: Strategies for Resolving the Environmental Crisis*. Reading, Massachusetts: Addison-Wesley.

Ewel, J. J., Mazzarino, M. J. & Berish, C. W. (1991) Tropical soil fertility changes under monocultures and successional communities of different structure. *Ecological Applications*, 1, 289–302.

Gause, G. F. (1934) *The Struggle for Existence*. Baltimore: Williams and Wilkins.

Grime, J. P. (1979) *Plant Strategies and Vegetation Processes*. Chichester: John Wiley & Sons.

Hairston, N. G. (1989) *Ecological Experiments: Purpose, Design, and Execution*. Cambridge University Press.

Hobbs, R. J. (1992a) Function of biodiversity in mediterranean ecosystems in Australia: definitions and background. In Hobbs, R. J. (ed.), *Biodiversity in Mediterranean Ecosystems in Australia*, pp. 1–25. Chipping Norton, NSW, Australia: Surrey Beatty & Sons.

Hobbs, R. J. (1992b) Is biodiversity important for ecosystem functioning? Implications for research and management. In Hobbs, R. J. (ed.), *Biodiversity in Mediterranean Ecosystems in Australia*, pp. 211–29. Chipping Norton, NSW, Australia: Surrey Beatty & Sons.

Lubchenco, J., Risser, P. G., Janetos, A. C., Gosz, J. R., Gold, B. D. & Holland, M. M. (1993) Priorities for an environmental science agenda in the Clinton-Gore adminstration: recommendations for transition planning. *Bulletin of the Ecological Society of America*, 74, 4–8.

McNaughton, S. J. (1993) Biodiversity and function of grazing ecosystems. In Schulze, E. D. & Mooney, H. A. (eds), *Biodiversity and Ecosystem Function*, pp. 361–83. Berlin: Springer-Verlag.

McNeely, J. A., Miller, K. R., Reid, W., Mittermeier, R. & Werner, T. (1990). *Conserving the World's Biological Diversity*. Washington, D.C.: IUCN, WRI, World Bank, WWF-US, Conservation International.

Mills, L. S., Soule, M. E. & Doak, D. F. (1993) The keystone-species concept in ecology and conservation. *BioScience*, 43, 219–24.

Mooney, H. A. (ed.) (1977) *Convergent evolution of Chile and California-Mediterranean Climate Ecosystems*. Stroudsburg, Pennsylvania: Dowden, Hutchinson and Ross.

Mooney, H. A., & Dunn, E. L. (1970) Convergent evolution of mediterranean-climate evergreen sclerophyll shrubs. *Evolution*, 24, 292–303.

Mooney, H. A. & Godron, M. (1983) Preface. In Mooney, H. A. & Godron, M. (eds), *Disturbance and Ecosystems*, pp. v-vi. Berlin: Springer-Verlag.

Mooney, H. A. & Gulmon, S. L. (1983) The determinants of plant productivity-natural versus man-modified communities. In Mooney, H. A. & Godron, M. (eds), *Disturbance and Ecosystems*, pp. 146–58. Berlin: Springer-Verlag.

Mooney, H. A., Medina, E., Schindler, D. W.,

Schulze, E. D. & Walker, B. H. (eds) (1991) *Ecosystem Experiments*. Berlin: Springer-Verlag.

Odum, E. P. (1963) *Ecology*. New York: Holt, Rinehart and Winston.

Orians, G. H. & Solbrig, O. T. (1977) A cost-income model of leaves and roots with special reference to arid and semiarid areas. *American Naturalist*, 111, 677–90.

Paine, R. T. (1966) Food web complexity and species diversity. *American Naturalist*, 100, 65–75.

Pimm, S. L. (1993) Biodiversity and the balance of nature. In Schulze, E. D. (ed.), *Biodiversity and Ecosystem Function*, pp. 347–59. Berlin: Springer-Verlag.

Raunkiaer, C. (1934) *The Life Forms of Plants and Statistical Plant Geography*. Oxford: Clarendon Press.

Schindler, D. W. (1991) Whole-lake experiments at the experimental lake area. In Mooney, H. A. *et al.* (eds), *Ecosystem Experiments*, pp. 127–39. Chichester: John Wiley.

Sellers, P. J., Heiser, M. D. & Hall, F. G. (1992) Relations between surface conductance and spectral vegetation indices at intermediate (100 m^2 to 15 km^2) length scales. *Journal of Geophysical Research*, 97, 19033–59.

Smith, T. M., Shugart, H. H., Woodward, F. I. & Burton, P. J. (1993) Plant functional types. In Solomon, A. M. & Shugart, H. H. (eds), *Vegetation Dynamics and Global Change*, pp. 272–92. New York: Chapman and Hall.

Springett, J. A. (1976) The effect of planting *Pinus pinaster* Aiat. on populations of soil microarthropods and on litter decomposition at Gnangara, Western Australia. *Australian Journal of Ecology*, 1, 83–7.

Swift, M. J., & Anderson, J. M. (1993) Biodiversity and ecosystem function in agricultural systems. In Schulze, D. & Mooney, H. A. (eds), *Biodiversity and Ecosystem Function*, pp. 15–41. Berlin: Springer-Verlag.

Vitousek, P. M. & Hooper, D. U. (1993) Biological diversity and terrestrial ecosystem biogeochemistry. In Schulze, E. D. & Mooney, H. A. (eds), *Biodiversity and Ecosystem Function*, pp. 3–14. Berlin: Springer-Verlag.

Walker, B. H. (1992) Biodiversity and ecological redundancy. *Conservation Biology*, 6, 18–23.

Warming, J. E. B. (1909) *Oecology of Plants*. London: Oxford University Press.

Woodwell, G. M. (1974) Success, succession and Adam Smith. *BioScience*, 24, 81–7.

18 Defining plant functional types: the end view

F. I. Woodward, T. M. Smith and H. H. Shugart

Introduction

The chapters in this book and the discussion that took place at the end of the workshop on Plant Functional Types have demonstrated clear individuality by scientists in their interpretations of the identity and utility of functional types. Much of the discussion used metaphors, e.g. *teams* and *players* of species and the *periodic table of species*, in attempts to clarify viewpoints. If anything, this approach exacerbated the problems of understanding but also reflected that the workshop and this volume are breaking new ground in actually grasping the nettle (or should it be a stinging functional type?) and pinning down the broad concept of functional types. There was unanimity in accepting the need for functional type classifications as a means of reducing taxonomic species diversity to manageable levels of understanding.

Natural functional types

The longest-established pedigree for defining functional types may have no *a priori* grounds for defining groups but awaits the accumulation of field information on (typically) life-cycle characteristics, e.g. early- and late-successional species. The field observations of life cycle and some environmental characteristics are subjected to a statistical technique such as ordination, and then groups of species with similar correlations with the environment can be excised. Classic examples are shade-tolerant and shade-intolerant species. More sophisticated and extensive approaches to recognizing these **natural functional types** are described in Chapters 9 (Bond), 7 (Grime), 4 (Hobbs), 12 (Lauenroth), 1 (Gitay & Noble), 11 (Sala), 13 (Scholes), 5 (Walker), and 6 (Westoby & Leishman).

There may be no aim in defining natural functional types beyond the natural human tendencies to classify nature, although sharply defined objectives can be addressed by the approach. Care may be required during the definition of these functional types to consider not only the environmental constraints, but also the evolutionary constraints. In a multispecies study, species may not represent independent points for assessing environmental correlates and groups of species may be closely associated in environmental space because they are taxonomically and genetically related (Woodward & Kelly, Chapter 3). So, for example, observed patterns in species associations with shade have been interpreted as a consequence of large seed size (Foster & Janson 1985). However, excluding the effect of taxonomic relatedness removes all significant correlation between seed size and shade (Kelly & Purvis 1993).

The initial emphasis on field observations requires a change of emphasis to include parallel studies in controlled, or manipulated, environments when considering the responses of natural functional types to novel environments, such as CO_2 enrichment and climatic warming (Grime *et al.* 1987; Hunt *et al.* 1991; Thompson 1990; Woodward 1992). Such a change in emphasis from the observational, with a limited range of environments, to the experimental, with a wider range of options, also presages a change in emphasis from natural functional types to **applied functional types**.

Applied functional types

A number of the chapters in this volume are concerned with defining functional types in order to predict dynamic plant or vegetation responses to changes in particular environmental driving forces, such as human land use patterns (Chapters 10, 12 and 13) or climate and CO_2 (Chapters 2, 5, 8, 10, 14, 15 and 16). The functional types that are necessary to achieve this end can be described as applied functional types. In some cases (e.g. Chapter 3) they bear no resemblance to the natural functional types because of the overriding emphasis on function and process, rather than structure and morphology.

It has always been hoped that readily recognizable structural types of plants could be readily translated into functional types (Box 1981). However, although there is a rich range of recognizable plant structural types, this does not translate into a rich suite of functional types (Woodward 1987; Smith *et al.* 1993). Therefore, it has been suggested that experimental studies, particularly based on threshold responses to the environment (Woodward 1987; Chapter 3, this volume) are an absolute necessity for defining the applied functional types which will predict the required dynamics and directions of vegetation change. Grime (Chapter 7) provides some hope that this definition can be realized from simple biochemical tests, without recourse to experiment; however, there is no evidence that this approach can provide the necessary breadth of information.

Attributes to be defined from functional types

The discussions at the end of the workshop were primarily concerned with applied functional types. A number of plant and vegetation processes need to be understood and predicted in order that future responses to climatic change can be addressed with some likelihood of success. The following list of attributes was judged to be central to predicting plant and vegetation responses (in no particular order of emphasis).

1. Vegetation feedbacks to climate, e.g. albedo, roughness length, canopy stomatal conductance and canopy carbon dioxide exchange.
2. Responses to the global suite of disturbance types, e.g. fire, wind blow, grazing, human land use, waterlogging.
3. Resource capture, e.g. CO_2, nutrients and water.
4. At the landscape and patch scale, dispersal mechanisms.
5. Life cycle responses to (a) CO_2 enrichment and (b) changes in temperature.

There was a realization that in order to achieve the goal of predicting vegetation responses to environmental change, and to extract the necessary attributes, there would need to be a mix of applied and natural functional types. The natural types have the historical and numerical precedent while the applied functional types are more directed to the research question, but are as yet insufficiently enumerated. A compromise scheme of the large-scale climatic filter approach (Chapter 3, this volume) overlaying the finer scale C-S-R design of Grime (Chapter 7), was considered to provide a profitable way forward to addressing the question of climate change.

Cross-scale conformity

It was recognized that neither functional types or attributes are likely to be defined in all vegetation types and their locations across the world. Therefore, functional types and their associated attributes need to be transferable between regions. This apparently unlikely event was demonstrated to be likely by Sala (Chapter 11, Argentina), Lauenroth and Reynolds (Chapter 12, USA), Bond (Chapter 9, South Africa) and Walker (Chapter 5, Australia). Such congruency of outcome suggests that functional types for garnering the essence of processes at the large regional scale are rather fewer than for smaller-scale local scales (Chapters 6 and 11).

Conclusion

The challenge that remains for applying functional typologies to global change issues is considerable. For a functional type cosmology to produce satisfactory results in predicting the change of vegetation in a world with altered climate (temperature, moisture, and CO_2 concentration in the atmosphere), the goals and the time and space scales of the application must be stated very clearly. The contents of the chapters that have preceded proscribe any *ad hoc* approach to developing functional types for global change. One important determination is the degree to which demographic aspects of plants (reproductive features, establishment, mortality rates) are tied to the functional aspects (metabolism, physiology, morphology). In several biomes, demography and function appear to be decoupled. Given the seed of a plant, the process of inferring the attributes of the plant from which it came (without a prior knowledge of the flora) is relatively uncertain. Barring the trivial cases (coconuts would not be expected on a herbaceous plant), a light-seeded plant could be a herb, tree or shrub.

If global change excites processes that involve demography (migration) as well as function, then the necessary functional typology will likely need to be two-dimensional (demography × function). It appears to be equally enigmatic to riddle the response of plants to CO_2 from their other attributes, implying an important third dimension to the typology.

The chapters in this volume clearly indicate that functional types are recognizable at all scales of study and can provide both static and dynamic understanding of plant and vegetation processes. The fact that functional types are recognizable goes against the work of, for example, Tilman (1988), where plants fit on continuous curves of responses to and trade-offs with resources. In this approach, there are no obviously discrete units, ready to be labelled functional types. However, the incomplete filling of the response surfaces by plants (Tilman 1988; Smith *et al.* 1993) indicates that, perhaps because of historical and genetical constraints, and/or because of very strong constraints of trade-offs, there is a natural tendency for the evolution of discrete functional types.

Although the development of functional types for global change remains a daunting task, the preceding chapters also point to the richness of approaches that can be brought to bear on the problem. Ecologists are not without ideas to address the problem, and this diversity sustains an optimism.

References

Box, E. O. (1981) *Microclimate and Plant Forms: An Introduction to Predictive Modeling in Phytogeography*. The Hague: Junk.

Foster, S. A. & Janson, C. H. (1985) The relationship between seed size and establishment conditions in tropical woody plants. *Ecology*, **66**, 773–80.

Grime, J. P., Mackey, J. M. L., Hillier, S. H. & Read, D. J. (1987) Floristic diversity in a model system using experimental microcosms. *Nature*, **328**, 420–2.

Hunt, R., Hand, D. W., Hannah, M. A. & Neal, A. M. (1991) Response to CO_2 enrichment in 27 herbaceous species. *Functional Ecology*, **5**, 410–21.

Kelly, C. K. & Purvis, A. (1993) Seed size and establishment conditions in tropical trees: on the use of taxonomic relatedness in determining ecological patterns. *Oecologia*, **94**, 356–60.

Smith, T. M., Shugart, H. H., Woodward, F. I. & Burton, P. J. (1993) Plant functional types. In Solomon, A. M. & Shugart, H. H. (eds), *Vegetation Dynamics and Global Change*, pp. 272–92. New York: Chapman & Hall.

Thompson, K. (1990) Genome size, seed size and germination temperature in herbaceous angiosperms. *Evolutionary Trends in Plants*, **4**, 113–6.

Tilman, D. (1988) *Plant Strategies and the Dynamics and Structure of Plant Communities*. Princeton University Press.

Woodward, F. I. (1987) *Climate and Plant Distribution*. Cambridge University Press.

Woodward, F. I. (1992) Predicting plant responses to global environmental change. *New Phytologist*, **122**, 329–51.

Index